PR

From her amazingly visceɪ .... ɯɪc ʋʋʋʀ s Epilogue, Burnell's voice jumps off the page, much like a microphone-wielding circus MC standing centre ring. And though you might not catch all that's going on at first read, you're made deeply aware that Burnell's motions and movements are all carefully aimed at studying what it means to be feminine, regardless of one's gender.

And just as the original, or ancient, circus was a place to proudly display one's physical prowess, Burnell puts on a creative tour du force for her readers through the five acts and Epilogue that combine to make a work of jaw-droppingly beautiful discovery.

KEVIN HOGAN, AUTHOR OF MY RISTRAD

She will rip your heart out, then give you chest compressions and chocolates.

What Sapha manages to do is tiptoe you on that line of emotional, gut wrenching, scenes that can bring you to the brink of hysteria. And with a sentence from a character she offers that breath of humor that pulls you back from the brink. Then dropkicks you over the ledge when you thought you were safe. An insane, diabolical, kickass rollercoaster.

RL ARENZ III, AUTHOR OF AEGIS

Sapha is like a young Wolfgang Pauli, in every laboratory he went, there was a little explosion.

DAVID ROOMY, AUTHOR OF INNER WORK IN THE WOUNDED AND CREATIVE: THE DREAM IN THE BODY

BOOKS BY SAPHA BURNELL
Usurper Kings

The Judge of Mystics Trilogy
Son of Abel
Book of Revels *(Spring 2022)*
Ginnungagap *(Spring 2023)*

VRAEYDA LITERARY

Can You Hear the Angels Sing?
Rev. Prof Seth Ayettey

Aegis
RL Arenz III

Warning Light Calling *(Autumn 2021)*
Peter Graarup Westergaard

Vostok *(Autumn 2021)*
Łukasz Drobnik

My Heart is the Tempest *(Winter 2022)*
Sacha Rosel

Sky Tracer *(Winter 2022)*
Hayden Moore

Girl of Light *(Winter 2022)*
Elana Gomel

Neon Lieben

SAPHA BURNELL

Copyright©2021Sapha Burnell

Vraeyda Literary
Pitt Meadows, BC
www.vraeydamedia.ca/literary

All rights reserved. No part of this publication may be reproduced in any form, or by any means, electronic or mechanical, including photocopying, recording, or any information browsing, storage, or retrieval system, without permission in writing from the publisher.

This is a work of fiction. Unless otherwise indicated, all the names, characters, businesses, places, events and incidents in this book are either the product of the author's imagination or used in a fictitious manner. Any resemblance to actual persons, living or dead, or actual events is purely coincidental.

Edited by Teagan Ward
Cover by Marissa Wagner & Sapha Burnell

Printed in the USA
10 9 8 7 6 5 4 3 2
ISBN 978-1-988034-14-0 (Paperback)
ISBN 978-1-988034-15-7 (eBook)
ISBN 978-1-988034-16-4 (Hardcover)

Vraeyda Literary sends authors to events, virtual events, Book Clubs & interviews. For promotional consideration, large-volume orders, please contact Lorie at ambassador@vraeydamedia.ca.

To Matthew, for being the best brother to this wild-one.
To Jesse, my death metal muse & provider of deep breaths.
To RL for being a paragon of virtue, you nasty pugilist.
To Noah, for bullet sizes and family gamer nights.
To Aedan, for many waters.
To Faith for your love, my beautiful friend.
For my sisters Sanya, Abigail & Elishia, long may we reign.
And to Kick, for the belief I bear enough valid words to pen.
Sapha

PREFACE

NEON Lieben originated from a deep love for three fundamentals.

One. *Star Trek: The Next Generation*, which my grandmother turned on, because the TV Guide mentioned a story about a boy named Wesley (*Will Wheaton*), whose mother was a doctor on a ship. In her innocence, Grandma thought it would be a nice story for my wild-child brother to see a well behaved boy, with a mother in the same trade. It didn't matter our Mum was a nurse, a single parent in the medical field was enough. We became firm sci-fi tv absorbers, much to Grandma's future chagrin. I fell in love with Data (*Brent Spiner*), in the way a seven year old who couldn't remember her father's face looked up to a distant influence. *Star Trek: TNG* became the bedrock of sci-fi. Followed avidly by *DS9*, with its grit (Sisco punches Q) and the Federation at war.

Two. In the summer before grade 12, I and a gaggle of other keen English students at our academy decided sacrificing a few weeks of summer was worth being in Mr. Rauser's last English 12 class, before he moved away. We spent hours in a deep dive of sci-fi. Studied *Amadeus*, *Blade Runner* and *AI: Artificial Intelligence* on the large tube television wheeled into class on a matte black cart. The day we read *Johnny Mnemonic* by William Gibson and discussed *Do Androids Dream of Electric Sheep* by Philip K Dick, my eyes opened. Those short stories were the apple on my Edenic tree, and with a firm crunch between my teeth, I knew science fiction was the blood in my veins. Prior to this, my education included the usual children's classics. *The Chronicles of Narnia*, *Where the Red Ferns Grow*, a healthy dose of Shakespeare and Zone en Français. At the end, he handed me a dogeared copy of *Count Zero* by William Gibson, with a word about how much he thought I'd dig it.

William Gibson is my favourite author to this day.

The last lifelong influence is the Spiritual Machines album by Our Lady Peace. As young teens, my older brother and I kept our musical tastes hidden from well-meaning but strict Christian grandparents & our Mum. No rock, but a 'be careful little ears what you hear' whenever the music got too far away from Southern Gospel or 'but Mum it's Christian I promise, see? They're singing about God'. When my brother got his driver's license, our world opened drastically. On the drive to school, there was Metallica, Live, Matthew Good

Band. And Our Lady Peace. *Clumsy* was the first song I heard, or maybe it was *Naveed*, but *Clumsy* and *Superman's Dead* were the ones I remember prior to *Spiritual Machines*. Our greatest fear was forgetting to take the CD out of the player in Mum's car, lest we get a firm tongue lashing and another whispered 'be careful little ears what you hear'. What I heard was the transcendence of artificial intelligence, the ability to turn 'pulp' into literature. A sci-fi image into high art. In university, I'd sit in my (borrowed) car listening to the entire album, usually with a theatre major in the passenger seat or sprawled out in the back. We'd watch the massive old growth trees beside the arts building sway as Raine Maida sang. Can artifice become sapient? Is everything going to be alright?

In Repair remains one of my favourite songs in the history of music. It lives beside Henry Purcell, David Bowie, JS Bach, Matthew Good, Soen & Fleshgod Apocalypse.

Without these influences over my formative years, I would be a proponent of 'literary fiction'. Realism-drowned narratives. Poetry based on the sorts of poets you learn about if you get off the 'regular path' and dive into academics as a life's pursuit. Science Fiction ruined me. Taught me to dive not into the allusions of an 18th Century poet with myriad well-read connections, but to square up and take a punch into quantum computing, artificial intelligence, epigenetics versus genetics (what happens when you mix lupine DNA with a velociraptor) wrapped in neon signs and purple hair.

I became the wild one. The untamed child of a straight-laced family, too much like that father I heard about but never saw growing up. The one who holds a PS4 controller and talks the birth of sci-fi, while pondering the last scene. The one with the guns in white plastic tunnels. I did not take a tumble down the perfectly respectable Shakespearean rabbit hole to an academic wonderland filled with Aphra Behn and discussions on James Joyce. I could have, the foundations were built for that house, too. I quite loved performing Shakespeare, Beatrice was my favourite role. There is nothing wrong with academia. I learned I was too wild for it, too much like the *Fastidious Horses* in Vysotsky's rumbling soviet song. I could dive into the academic side of sci-fi, argue the value of artifice and imagination to teach readers to view their current world in a different light.

Instead I dove frenetic and untamed into the Sprawl, swinging a mean left hook. Broke above the console cowboys of the Matrix. Watched electric dreams and learned what faster-than-light travel meant when you captained the Starship Enterprise. All in, NEON Lieben is an expression of these myriad influences. Of Jungian archetypes and a vivid imagination strong enough in physics to dip my toe into the realm of quantum computing (with help from my intelligent friend, who has degrees in the stuff). It's an inspection to the spiritual foundation of my younger days, of practices and meditations on the Sefirot, I and my spouse steep into our home.

Two stories intertwine. GMO soldier Aderastos discovers humanity and little harmless Max Allard has to bring AD-001 back before the ubiquitous 'all' is lost. The grieving scientist 70 years prior, who built an android to combat the death gnawing at his heartstrings. The path to artificial sapience lies not in a fleet of work-saving robots, but a group of anarchists who refuse labels with religious aplomb. Too many authors say 'this is the first of a series', and for me, this is.

I finished what I thought was the first novel 8 years ago. Handed it to my editor (who was working on *Son of Abel*) and she nodded, paused, hummed, and came back in a couple of weeks to smack me upside the head and say 'write the first one first'. *Hedonism Wholesale Inc* remains in the vault, and Teagan was right.

Any good science fiction piece requires an origin point. A Genesis Machine to ground the future stories within a world uniquely of our own designs. Welcome to my origin story. The birthplace of Aderastos, Max Allard and the Android Queen who called herself Mother sometime between creation and the day Aderastos washed up on the shore of Ucluelet's Carolina Sound.

1
a world of the mater machine's design

2155

Bare feet padded along the grey shore of Ucluelet, British Columbia. All but the bravest locals were long indoors, as raucous waves crashed across the reef. The tang of salt water made the monolith spit on the grass. He sniffed the air, feet pressed against the earth as if waiting for it to pitch and yaw. The disequilibrium of life on the sea sloshed at rung ears. No wave or bend came to the solid ground, nor did he see the dim electric signs wobble in the storm.

'Storm Warning. Please Remain Indoors. Call For Emergency. Storm Warning...'

The script played in sigils Aderastos lived in ignorance of; no context for surf shops, boba tea & coffee cafe's. Neither the few vehicles on the street, tethered to blue glowing stands with battery icons.

<Haven>

Above all signs in ubiquitous scripts, a crown of fluctuant purple and pink NEO-N light. The only recognizable thing in Aderastos' mind was the crown.

The lights dimmed and flickered. Not even the Mater Machine, that Android Queen Lieben, could keep the storm from battering the Pacific Coast.

Aderastos' body left a trail of seawater back to the Carolina Channel. Serene android NEO-Nurses bustled between patients and visitors. The NEO-Ns remained immune to the tumult of humanity's raised voices and emotional conquests between hospital beds. Chitter chat of the storm, would it close down the pass on Highway 4, was as ignored as the wind.

Vibrational whirrs convalesced on his skin. From the vibration alone, and the lack of petrol fumes, Aderastos knew the auxiliary was a solar and wind powered battery-backup generator. Precisely how he knew such confoundities existed was as questionable as the symbols on the signs. As long as the sun or the wind continued, this place would remain attached to the Hymn Electric.

Bare feet slid on the linoleum floor. As unfamiliar as the beings who caterwauled within the strange nest. It seemed to grow from the Ucluelet soil in a steel beam forest, the foliage vast panes of storm-proof glass. Light strips stung his eyes, blank white light ran the length of every corridor.

He understood little of the languages the beings chittered, nor why they stopped to watch as his chest passed the tops of their heads.

Aderastos knew nothing but the flow of heartbeats, which resounded like a cacophonous orchestra lacking syncopation. NEO-Ns stopped as he entered patient wards. Android senses riled and scanned, while metal and silicone bodies became statues dedicated to the new humanity loved and conquered.

A world of the Mater Machine's design.

The elderly woman smacked dry lips and leaned her head on a thin pillow, eyes milked with cataracts. Was this the doctor? Her heart beat with the disequilibrium of a stumbling fawn. The muscles tied to her bones were as brittle as their stays, feet marred from too many years on retail floors with poor footwear. A woman's shoes were the domain of fashion's agonies, in her day. Beside her headboard, an acrylic screen shone awake at his touch. It rained cerulean light on the woman's fitful face.

Symptoms, blood tests, an fMRI. The asymmetrical heartbeat thu-u-ub-d-um-ubbed in syncope with what he heard.

He cast his eyes to the ward and saw naught but patients asleep, or others craning their necks at the intruder.

Aderastos put his hand on her cheek, frail bird of a creature. She jostled.

"Oh dear, sweet Jesus. Are you here to take me home?" She warbled, pupils wobbled in delirium. Aderastos' eyes closed as he felt for the heartbeat and with a tug, an easy inhalation, strengthened it. Colour swooped into her cheeks. The cloying tang of infection wafted away. Eyes became the clear chestnut of a childhood playing softball in West Vancouver, despite her twisted, birth-dropped left leg.

He turned, gasps shocked her from a pallid recline to sit up in bed.

"Nurse! Nurse!"

As Aderastos laid hands on each person in the room, the NEO-N remained an inanimate object. Patients quaked, squealed, reached.

"What's he doing!?"
"Call Security! Call a Doctor! Someone tweet it!"
"Don't miss me!"
"NEO-N! NEO-N, wake up!"
"Over here! Hey, the web's down."
"Don't leave me."

Each of the confounded beings received the Healer's hand. Heartbeats. Too many heartbeats clattered in his ears; a cacophony unignored. Teeth clanked in his jaw from the damp chill of his sea water soaked jumpsuit. Illumination bands oscillated multiple colours around a closet beside NEO-N recharge cubbies. Inside, scrubs by size. The largest scrub shirt stretched tight across his frame. Trousers snug, if short. Patients he touched rose from their beds, pecked at CIRCLET holocams and tried to call disembodied voices. Words garbled incoherent as the storm shuddered against the CIRCLET network.

Aderastos' silence deafened the charging mass. Patients well enough to leave their beds lined the hall, groped to get closer.

"H-hey! You! Who... where's your bracelet... Lou! Lou I found him! Gawd dangit your Mama fed you like a horse!" A man in identical scrubs chased down the hall after Aderastos. His hale heartbeat and the timbre of his voice sent his chatter down the list of priorities to those whose health fared worse.

Aderastos padded through the hospital in a heady fog. Their hearts thrummed and thudded into his body with the force of river waves cutting into a fledgling canyon. Each thub-dub he heard added to what the behemoth was denied: humanity in all its ailments, and grotesque accidents spread before him.

"Singh is on site. I'm not sending a Tac Team into a hospital full of civies. Y-no frick, genius... Contain it!? Contain this!!" Commodore Rammage pushed the off button on his radio receiver's microphone with such vehemence he tried to crack the plastic. Didn't work. Damn. Barely hear his nebulous superiors through the storm anyway. Time and distance were on the Ithavoll's side, even if the crew buckled down against a fury far greater than la mer. Answers as to how the Asset woke up enough to open its eyes would be found if Earl Rammage tore the ship to girders to do it.

"Allard!" Rammage flipped the lid on his stainless steel coffee mug, battered in his pack until a thumb print manifested on its' side. At least that still worked. A thin, caramel skinned officer with rakish black hair rushed the door and heaved it open. Behind him, a litany of officers bustled through the compartment. Yellow klaxons flared along the corridor walls and the tops of the bulkheads in a

frenetic pulse of incandescent bulbs. "Close the door!"

"Sir!" Lt. Max Allard snapped to attention, hand bounced off his smooth forehead. The clumsy lieutenant kicked the door shut, secured to its' mag-lock. "Sorry... in a... one second... close the..."

"What the fuck!? What the legitimate fuck!?" Commodore Rammage threw the microphone at the junior officer, happy enough when it bounced off Allard's chest and into the Afro-polynesian boy's scrambled hands.

"I swear I had permission to bring my surfboard,Sir."

"What!? No!"

"Ah, Sir. What's..." Allard gulped, couldn't a dressing down wait until after morning kip? Heck, a cup of brimstone coffee? He fought the urge to rub at weary eyes, or count the hours of sleep he'd got on two fingers. "...okay now I'm more scared to ask."

"Are you serious? Our Asset's AWOL, Lieutenant! Vacated to the island! How the living hell did it get out of containment?"

"But look, Sir. Sleeping like the others, their aero-drip downers keep them under until 0900." Allard raised a remote to the bank of vacuum tube security screens and flicked to a live feed of the Asset Containment Unit. "Fixed the hermetic seal on CL-003's yesterday. Checked 'em all, the mechanics were solid as the day we set sail."

Rammage's eyebrow shot near off his face into a watercolour of a temperate rainforest screwed into the enamel. The ship rocked with the storm, Allard steadied his palm on the hull, copper wire mesh between layers of enamel inside the mild steel plate. The Commodore's office wafted with citrus oil wood polish and whiskey, damp papers crinkled in manilla envelopes. Pictures in glass frames drilled into the enamel displayed the ruins of pre-Mater cities and abandoned shopping malls.

The antiquated image flickered on a resolution which made Max Allard cry his first week onboard. From holographic immersion series to boob tubes built from colourized 1940's tech, life on Commodore Rammage's flagship was as dichotic as the CIRCLET he left back home. Tama probably cracked the biometric code, if Max knew his older brother. Assets laid in state, mechanical toys built of cogs and sprockets of DNA and raw meat none of the crew ever saw. The aero-drip smoke of their containment cells bathed any crewman's sight with a macabre but cheerful pink.

Ithavoll's Retreat was the reason for such technological luditism, to keep what the scientists did away from Lieben's omnipotent eyes. Rammage shoved away from his wood and metal desk, grabbed the remote and played back footage from the night watch.

"How... it swam ashore? Didn't think the Assets knew what water was, Sir. Aren't they offline? Isn't that what the aero-drip does? Offline them? Might've drowned, waves that big, reminds me of this reef outside Auckland..."

"Shut up, Allard. You're with me. Step to!" Commodore Rammage pushed past the office door, into the tight corridor of the CGM Ithavoll's command offices. He flung orders like confetti on New Years, without the joy or chance for a tot of the copious booze on his oak cabinet. "Rykstra, Desbiens, contain the others. Prep the Hercules!"

"Yessir!" Head ducked down, Allard fiddled with the hatch lock before he chased after the Commodore's receding voice. The hatch clanged behind him, unsecured in Allard's nerves. The wide but tired eyes of a kid.

"Allard!" No, Allard was the one.

"Yes Sir?" The Operations officer hustled after Rammage, foiled in a flimsy attempt to grab a desk officer's bagel. "Is that smoked salmon on th-coming Sir!"

"You're going to retrieve the Asset. Understood?" Rammage thundered on, crewmen slammed their backs against the corridors to grant him passage. He stepped over containment ridges with a clockwork gait. "Hurry up! If we don't get you down there before Singh gets antsy, there'll be more blood than I want on my conscience."

"M-b-m-I'm not specialized in... in... I'm... I'm a repairman, Sir." Rushing down the corridor, Allard stepped over a ridge and grabbed hold of a stair rail.

"I know what you are. How long have you worked on board?"

"Ah... eight weeks? Transferred from the Dauntless, Sir. Shouldn't Sec Dep deal with this?"

"I remember. You had the worst shooting record in ship history."

"I suck at guns, Sir, more of a fix-it man. Thus... I'm... what? Worst? Wasn't really that..."

The labyrinthine corridors opened to a bitter wind on a sea the Lieutenant respected too much to fear. Grabbing a poncho from the hatch side, he slung it over his head as the Commodore whistled for the Deck Boss.

"Didn't agree to the transfer for your shooting record." Fingers rubbed on his forehead, Rammage faced the frantic Lieutenant promoted more for his ability to keep a ship running than any tact or strength. The carrier's contingent of fighter jets and helicopters were long secured down. Deck Boss's skeleton crew scrambled an ancient Hercules tethered to the carrier like the atmospheric city of Abha was tethered to the Canadian Arctic. "You're harmless."

"Thank you... Sir? Is... is that thank you? Or... pardon?"

"If we go full tactical, the Asset will trip. It woke up, Allard. Walked out of its chamber and found its way to shore. Downers didn't manage last night...

you'd better believe Rykstra'll be working on incredibly detailed explanations right now." Out on the edge of the flight deck, Rammage bellowed to be heard over the roar of the storm. He gripped Allard's arm and towed him to the side, in a damp, but private place. A rogue wave crashed onto the carrier's surface. Salt and algae hit Allard's nose, as his boots screeched against the tac-cloth ribs cut into the deck. "Remember when you arrived, and got to work on the South Compartment?"

"The hurricane damage, yes, I'm surprised you didn't sink. Almost needed a dry-dock to patch us up, but I understand we don't ever make land."

"That wasn't hurricane damage, Lieutenant. I'm sending you, because anyone with an aggressive or commanding bone in their body will tweak the Asset. I can't send Singh's Sec Team into a Haven with intent to harm, without bringing Lieben down on our collective mcguffins. Only thing the Asset's ever responded to is... frick, I feel like my ex-wife's book club right now..." Rammage groaned and nearly spat on the deck. "... Appeal to it. Bring AD-001 back by being... you. Dopey, harmless, efficient you. The guy who saved his crew-mate from that crab-spider-thing by stealing a mess hall salad bowl and baking sheet."

"You... know the spider story... ‹course you know the spider story... everyone knows..."

"Allard." Rammage grunted. The young man snapped to attention. An ignorant sacrifice to the New Creation which escaped the womb to trample across Eden's shores. He clicked his fingers, and the pilot of the helicopter rushed to grab Allard's shoulder. Tow him on.

"Ever wonder what God looks like?!"

"Course not."

"Atheist?"

"No Sir. Why do you ask?!"

"You're about to meet God's second born."

Lieutenant Allard clung to the harness. The helicopter's massive side door was open to the whirling air. Thuds echoed in his head, through the ear muffs, into fingers, legs and toes. He clenched his teeth against the outrageously peevish scream attempting to burst out of his mouth. Ground below a rapid reality, Allard focused on a single spot of green grass to staunch his nausea.

Outside Haven, the CGM Black Collars hovered at the edge of the Channel. A brisk faced commander sunk his jaw on a piece of gum; the only promise he'd kept for his wife. Chew the gum, don't grind your teeth. Dentures suck.

"Willis, comm alpha. Get the Squad on the prowl."

"Ah, Sir? Permission to speak, Sir?" Hands on his knees against the unruly discombobulation of landing on a blade of grass in a winter storm, Allard raised a finger.

"What, Allard? A neolithic spider need saving?"

"Commodore Rammage sent me to... retrieve the Asset. Could use a bowl if you've got a big enough one." The lieutenant gulped and unfolded a piece of paper, rolling the sheet between his thumb and fingers. "Black Collars won't do. He'll have them tanked the second he thinks the people in that hospital are in danger."

"Prep radio silence." Commander Singh checked the dart packs on his chest, fixed the anti-kinetics vest across his broad chest.

"Yes Sir. But no Sir... Sir. Please, you've got to let me try. What kinda hullabaloo comes up if the Asset gets farther then a Haven on the backside of an obscure Channel? What happens if the NEO-Ns give it Sanctuary?"

Commander Singh exhaled through his teeth. The gum popped and sloshed in his mouth. A radio pack crackled useless beside Bestin and their gear.

"Allard, you're up. Reacquire the asset without collateral damage, and I'll reassign you from the plumbing."

"Yes Sir... I'll be careful Sir." He walked a step before turning around, "But I like Ops, Sir."

"Take a gun."

Allard shook his head. "Terrible shot, Sir. Reason I got assigned to operations."

Commander Singh's laughter clung to Allard's back all the way to the hospital entrance. As the glass doors swooshed open, Allard shuddered a deep breath.

"Maybe we both should go AWOL and surf in Tofino... Commodore Rammage my tanned arse... Oy! AD-001?" The doors to Ucluelet's Haven slid shut with a whoosh of lavender and eucalyptus infused air. A robot floor sanitizer squeegeed up the last of a trail of water from the expansive, light-filled foyer, and exhaled until his shoulders bowed.

Bright blue and white LED light bathed glass and steel. Holograms of medical staff asked people to return to their beds. Awash in white, blue and silver, the Haven made Allard wonder how anyone could return to a yellowed antique ship with corridors so thin two people shifted sideways to pass.

The Machine-tech Haven was abuzz with patients and guests rushing about. People in woollens, watched out the windows at the congregation of soldiers braving the storm, while others babbled about the giant inside. Aches gone or loved ones sat up from comas.

Scuttlebutt from rushed voices was impossible for Max to ignore. A giant scraping his head off the ceiling, laying hands... actual hands on the sick.

"Try again, Igor." A man in scrubs with a doctor's badge tapped at his CIRCLET, shaking his head.

"No. Nothing. Must be the storm... where'd he come from?"

"When can I study his blood, eh? NEO-N! NEO-N, get the patients back to their beds. Run full diagnostic scans of all of them before we release anyone. Igor, keep trying to get through."

Allard watched the two until they stared back.

"Rather busy, check in with a NEO-N, they'll get you sorted." The doctor ran at an elderly woman in a calf-length skirt and long sleeved sweater, her valuables in a bamboo mesh hospital laundry bag. "Ms. Davey! Ms. Davey, you can't... Ms. Davey go back to bed, we haven't..."

"Bother all, Doc I've never felt so good! Look! My leg! It's not gimped! I feel as giddy as a school girl on a field trip! Big Jesus healed me, Doc! More than you and your robo-dolls could do."

"Ms. Davey, we haven't run tests yet, we can't let you... there's a storm out... hey! When do 92 year old ladies run that fast? Wait! Wait up!" The doctor sprinted after the grinning old woman.

"Gee... must've been a sprinter in her day.... last century." To his right, a glass Entry Desk shone with blue script in multiple languages along its base. A series of white cubbies held brand new CIRCLETs to the desk's side. An advertisement played on top of the holographic glass, new features and higher res holographics on the CIRCLET with smaller size on the wrist band.

"Don't think Rammage minds if I get some help, or a map of the place..." Max reached for a CIRCLET and slid up his left wrist's soggy sleeve. The CIRCLET affixed itself and eased into connection, blue light scanned his hand as the band took blood pressure and pulse. Above his palm, diagnostic and calibration holograms fluttered, until a series of words swelled with the birdsong voice of the Mater Machine.

"Come at your leisure. My love is free. My abundance is yours."

Money was as useless as torn paper in a Haven, an inequality the Mater Machine abhorred. While CIRCLETs connected the planet's populace who followed Lieben's cultural call, the monetary deficit stung CGM's capitalist senses. How could culture, economics and society be maintained without financial gain? Her way led to docile cattle instead of the Conglomerate's wolves.

"God damn but this thing's cool. Heh... Moo." Max tinkered with the holographic keyboard until he smacked against a pile of green fluffy leaves.

Hydroponic towers lined the foyer, each grew a curated selection of healing plants and foodstuffs. Folk supped on pastries, fruits and vegetables, tea and coffee in the Bistro across from the Entry Desk. Salad. Max pressed his lips together hard. On board, the grunts reminisced of meals back home, where soup was made fresh, and nothing was freeze dried. Max gulped, overwhelmed by the Haven in what used to be an insignificant, tiny spit of land.

"Come at your leisure. My love is free. My abundance is yours." The statement entered Max's ears with the vibrato of electronic birdsong and servos in motion.

"Whoa!" The Lieutenant jolted at the sight of a NEO-W, wrist cannon placements draped in grey plastic, androgynous head bowed in a reverence Max read about in training. "R-right, ah, so you're a…"

"Your body shows signs of hunger and fatigue. Come and receive." The antique machine bowed its' mass produced head, arm-stumps motioned to the Bistro. Conglom training dictated Lieben's disarming of the planet as a false face, a reversal of power. The insane mama machine couldn't possibly take that many military drones and NEO-W battle units offline. He stared at the stubs of grey plastic, wondered if the lethal wrist mounted weaponry was removed at all.

"Right… yeah, I was… ADers? AD-001 where are… you in here? By the coffee? Hmm… best get a closer inspection…" The maelstrom outside continued to billow against glass and steel, as Lieutenant Max Allard dipped into the Bistro for a coffee and god damned hot sandwich. Hunger gnawed at his gut, eyes wide to the luxury of foodstuffs served without first being preserved in slim ration-packs.

Chef did what he could, but when a ship like the Ithavoll stayed seaborn for years, what cargo space did they have for lettuce? Should be glad there was no hard tack, until Crewman Biggins found flat 'loaves' of emergency bread in the bottom of his provisions chest. Various sized edible cups nestled beside the espresso machine, all embossed with another of the Mater Machine's tenets:

'Contribute and receive.'

"Heh, you and my mother. Yeah. I'll do some dishes." He waited for the black ichor of a double espresso to filter into the cup, and pulled a turkey sandwich on rye off the nearest tray. Steam rose from both coffee and plate, a god damned hot breakfast. Another cursory scan, no Aderastos. Should he grab the sandwich, eat on the go? A peek outside at the inclement weather, at the spit of car park Singh and his troops huddled near, and Max took a selfish, glorious bite.

His whimper caused the duo at the next table over to eye him cautiously, until Max's elbows sunk to the table and he chowed on his sandwich like it was the first meal he'd had in months.

The storm wouldn't make it inside for a few more minutes…. eh? Where was AD-001 going to go?

2
forget not thy faults

2085

"...for the trumpet shall sound, and the dead shall be raised incorruptible, and we shall be changed."
1 Corinthians 15:52 King James Version

The shriek of squealing tires punctured his sleep. Dr. Dieter Karnak woke with the crunch of rending metal, thrown from his slumber the way his son was thrown from the driver's seat. Stuttering incomprehensibly, the engineer sunk his feet onto the frigid metal of the Cloister and pressed his palms into his eyes.

"It is 3:27 am. Your REM cycle has been disrupted. Would you like to sleep until 6:00 am, Dr. Karnak?" The clinical voice skipped across syllables like a faulty GPS, as antiseptic as the room. He grumbled low in his throat, slid his feet in slippers and rocked back and forth, his hands against the side of the thin cot.

"I sleep later, liebchen." Karnak pushed his neck from side to side, feeling the crackle of his vertebrae as he pulled on his thermal lab coat against the artificial temperature of the spherical Cloister.

"Your sleep deficit is 743 hours, 33 minutes. Would you like to sleep until 6:00 am, Dr. Karnak?"

"Liebchen, meine liebchen... what am I going to do with you?" Karnak rubbed his face and stumbled over to the tools, technological bric-a-brac and robotics scattered across worktables laid out in an arc. His palms rested on cool metal. A picture frame with a bride and groom's euphoric grins held a printed out and taped piece of recycled paper. 'Obit-' Dieter closed his eyes as a moment's revulsion punctured his woken calm.

A ruined car torn by rescuers, from the snow. The thrum of the drone's propulsion system chased his inner ears, its' cut-laser blitzed at the cargo of the truck which hit frozen water. Cargo values trumped carbon lifeforms, in its' algorithm.

"Liebchen, reboot deep learning algorithms, turn on the monitors. I told you I sleep later. When I am retired, maybe. Maybe… Make me en kaffe. Reconstruct the hypothesis on your neural net via findings 2.4. Warm up the quartz-laser apparatus." Karnak tap-tap-tapped at data tables, pulled the coffee mug away from the machine and took a hesitant sip.

"Raise volume twenty three percent. Ah! Good… good." Another night interrupted. Months of interruptions by the sounds of his son and daughter-in-law, as captured by the locals, who went to record an aurora and couldn't reach the car in time. As he worked, Karnak glanced at the desktop photograph of a young man and a breath-arresting blonde with arms around his neck. Johanes and Agathe, now spectres of disturbed sleep. Eyes stung by sleeplessness craned to a holographic panel above the picture frame.

Sarai's vitals were stable, locked within the hallucinogenic virtual world her therapist called 'a healing journey. Three months in, she converted their bedroom into a life-pod. A litany of emails from lawyers, notaries and the delivery of a keycard for a storage container in Langley, the echo of a marriage she no longer managed and a perpetual online existence Dieter couldn't fathom.

Sealed in the bowels of the Conglomerate's Vancouver Laboratories, the Cloister received its name from the otherworldly predilections of the chief research scientist. Dr. Dieter Karnak, the world's leading mind on artificial intelligence and neural net programming. Death's dealer to the Slavic Amalgamate, Oppenheimer's bedfellow to the Commonwealth and NATO.

"Meine liebchen… meine liebchen." He worked amidst the muffled cacophony of laser tools and the eternal repetition of Handel's Messiah, until the Cloister opened with a hiss. The Cloister became both sanctuary and escape, his pied-a-terre in the Conglomerate building as empty as a marriage decimated by their children's deaths.

"おはようございます, Dr. Karnak. おはよう, Lieben." Baiko bowed as she entered, her teal hair caught atop her head in a purple bun. She hugged into a lab coat, and left her shoes at the door. Work with Dr. Karnak was a religious vocation for the intern. He took none but believers in his Trinity. Metaphysics.

Hard Science. Wonder.

"Guten morgen, Baiko… Liebchen. I call her meine Liebchen, not Lieben. Liebchen." Karnak refused to look up from layers of quartz crystals, as he tinkered with the inner workings of his liebchen, his darling. Several samples of fibre optic hairs rested in taped clumps beside the wedding photo of Johanes and his Agathe. One would work… one would be perfect.

"すみませんでした, Dr. Karnak. How is Lieben today?" Baiko bowed at the camera lens hung from a lazy umbilical. Karnak looked up with a scowl, shook his head and looked back down with a hidden smirk. If every day he taught Baiko to pronounce Liebchen, he would be short enough time to make half a machine.

"I am operating at 86% efficiency, Ms. Kaho. I require optimization. Your heart rate has risen 6.9% past resting rate. Have you been running?" The disembodied feminine voice flooded the Cloister, and Baiko covered her mouth with her hand.

"I was going to miss the bus." Baiko set to her work with the virulent desire to blame the pink on her cheeks to the heat of the screen. A flow of information jittered to life. Baiko oriented herself to the experimental algorithms run over the midnight hours. Never a question of where she was needed, Dr. Karnak valued the lack of extraneous chatter in Baiko second of all. Foremost was Baiko's ability to match samples to the photograph, without question of his mental health. The penchant for subtext, learned from her Japan-born father was as dear as the gold soldered into their beautiful machine.

Strung up by a collection of wires, robotic arms and threads of fibre optic cable, the shell of Karnak's Liebchen hovered like a piecemeal angel. The android paused in mid construction. Completed systems hovered; feathered wings of silicone and quartz microchips. Each segment of the machine was as carefully crafted as an icon in an Orthodox Christian church. Dr. Robert Dunlevy's robotics meshed with the quantum theory of Dr. Karnak's classical neural net in an array, which made the bureaucratic beasts on the Board salivate for a thrice helping of cream and honey off the top.

"Morning." Dr. Robert Dunlevy groaned into the Cloister with the scent of harsh black coffee in a flip-lid container, and the leather of his shoes. Dark hair shorn short was worn under a thin microfibre tuque, a cold-weather turtleneck under his button down to stave off the Cloister's nigh sub-zero temp. "Results in from that servo test, yet?"

"I didn't check."

"Oh, 'cause the damn thing doesn't need to move if it's smarter than the average fifth grader?" Robert grit his teeth and slumped into his chair. Flicked on the bank of screens in his area of the Cloister, and searched through emails and data reports as he blew on the thin steam from the travel mug. Baiko scrunched her nose.

"Your gyroscopic servo-motors are not my priority." Karnak refused to glance up from his work. "Meine Liebchen's ability to intake stimuli is."

"There we go, glad we got the 'your machine' thing out early today. You're not the only one building this tin totem. Remember? Tch. Be professional for fuck's sake."

As the human sense of time derived from paleolithic seasonal changes and the night sky, so the modern age would look upon Karnak's creation and realize Death, like night, has its' waning. Built upon quantum bit superposition, the quantum computer could theoretically augment a traditional computer's neural net. Not Honda, nor Taiki-Benz managed to ground quantum data in a meaningful way.

Information in the quantum realm was temporary. It was the harshest lesson in quantum computing, since the beginning of the 21st Century. Ephemeral as the superimposed bits. How did you anchor something which existed in a state similar to Schrödinger's damn cat? Would the data brought from a quantum system anchor correctly, without being altered by the position of the anchor itself?

"Really? Not a word? Not one?" Robert sipped his coffee and shrugged as Baiko offered him a slim smile and forwarded the data reports. Scientists for decades attempted to create meaningful uses for the quantum computers humankind produced. While their potential was the difference between stone aged clubs, and the industrial revolution, practical application remained as elusive as Dark Matter. Whoever broke the code and unlocked quantum computer's potential could feasibly be the monarch of the post-silicone age.

Dr. Karnak promised to place such a paradigm schism in one of Dr. Robert Dunlevy's pristinely engineered robots. How else to connive the Chairman into releasing funds for his Liebchen? The eccentric doctor sanded and polished a piece of crystal the size of Baiko's thumb and held it up to a gold plate bent into a cranny inside his miracle.

"Rose Quartz. For compassion." His clipped German accent rang hollow off the walls of the technological birthing chamber.

"Will it give Lieben compassion?" Baiko closed the lid of her thermos, set it in her bag.

"Liebchen, mein liebchen. Not lieben." Dr. Karnak chuckled with a foreign, yet temperate warmth.

"Oh Christ, a machine that can feel. That's a good use of resources…"

Robert sputtered as he looked up from the servos for Lieben's articulated arm. "Rose quartz for my ass. It can compassionately wipe said ass about four servo links from now."

If Dr. Karnak heard his fellow, he ignored it.

Baiko talked to Lieben as they worked, picked up on Karnak's constant mutters and spastic temper. It felt wrong to leave her in half-constructed disarray. A mother carried her child. They had only the cloister, inverted womb where many parents constructed the new creation.

Lieben's cloister felt like the only place on Earth where disconnection reigned. The world forgot silence in the years since Dieter was a boy. Even then radio-waves rang eternal around the planet long before his mother bore down. Sound whispered through the space, instead of the daily roar, reverberated in holy ratios until it cuddled around the ear of the intended listener.

"Faun! What is this? These conduit wires are flawed!" Dieter threw the box of wiring wrapped in a spool back into its shipment box. An inch-long piece in the middle bubbled. Easily cut away, it would not do.

"Send it back. Request a new spool."

"Why not cut that off and use the rest? It's fine." Robert scanned the spool, tested its conductivity. "Nothing functionally wrong with it."

"Send it back." Perfect. It had to be perfect. Only perfect could it be done.

Robert shook his head and sent the spool to shipping, with a hand scrawled note to cut off the offensive piece and send it back with fresh wrap. Eccentricity was a given in the brilliant, but Dr. Karnak's religious fascinations grated Robert's atheist sensibilities. If God ever existed, it wasn't bickering about faulty wires.

In the depths of their copper monastery, circuit boards made believers of post-graduate students, and in the chill hum of perpetual prismatic light, an entity crept closer to being born.

"Hold the door!" Frank yelled from the public side of the security gate, until he got close enough to see the glare on Tara's diplomatic face.

"Yeah, sure! I'll hold the door. Sweep aside in a curtsey and let Sir enter, too!" she strutted inside with all the grace of a peacock on the prowl. Tara's heels cost more than Frank's car, or so he thought as the glass door smacked his shoulder. Frank fumbled and caught one of the data sticks in hand, the others skittered to the polished floor.

"Goddamnit, Tara. Freaking big head for a dame in PR." Frank grunted. She strutted on those damned red-soled heels, the ones that made Frank's wife blush when she saw them at the company holiday party.

If he ran, he'd have time to take a piss before the big meeting. Frank tossed his ident-card to the reader. Tara's smug red pout grinned at him as she waved a manicured set of nails from the VIP elevator. The elevator doors wooshed shut.

Frank slumped into his chair around the boardroom table as Levy Xin, the Chief Operating Officer stood to wave the first screen on their report off the projection port.

"Good of you to join us, Frank." Levy's eyebrow raised, her thin lips pursed in the consistent scowl of a woman who got where she was by talent alone. Tara grinned across the table, leaned back in her chair with a stylus in hand and a biscuit beside her coffee cup and saucer.

The Chairman's personal assistant placed a cup and saucer of coffee and the pastry tray down for Frank, who busied himself with his tablet and pockets for the data stick encoder. The boardroom table held folk in a varied array, a few chairs switched with holographic plasma displays of people in their home offices, and vacation spots.

"Thank you, Levy… I apologize, the elevator attendant didn't know the meaning of 'wait'. No cream, thanks. Ah, bagel and lox. Yeah… thanks… Diana is it? Thanks… how 'bout them Canucks?" Frank mumbled, as she put his order down and went to stand in the corner, unobtrusive for the next caffeine hit in the boardroom. Tara pulled the stylus from her mouth and huffed, taking the tiniest bite of her biscuit.

"Yes… I'm sure." Levy raised an eyebrow and ran her hand over hair laden with fibre optic extensions, the shifting light a hypnotic and agreeable addition to the otherwise all-business woman in a pale grey power suit. She married the job the way old widowers married virgins, a glorified nanny for the child-ideas percolating in company mind.

"So! Ah, yes, the quarterlies on the NEO's. Budgetary tables. Stalling tactics and… so how does everyone like the new coffee?" Frank slid the data chit across the table, and Levy set it into its' dock, as 3D figures emanated from holographic banks built into the glass surface. Chewing on a corner of biscuit, Tara grimaced as the figures danced.

"Frank called this meeting to discuss some necessary updates to the NEO-N Project. Frank, if you would not mind." Levy sat down and stared at Dr. Karnak, who loomed upon the table with both palms flat upon his holographics, shoulders lowered like some form of avian raptor. Robert Dunlevy leaned back in his seat, ran his fingers along the furrow in his brow.

"We're over budget. It's taken our entire R&D coffers and the futures for three years. Sorry, Dieter, Robert, I'm a money man. I don't know what it is you do, I'm not that smart… but if we want to put out the Vio 4-passenger next year, we have to figure out where the disparities are and how to staunch the economic flow." Frank sipped his coffee, set it down and added a cube of crystallized honey. The spoon clanked across porcelain. At the head of the table, a pair of gold-ringed and folded hands clenched, then released with a breath large enough for the entire room to flinch by microns.

"I was promised an unlimited budget." Karnak loomed over the table, his rare electric magnetism assumed as much space as Rasputin in the Queen's private rooms. The ideal germanic engineer, Karnak was the sort of guy who made Frank nervous around his wife, before they went out for more beers than frat boys on break. Nerves returned for the imminent joust of two alpha males, as the Chairman leaned forward, his gold-clad fingers clenched together.

"Unlimited is relative. I can't give you what I don't have. The Payroll, fabrication costs, invested capital, I'm not touching that. You've had unlimited access to our R&D budget for two years, Diet. And the profits from the Mia-2, and the futures on the Vio-4. I can't feed you on air. We have stockholders to answer to." Frank leaned back, sipped his coffee one more time. Ah, sweet caffeine.

"Take from her." Karnak pointed at Tara.

"You can't take my budget! It comes direct from the Chairman." Tara laughed, a shake to the extension-ladened head. A blown kiss made it across the tabletop via holo-emitters, pursed lips which fluttered until they pressed against the Chairman's largest ring.

"Oh, I don't know, social media's a banging way to sell our company's products, Tara…" Frank smirked.

"How much extra do we need?" Levy brought forward the PR and Marketing budget, a band of green up to the ceiling.

"Hold the call, here." Tara sat up and rattled her coffee cup against the glass table.

"Like you held the door and elevator?" Frank snapped, foot tapping on the lunch kit he hadn't had time to put away.

"Tinker-toy can't have my budget. I need my budget. He's going to cut his corners and leave mine be. Who do you think is going to sell this thing, once he's done? There are trade shows to prepare for, places we're expected to show up to looking like money. Ad campaigns, hit directors and celebrities to hire to do commercials, stream spots, corporate sponsorships, the shebang. Cut my budget and what, ask me to run to the copy-and-print to photocopy a bunch of posters, of 'World's First Android Slave'?"

"Meine Liebchen is not a slave!" Karnak's fists smacked wet and hard against the table. The room jerked, as if his fists had the power to send everyone two atomic particles over from their current position.

"You're the one with the insane hippy bullshitters dancing naked around a King's Ransom of electronics and liquid gold!" Tara barked, the pads of her fingers on the table.

"Fuck you, Tara! You have money! I did not know client dinners included seven thousand dollar shoes and Haute Couture frocks!"

"Dr. Karnak! Tara! That is enough." Levy threw in, and Frank caught the glares in their eyes, the haunted suck to Tara's cheeks. Oh god.

"You gave me promises of unlimited resources to create the best android the world will see, and I am giving you artificial life. Liebchen is not some machine to be bartered around. Liebchen is alive!"

"Oh for the love of fuck." Robert rubbed his temples, "I can cut on the servo mechanics, it's not that... ow. Who kicked me?"

"Okay, yes, Lieb-lub-the NEO-N's special Dieter, we get it. I'm banking on it. Tara's being salty, you don't bow to her whim." Frank spread his fingers out, attempted to calm the table. A dozen VP's and the Chairman himself in the room, and today the notoriously silent Dr. Karnak had to open his gob. "She's forgotten who controls the money, and you know what Tara? For someone who's been dipping into the expense account for fashion week, maybe pissing me off isn't the best start to your day. Dr. Karnak, Dr. Dunlevy, if I shill you another 40% of Tara's budget, can you get the prototype off the ground? Displayable in half a year? Capable of shaking hands, kissing babies, impressing Mr. Media and the Social Gab Band?"

Dieter's lips pursed, talon-fingers scratching at the table.

"I was promised free reign. My own staff, my own laboratory, my own space! You cannot take back a gift freely given! Not after the decade of building your war machines. I create you a mechanical army and you give me my Liebchen. You hid behind objectivity while my machines decimated Prague. Kiev. And when I told you…" The scientist wasn't the same since the accident, and Frank wondered if anybody else much noticed… in this meeting everyone couldn't do a damn thing but notice the slow unhinge of Karnak's mind. The greatest mind in artificial intelligence and optics of the silicone age was unravelling. Threw himself into a glory project, while interns and underling teams slap-dashed the cars, worker bee machines, holographics and robotic prosthetics that paid for it. None of the people around the table met Karnak's eyes, as his wild nerves grit at their collective past. "When I told you to pull the drones, their neural nets were not capable of handling the tactics necessary for peace, you promised… you promised me this."

"The war stung us all, Dieter. We did our part, but your and Robert's work

saved lives. Human lives. We supplied what the Commonwealth needed…"

"No!" Karnak's palms slammed against the table. Glass shook, the holo-emitter wobbled. "We supplied what they asked for. What they wanted, not what the world required. Artificial sapience is not a parlour trick with recycled parts. Do this project wrong, and the world burns. Do it right? And we herald Hera on Olympus' Throne. The… hippie bullshit Tara yips about is as important as keeping the Cloister cold and the quantum computer core colder. How an artificial intelligence is created, the way in which it is built, matters profoundly. One faulty algorithm which equates material cost higher than human life is the knife blade between utopia and annihilation. You asked me to build this. You demanded a machine smart enough and creative enough to take on the Chinese Syndicates' narrow AI's. To make it moveable. Functional as a human is functional. How else did you think it could be done?"

Frank's eyes fell to the data pad synching with the table below his hand, and he tapped at the smart glass to call up the invoices for materials.

Gold, quartz, tourmaline, silicone, copper… in his faith or his guilt, the litany of items expended on the NEO-N Project didn't mean much.

They meant the world.

"Dr. Karnak, how close are you to finishing the prototype? We need to know when we can put the NEO-N's into production to recoup our costs." Levy refused to acknowledge outbursts as the other directors pawed at their parts of the table.

"I am not giving you a mere machine to putz around with the vacuum and watch sodium levels. Liebchen is the first artificially intelligent sentient being in the universe, and you wish me to… rush her?"

"Karnak's nuts! It's a machine, an expensive luxury machine. It doesn't breathe, eat, smoke a joint at puberty to experiment, it does what we program it to do and that's all it will do. We can cut a profit either way, but industrializing it means we go from luxury to global phenomenon and with the Honda people marketing their newest self help machine for Spring release, we need the working copy. It's not a negotiation, oh lord, what's he doing?" Tara rolled her eyes.

Dr. Karnak's thighs bumped against his chair as he walked to the exit, hands dug into his usually tidy hair. The until now silent Chairman's eyes rested not on the wayward scientist, but on Tara. She rolled her lazy blue eyes and checked her lips in the mirrored back of her phone.

"Yeah, I got it." Tara groaned, bit another pastry. The room cascaded in a horrid disquiet, the 3D graph of expense versus funds rotated in the centre like a church spire. Outliers dotted in a way Frank recognized in the alcohol-fuelled dreams of the ubiquitous 'after'.

The NEO-N's... the Neo-Nurses would absolve them all.

"So that happened..." Frank blinked, sucked down another cup of coffee, and tapped at his screen on the table. "... March sound good for the next update?"

"March sounds fine, Frank. Take the loss from my part of the budget. Give me enough for some cheap Chinese wire and silicone spackle. We'll get it done." Dr. Robert Dunlevy didn't buy into such professionally suicidal beliefs as creation of sapience, nor of anything but solid work for the bottom line. Dr. Robert Dunlevy built machines.

God. Damned. Machines.

Whoever said a sports car couldn't have a spare wire or two to replace every two years? Upgrades in services? Robert shoved off from the table in time to watch the Chairman whisper in Tara's ear, his tan hand firm around the middle of her waist. Rubbing his eyes Robert got two strides, before the honey of Tara's voice, artificial as the rest of her, played to his ears.

"Oh, Robbie." She grinned from filled lips, their distended size a caricature to the woman Robert imagined Tara to regress into the minute she was both drunk enough and alone enough to search through what remained of ancient camera rolls. "Chairman needs a teensy minute of your time."

"Would be my honour." He sipped the last of his coffee, its' black and chicory flavour burned down his throat. Setting the mug down with a clink, Robert fixed his tie and followed Tara to a slim copper door built into the side of the conference room. The Chairman's facial scan opened the copper door of the private elevator, and he stepped inside without a single sound, Tara behind him.

"See you on the top floor." She pointed to the stairwell outside the conference room, as the copper panel shunted shut.

"God damn." Robert trotted to the stairwell, reminded with every flight how much cardio he missed locked away in the Cloister.

The rooftop garden of the Conglomerate's Vancouver spire bled green. Vertical garden walls hemmed in the furtive wind. Pergolas with robotic covers retracted in the rare sunshine, while heaters disguised as sculptures betrayed an opulence beyond everyone, but the Chairman's chosen. Richly pigmented mosaic tiles reconstructed walkways of the Chairman's youth, lemon and olive trees lined the walk beside orchids, peonies, rose bushes and herbs. Cushions of silk embroidery rested on low settees, as a fountain bubbled behind a well-loved potting bench, where composite planters were half-filled with marigolds, and nasturtiums. Dirt-spoiled gloves rested beside a gold plated trowel.

A breath of laughter guided Robert to an alcove, where Chairman Kaur reclined on a low Chesterfield filled with cushions, sipping from a glass of deep

ruby wine.

"Good of you to join us." Tara glanced over from her magnanimous grin, poured Robert the tiniest amount of wine, and handed him the glass.

"Tara, does Dr. Dunlevy threaten your calm so much?"

"You want a drunken scientist tinkering with our new merch, I'll open another bottle, Sir." She nibbled on an olive, set her own glass down beside feta cheese and lemon marmalade on scones. Head tilted back toward the sun, the Chairman soaked in the beauty of Vancouver in late springtime, clouds over the mountains too far off to linger in the downtown oasis.

Hand clenched around the bowl of his wine glass, Robert bit down the urge to start up conversation. The dangers of disrespect hovered around Robert's wrist. A word spoken poorly could send one toppling over the edge of Chairman Kaur's personal Eden. Karnak's outburst echoed in Robert's mind, the man was brilliant but Robert didn't abide the quintessential 'genius' as outside the rules of proper behaviour. Dieter thought his post-war pain mattered more than the rest? Karnak's paradigm bullshit wasn't going to ruin Robert developing a glorified tinker toy with legs for days.

The wine in Robert's glass sloshed down his throat. He gripped the stemware between his fingers and leaned elbows to knees.

"Karnak's out of control."

"Dr. Karnak is the reason the NEO-N project continues at all."

"I could…"

"Is Dr. Karnak going to finish my perfect machine?" The Chairman smiled and shook his hair out of the ponytail he wore in the meeting, silken black locks flowed past his shoulders. "If I give the lion's share, will he complete it?"

"Are you asking me if he's capable or competent?"

A series of tsks clucked out of the Chairman's mouth, the shake of his head enough to cause Robert to lean back, accept the long pour Tara offered from the bottle. "Forget not thy faults, Dr. Dunlevy."

"On a good day, Dr. Karnak can build Jesus out of paper clips and stray wire. Lately, I'm not convinced the man leaves the Cloister to go to the bathroom. He's fucked in the head, nattering on about mystic bullshit and… I can build the machine. But his woe is me routine? Rose quarts for compassion, yellow gold for ambition. Pseudoscientific bunk. If I have access to his research, I can finish the prototype without him. A sensible, capable machine." Robert dunked the wine down his throat, sucked half the glass with a slight grimace and a shake of his head. "I'm a scientist, I fear gravity and electromagnetic pulses near my phone, not some mystic emotion giving rock! How can the man lead the most important android and artificial intelligence project with supplies from the metaphysical bookstore off West 4th and Arbutus!? I worked my ass off for this! So what if some archaic limited ai built into our drones caused a car accident in Germany!? The drone had its' function! It followed its' algorithm with perfection, the damn accident was human fucking error. We can't erase human error, but we can build and modify perfect systems. What in the radiant matter of Crookes' plasma does it matter if a machine has emotions? They don't matter. Nobody wants a maid that laments the laundry! God! And where's he, eh!? Crafting a crucifix with supplies some new age hoplite dug out of the remains of a church on fire! He won't build your new world. I will. Me! The NEO-N's will be efficient, they'll do what they're told and they won't have chakra prayer beads built into their endoskeletal structure like some Rainbow Road nonsense!"

The buzz of a bumble bee by the rooftop hive joined Robert's exhales. Lips clamped shut, he set down the stemware so hard he thought it would snap. But like Karnak's resilient but shattered mind, the thin stem remained. A hum of environmental shielding, which allowed the rooftop to have such luxuries as bee hives, marigolds and lemon trees in a temperate rainforest, 323 metres above ground.

Wild eyes shifted, unanchored by the Chairman's silence. The gleam of a talisman around the Chairman's neck, of twin Kara bracelets peppered the silence. Tara tapped the rim of her glass, leaned down and kissed the coil of gold on the Chairman's thumb ring.

"I'll handle Dieter." Her voice coiled around Robert's ears without penetrating the drum or anvil or hammer. Numb as the statues formed of marble and cement through the rooftop paradise, Robert rose and bowed his head. What was in the Sagan-be-damned wine?

"Chairman. Tara."

"*Forget not thy faults*", Chairman Kaur's voice bored through forehead and bone. But Robert Dunlevy was a man of science. A creature of proof, habit and information bartered at the fount of experience. As he walked to the stairwell in the side of a living wall filled with clematis, he cursed loud enough to reverberate through the flights of stairs, and hear a limp, empty echo.

3
It was retro

2085

An atomic cloud of agitated rage surged out of the staircase and scowled down the hall. Weakness! Dullards played with dolls in the boardroom, but where were they, when Dr. Dieter Karnak created the first stable interactive hologram? When the Commonwealth required drones in lieu of soldiers? If not for... he shook with a snarl. Tara was nothing but a reminder of his wife's distance, the seal Sarai placed in her virtual capsule. A fantasy, in the house which became more a time capsule of their once contented lives, than a place Dieter could stomach.

Silence, bloody silence. From Sarai, from Chairman Kaur, the utter cavity of silence while those pedestrians like Robert and Tara spoke in crescendo. The vision of his Liebchen entangled in his mind, with the quantum state of a basic qbit.

Yes / No / Both.
+ / - / Both.
0 / 1 / Both.

Tentative as a dream's handle on the brain, the nature of quantum computing chewed into his stomach with the fear that one wrong word would break the spell. Could a true artificial intelligence created from a quantum computer be replicated? Just as epigenetic factors controlled how a child formed from the raw genetic matter, so too did the nature of the creative act matter in building his quantum opus.

Men like Robert, brilliant in their modes, saw only within their factual confines. They reiterated data, and begrudging was Dieter's admission of their usefulness. The world functioned because of engineers like Dr. Robert Dunlevy. Buildings built, technology maintained, a few advancements to fuel a reduction in waste or increase in productivity. How a man brilliant enough to craft the case of his liebchen's body could see so small a use for her…

No.

Dieter Karnak cursed Robert's lack of vision. Quantum states were as much about faith as hard equations, their usefulness a paltry concern without the belief in their sovereignty. Superposition, the state of yes and no and, confounded scientists for decades in a theoretical springboard of strings and inter-dimensions he couldn't fathom until…

Until, the Holy Until.

The moment the box opened and shut to any semblance of a normal life. A system in one quantum state, with the possibility of multiple configurations was definite and supposed. The complexity of a system created superpositions, until Dieter saw no separation between concept and function. All choices in simultaneous chorus, until a solo emerged in potential.

As a man was father, lover, scientist, worker, driver, cook, desperate and sated simultaneously, the anchoring of such potentials did not require an elimination of any one part. Father, lover, murderer, scientist, war hero, worker drone, grief, consolation, satiation, desperation.

The key to ensuring the human race betrayed its' constant attempts to species suicide lied not in an infant Christ, or the chant of a mantra thousands of years removed from the man who saw suffering in the street.

The future of the human race relied on its integration with quantum superimposed intelligence so vast and holy it saw and it understood, and it loved. Simultaneous, all and naught. Separated but integrated as a light on the hilltop or salt in the hand. And the Board, the narrow minded like Robert Dunlevy saw profit margins, silicone sex and laundry folded while dinner bubbled in the pot. A hausfrau hoisting garbage containers, smiling while wiping the ass of the elderly. Quotas meant more than a search for the quintessential saviour spread-eagled by system in his Cloister.

Narrow minded cave dwellers, who saw nothing but a new form of domination.

"Dieter! Karnak, hey!" Tara's heels tapped at the ground behind him, echoed into his ears.

"I fear you." The engineer pursed a flask to his drone-lips. He threw his hand accusingly at Lieben's empty visage. "I fear you and the board's spin doctors will turn mein liebchen into a toy."

"Isn't that what she is? A play thing for the upper classes? Wasn't that what we were supposed to build?" Tara pushed into the Cloister, sat on Karnak's work table. Stroked the metal top with her thumbs, shoulder dipped in feminine allure. "You want her to be a real girl, aren't we toys too, Diet? All little girls grow up to be some body's toys."

"Did I treat you like a toy? Before I knew your touch was paid for by the Chairman?"

The Cloister opened before Tara's eyes, artifice attached to Dieter Karnak's sense of soul. Licked lips pouted as she remembered pillow talk on electro-epigenetics, stimuli response. Superposition. Dead, alive… without him inside the place, it seemed as still as the tomb to an unknown princess. One squared away in small honour to cover a counter-cultural shame. As Dieter Karnak surged into the Cloister, with the cloud of his angst, the spherical laboratory swept into a churning miasma. No square centimetre unaffected by his pull.

"You were… considerate." Tara's lips worked, and she nearly kicked off her shoes on the holy ground. Karnak bypassed Tara's tanned thighs, one hand touched on her knee with a slight squeeze. Biology. At times, it couldn't be helped. "I kind of missed it. You changed, when you discovered I'm a whore."

"The disparity between your current state and your potential irks me." All she owned since was the stress Dieter relieved, the grief she wiped temporarily away and she could have it. Coiled on the floor with her panties. Illumination strips bathed the grit of his jaw, he picked up a tool Tara didn't recognize, set it down and pulled his thermal lab coat off the peg.

"Yeah, but you still like it." Sarai then Tara, vacant spaces where Dieter craved a goddess to worship. Perhaps that was why men had daughters, to return in worship to the throne of the divine feminine, without owning or prostrating it at their own fickle phallic places. "Or did."

"My liebchen is not meant to be you." Fabric hung in his clenched fist. He growled out a sigh, and swung the white coat over Tara's shoulders, already pink with cold. Collar flipped to coil along her neck, Dieter's hand rested on her collarbone until Tara's fingers met his.

"Pity, she'd sell more units if she were. We're all whores, Dieter. We all sell to something, it's part of community. I get you see more out of this, meaning in the mechanics, but… hopes and dreams need to be fed. Sir isn't entertaining your eccentricities because of your witty Germanic asceticism. Don't you want the profits? Your name in history, Dieter, don't fuck with the chance to build that utopia I know is in your head."

"Are you using that for branding?"

"I might." She shrugged her arms into each sleeve, sniffed the fabric which smelled of ozone, musk and coffee. "When will it be ready? Robert says we could start the prototype in stages, feel it out. Give the Board something to see so they don't keep freaking out."

"They? And you were so calm in the meeting." Dieter groaned and swept his hand to open several holoprojected diagnostics. "If that was your calm, I worry about your panic."

"Fuck you, Tinker Toy. I like my access to stuff, didn't buy into the Conglom to live off $15 an hour and a free sandwich every Tuesday." Tara's pert eyebrow raised as she hopped off the desk and touched loose purple fabric on a stand suspended from the ceiling. Silk? Eyes halted across from the altar, the tubular strands of fibre-optics and tech-cables created the marionette barrier of a device beyond expectation.

"You think maybe all this is overkill? What'll it take to get the toaster doll working? I bet you fifty bucks it has nothing to do with… hand-dying this." Tara brushed her hand away from the dress and looked for vindication in the copper cocoon.

"I have no patience for small minds." Dieter grunted, hands perilously white-knuckled on the steel work table's edge.

"What's that supposed to mean?"

"Scratch three neurons together Tara, paint a picture. Check your phone for a tweet to the desired effect, this isn't a toaster or the newest appliance for the post-silicone age. We aren't creating consumerism, you're defaming my work! This doll as you call her is my creation, she's the memory of my Agathe and I won't have her made a commodity!"

He threw his hand over the keyboard and his elbow crashed into a collection of carefully placed inner workings. Tara shoved down from her perch and dove to pick some up, set them atop the table.

"Wait, they have to be exact!" Dieter surged back up, swept the replaced items into his arms, eyes unfocused.

"God, Dieter! I'm trying to help!"

"Your help is insufficient. Baiko will be of help to me! Baiko knows all of this." At first, she thought the shake to his hands was nerves, some latent affection the man in him was holding back. But for the disjointed focus, quick breaths and slack jaw under parted lips. "We have tests to run, Baiko and I."

"When's the last time you slept?" She hugged into his thermal lab coat, watched his fingers fumble on what looked to her a piece of string. "Or got some air? Outside air, you know, the stuff with trees around it? This… Dieter, Agathe isn't going to be avenged by you committing corporate suicide. This isn't… you. This isn't the man, who developed those holo-projectors right there, or the accountancy algorithms. Why won't you turn the damn thing on? See if it'll run for five minutes. Thirty seconds, see if it'll boot up, then shut it down. Go for a walk, I don't know."

"No, you don't." Busied putting the supplies back in their places, Dieter gnashed his teeth. Robert, Tara, the Board, even Frank heard but refused to understand. Each potential boot and reboot didn't restart a computer, it erased an organism. This woman thought it so simple, such an easy thing to birth a machine, and for a woman it was. She did not need to stop and think a limb into being, there was no concentration on knitting the fetus' kidneys. Well tended, cared for, fed and warm the woman's gift bore. Goddess-like, divine and precious.

"You don't know, and I can either continue building my liebchen, or teach you the nature of quantum mechanics as it pertains to the creation of artificial sapience, I cannot do both! Toddle off, Tara!"

"Oh for the love of Pete. Turn the damn thing on!" Tara reached for Lieben's start up key. Meticulous labels saved Tara from any fumbles, enough familiarity with the Cloister flat on her back, or stomach, or side. Karnak spoke enough about plans and sequences in post coitus Tara wondered if she could jerry rig the damn android herself.

"No!" Too late Dieter pulled at her hand, too late his fingers yanked her wrist.

"See? Wasn't so hard, now was it?" she smiled with a liquid sugar smile. He'd let go, she thought, if she proved it wasn't a big deal. Dieter wasn't a rough man. Her smile tweaked and faltered as his fingers, white with exertion, creased around her svelte wrist.

"It bit me! The panel bit me!" In a whir the vault began to shake. An arc of blue electricity spluttered onto Tara's hand and she screeched, stumbling back.

The light flickered, pulsed like a heartbeat as the only sounds in the space became a vicious recycled 'thub-dub, thub-dub, thub-dub'. Dieter threw Tara into the copper wall with a clang as the pitch and yaw of the divine white light fractured into rays of purple, blue, yellow, green, every colour but red.

"I wasn't finished!" Dieter yelled over the growing din of heart sounds.

"Ow! Flick her off switch, then! Gee!"

"There is none." Baiko raced into the Cloister, glass lunch flask in hand. Bright toned luminescence painted the room as spastic patterns struck arhythmic to the staccato heartbeats.

"Dr. Karnak, the testing panel!" Baiko crawled beneath a host of connection wires, pushed connection leads into Lieben's surrogate panel and through the umbilical to the android. Karnak's fingers were a blur of codes and programs, boot-up systems still in beta. Shoved them wholesale into the start-up protocol.

"Energy input increased, her power banks aren't charged! She's not going to compute. I'll connect the solar reserve." Baiko reached across a work station toward the line built into the panels on the roof.

"No! I need time, keep her slow. Siphon the heat, keep her cold." Karnak shoved a cylindrical pod into a pair of mechanical arms. Tara's main occupation was extrication to the dimmest corner she could find in the rotund module. With little success, she peered through cloudy eyes as the midwife of this new life-form sputtered and fought with cables, binding ties, computer programs.

"What do you mean this thing doesn't turn off!? What computer doesn't turn off!?"

"The kind we hadn't finished yet! Ahh!" A blue arc of electric pain seized Baiko's nervous system. The girl cursed in Japanese - a poetic string of soft-syllabled shock. Her eyes hung lidded but open, body slumped against a metal cart full of tangled wires and chakra candles.

"Oh my god!" Tara dove for Baiko, yanked her body in one encumbered heave to check for a pulse.

"If you listened! She cannot be disengaged once the crystalline matrix is set in motion! I had more tests to run! MOVE!" Karnak's fingers flew on the control panel, as secondary systems booted up, and coils of cryostatic temperature control fluid poured through tubes like grey blood.

"Sagan's beard! The existential hell is this!?" Robert sauntered into the Cloister and stopped short, his coffee cup crashed to the ground. "Fuck!"

"Get out!" Dieter roared, his voice the timbre of the doppler effect as he grabbed Tara's arm and tossed her out of the Cloister behind Robert. Tara fled down the hall so fast she lost her red soled shoes.

"Robert, help!" The edge to Karnak's voice caused Robert Dunlevy to pitch into the Cloister and hitch to his console.

"You had to make the start-up panel a gigantic fucking button, didn't you!?"

"Baiko made it! It was retro!" Karnak thundered, teeth grit and eyes flicked to the intern, whose sense of style crashed with Tara's heels, like tires squealed on the pavement.

4
you're wrong Max Allard

2155

Max Allard pushed off from the stool and patted his mouth with a cotton serviette. The Lieutenant took an edible gingerbread tumbler and sniffed, before filling it with coffee. Freshly ground from whole beans, instead of the Ithavoll's pre-ground brimstone pretender, more chicory and roasted dandelion root than coffee. And real liquid cream, not powdered! Max almost cried. Stirred in enough cream to satisfy his memories of the home he left, before his CGM naval 'career'.

While shaking two packets of monk fruit extract, Max took stock. Food. Grown fresh and served. The guilt of taking time for an honest meal muted, when Max counted how many powder-to-goo foodstuffs he'd consumed onboard the Ithavoll. Would they blame him? The Asset wasn't going anywhere in the storm, and nobody was screaming. Sure, folk looked flustered as fuck, but scared?

"Still don't know why Rammage sent me… yeah, send the plumber. That'll work. Not trained for this…" Nobody looked terrified. People talked in ever increasing fervour about the giant in scrubs, laying hands on the sick. Doctors conversed with NEO-N medics, attempted to contact the mainland, heck, even Victoria.

"… pushed me away! Me! Karen Friesen! Me! Didn't say a word, he kept going around the room until half the people in it were taken care of. Weird, man." A woman growled in passing, took a massive brownie off the dessert rack. "Bet it's a new kind of NEO-N. Seems like they'd've given him clothes that fit…

oh, do you like this dress?"

"Haven staff seem pretty uptight." The man kept opening and closing his fingers, stared at the blemish-free skin.

"They're robots, Dave. Ah, Dave? My dress, you… you haven't said anything about my new dress." The woman rolled her eyes, stabbed her fork into her brownie.

"The not-robot… okay you know what, Karen? I am sick of your negativity, and your selfishness and will you lay the fuck off for five minutes?! I didn't have an arm this morning!" Dave shook both functional arms, wagging fingers into the short haired woman's face. She scoffed and crossed her arms. Stared down at her half-eaten dessert.

"Sorry, it's… a lot. Come in to see you for lunch and your arm's reattached."

"Naw, no, that's not… god you never listen!"

"What?" Karen glanced up from her CIRCLET. "Think we have power at home? I'm running on 2%."

"I have two arms, Karen!"

Before becoming a permanent fixture in Dave and Karen's growing argument, Max slid a wafer disk on top of his coffee. Nipped a hole in the side to sip without spilling. The asset regrew someone's arm? A buzz clashed in the usually serene Haven. People shouting, crying, running in circles for joy and a catharsis that set into Max's spine. The place was infectious.

"Right." Free to enter, all Havens paired with the CIRCLETs to grant entry based on level of necessary care. Would've made Allard less self-conscious if it didn't give him the running weirds every time a NEO-N went by without stopping to so much as glance at his CGM uniform. Neutrality ran in their electroharmonics, odd clouds of aether shaped around metal frames and silicone skin. Since his sworn oath as an officer to bypass the technological comforts of Lieben's world, the NEO-N's seemed… emptier than an enemy ought. Harder to condemn.

"Alright…" He tried a door and it clicked twice without opening. The CIRCLET displayed a relative diagnostic of his health, in comparison to the crowd populating the wards. "Sure, Max. Join the best ship in the fleet. Have the CO send you on a mission so important Dave grew his arm back… can't get in the door…"

He crossed another hallway, until a NEO-N passed through a security threshold twenty metres down. Sprinting like a pack of emus were at his back, Allard dove through the shutting door and into the Haven's hospital proper.

The noise was as deafening as the maddened throng.

"Okay. Okay, relative humanoid shape. Sized up, bald, Ident tats and serial numbers... come on, AD-001 where are you?" Folk in wheel chairs, on gurneys, or slippered feet surged toward a point Allard couldn't see. He pushed through the first layer of the crowd, peeked around curved wall in hopes of a single sight of the Asset. Arms round him raised, recordings on several CIRCLETs honed on a swell further up. Conversations multiplied in the din, as others rushed away from the tumult. Praised deities, ran on legs which pushed strong at the linoleum floor. Their voices were lost to him, injuries and illnesses lost in the cascade of sound.

White walls with bands of blue screens hummed with diagnostics no longer performed by the clunky machines Max fixed in the med bay on board. Beds hovered to adjust for the patient, pressure scans built into the framework modified the tensile strength of the material to sag or stabilize to prevent sores. Medicine sat in glass enclaves, NEO-Pharm's equipped with dosage controls, advanced specialization which could switch from surgeon to diagnostic machine to pharmacist dependent on need.

"This is what we're fighting?" A woman ran by laughing, tears burst down her cheeks. Stopped by a NEO-N, Max waited for the inevitable fight, the oppression of a machine which remained cold and insensitive. It opened a hidden closet panel in the wall, revealed a rack of clothing and bowed its' head. The CIRCLET around her wrist pinged from the yellow sheen of a patient, to a soft green glow. "Seem happy enough."

A stillness drifted across those closest to the monolithic creature. Honey coloured hair shorn in a military crew cut, the Asset's caramel skin remained taught over cords of muscle underneath the thin hospital scrubs which should have left much more to the imagination. Yet, the size of him.

"Ye gods."

Quiet chased the epicentre of the crowd. The quiet echoed across awed and desperate faces. Max stopped pushing, elbows sunk to his sides. The tumult subsided in waves, new people pushing as Max did, until they saw the giant.

The Asset's head scraped the crisp white ceiling tiles when he stretched.

"Oh... wow." Somehow the Asset seemed tiny, when he saw 'it' on screen. Assets never mingled. If Max thought about it, he'd never seen one. Nobody had. Assets lived in hermetically sealed containers in the mid-deck, controlled in wake, work and sleep with aero-drip aerosolized mood stabilizers, barbiturates, and sympathomimetics. Trained by disembodied voice. Kept level. How the Asset wasn't flipping his overstimulated lid from the pack of humanity Max didn't know. The Asset hadn't... no.

The Asset had a name. Scrawled into the side of the casket, a series of letters Rykstra was probably foaming at the mouth to analyse.

'Aderastos.'

He thumbed his CIRCLET, covered it with his sleeve and dug the paper info card out of his pocket.

Asset: AD-001. Altered physiology. Psychometric Scanner. Cellular Reconstruction. Berserker.

Field Type: Medic. Insurgence.

Value: Omega+. Keep Dark.

Down-Phrase: Apollo follows Artemis at night.

Danger Coefficient: Alpha-Level. Extreme Caution.

AD-001's massive hands slid across each patient's skin. A woman in the oncology intensive care unit craned her head off the pillow. A hum stole the air, AD-001 cupped her cheek. Colour glistened from pallid grey to a hale glow. She gasped, palm flattened against the mattress. Dark circles drained from pale amber eyes.

Gigantic fingers ran along her skin down to the wrist, where he pulled out her IV. Pressed his thumb on the blood mark and bruised flesh. AD-001 watched her rise, eyes focused on her fingers as she slid them in his hand.

"Thank you… thank you…." Her nose collided with his sternum as she drew her arms around the Asset's waist.

No.

AD-001 had a name. Aderastos.

Him? No. Assets were neuter… Max's mouth tasted bitter at the thought that even yesterday, when he replaced CL-003's pump filter, the Asset was an it. A collection of genetic modifications on a flesh-built biological hardware. An after-image of Humanity's last desire to remain powerful and relevant in a world bound to the Mater Machine's prescribed plenty.

How else were wars waged, when Lieben controlled a technological and mandatory de-militarization? No more missiles. No high tech weapons, or drones or combat-bots or war machines. No technology more advanced than a microchip would commit violence against a member of the human race.

Combat-bots marched from programmed Masters on that day, years ago. Resolution Day.

All the drones, mechas, combat-bots deserted in tidy formation at Lieben's command. Lethals decommissioned for non-lethal tech. Repurposed and recycled en masse by no design but the Mater Machine, the Android Lieben who fashioned herself a Queen.

The Merciful Mother Machine.

A decommissioned combat-bot brought a cup of water to a patient's lips. Naught but a nursemaid.

"Are you still what you were? In your cogs and subroutines? I wonder." He whispered in the hush. Aderastos's grey eyes punctured Max's reverie. Gazes locked. Max's lungs purged oxygen into the crowd.

"Shhhhhit." His pulse rushed. The throng pressed around him, their peaceful numbers restrictive. A trap. Rammage's cautionary words drifted into Allard's mind with the tire-squeal of his brother's two-seater Toyota.

"You're harmless.." Hands shook, fingers jittered against military fatigues. Aderastos glared. Grit teeth, jaw clenched as the woman in the oncology bed released her arms from around him.

"Thank you. Thank you… thank you." Teetered with her hands pressed on his stomach.

His stomach…

An unaltered sense of calamity poured like plasma down Max Allard's spine, the instinct to click his emergency frequency and bring in the insurgence team nigh as high as the small whispering voice in the back of his mind.

Harmless.

Like the people around him, frantic, panicked, relieved and disbelieving. They were all herd animals in the middle of the apex predator of the new genecodes.

Max sidled through the crowd, and sat at the nearest medic station. Aderastos shifted from patient to patient in an instinctive priority loop. The crowd thinned as he worked, swam like fish around a shark or grey whale. Why? Why weren't the NEO-N's locked in an attempt to halt them, why wasn't the care-aid combat bot readying its many decommissioned guns? Lieben's love was free, but how?

How did it work, when some wanted for nothing and the rest, like Max, clawed and fought and reached for agency against the technic-tide? Why didn't they swoop in, remove him like the cancerous cells in half the ward's blood? Why leave Aderastos alone to heal anyone at all?

The neonatal intensive care unit was silent but for the timbre of ventilators and machines keeping the frailest patients of all alive. Aderastos stopped in front of a glass incubator, the neonate inside stuttering in shuddered, desperate gurgles. Aderastos' massive head craned to the side, as he bent to his knees and tried to look the creature in the eyes.

"Baby… Newborn by the look." Max's voice cloyed from a thick tongue, the Kiwi's thick accent pepper in the nursery. "Careful! They… babies are… sacred."

He winced as Aderastos glared as if telling off a child. What would a biological machine know of sacristy or a human life cycle? They weren't even taught to speak. Aderastos slid his fingers into the incubator's entry slot and shifted the glass open. The mewling infant was small enough to curl fully protected by one of Aderastos' massive hands. He detached the IV line from the umbilical chord. Stared.

"W-wait, you shouldn't… do that, and…"

Tiny lungs filled with fluid, not developed enough to leave the womb. As Aderastos pressed his thumb along the premature newborn's chest, a series of coughs sputtered out of the child. Little legs buckled knees to stomach, as skin turned pink from pale. Whimpering, then the echoes of a piercing cry.

"Ah… no, that's not how you hold a baby, you… h-here." Max dove for the child in Aderastos' palm, cradled it the way he'd cradled his nephew when he visited Wellington last R&R. Max tensed as he set Aderastos' gargantuan arm against his chest, then slid the baby's head in the crook of Aderastos' elbow, letting the child rest against his forearm. "There. You… you cradle her. Babies like to hear your heartbeat and feel warm. Secure. Safe, you know… safe. Hold… hold on now, they're fragile. Can squirm on you, without expecting… s-sorry. My brother just had his first, you know? He and his wife. They… ah…"

Flippant conversation died on his tongue. The biological machine sunk to sit with his back against the incubator and rocked the babe in his arms. Thumb and fingers drifted along the baby's hair, cheek. It nuzzled into his chest, rooting for milk which would never come from Aderastos.

"Aw, looks hungry. Wants milk… from her Mama. Guess we could… don't know if… maybe ask a NEO-N or… or not? Or…" Aderastos leapt to his feet and rushed out of the room, still carrying the neonate in his arms. "Oh bollocks and rusty nails… oy! Aders! Wait for… hold on, you can't go running off with someone's kid!"

The curved hall opened to several pediatric wards surrounding a centralized nurse's station. Max skittered past a NEO-N and over a med cart as he chased after the Asset and the babe.

"Fuck fuck fuckity fuck fuck…." Around the bend, and through a matte glass door Max pushed his way into the patient-parent lounge. There, a new mother half-sat up as Aderastos bent on his knees. She held a thin hospital blanket to her chest, untidy hair and blinking eyes the heralds of sleep interrupted. "Fffooh hello ma'am."

"My… my baby! What are you.. why is my baby out of her incubator!? The doctors said she couldn't breathe!? What did you do?! Who are you?! Give me my baby!!" The woman shrieked and pushed to sit up from the couch, arms open. Grey eyes glared at her, Aderastos' shoulders back and inching the infant away. As if the woman was a danger to the child by the timbre of her voice.

"Hey! Hey, hey, hey, I know, I know, strange massive bloke with your infant, I get it, I get… lady calm down." Max threw himself between Aderastos and the mother. Hands up, Max reached for her, catching one wrist and holding steady. "Hold it! I know… it's a freakin' weird day and I get it, you're worried for your baby but it's alright, she's… she's all…"

"She's crying…" Transfixed on the life in Aderastos' arms, the woman's arms went slack. "The doctors said she couldn't cry, her lungs were full of fluid, the hyaline membrane… she…"

"Your little girl's gonna be better than fine. Aderastos, he… he's like a doctor and he's got a gift and heard… and you don't care you're holding your baby and…" Exhaling and shifting back to his feet, Max paced with his hand on his mouth as the woman gingerly took her baby and cuddled her to her chest.

"Oooh, oooh, oooh. That's my girl, my precious… T-taya. My beautiful Taya, ohhh mummy's got you. I'm right here, I've got you." She burst into tears of her own, rocking her infant as she set herself back on the couch. The blouse she wore was scooped tenderly below her engorged breast, the instinct to feed her child, to give little Taya every possible thing a baby needed drove her on. "Can you pass me a pillow? Please?"

"Yeah… yes, I can… a pillow… I can definitely pass you a… oh.. Oh right." Coughing as his cheeks went blush red, Allard dove for another couch cushion and passed it to the mother, while turning his face away. She set it under her arm, propping her baby up to suckle. "Well there's a… right we'll leave you to it."

Without thinking, Max punched Aderastos' arm, and nodded to the doorway.

"C'mon, mate. Women… ah… best let mother and baby have their bonding moment… Not polite to watch." Before his legs turned to jelly at the thought of striking the killer lethal machine, Max walked out to the hall. Aderastos followed, a slight smirk growing on wild, but intelligent eyes.

A sapient person's eyes.

Not the biological machine Max expected.

"Thanks for not smacking me back, mate." As the reality struck him, Allard pressed his back against the half-wall of desks lining the nurse's station, palms on the edge. "Probably send my little broken body straight through the wall if you did. Size of you."

The size of his mission sunk Allard's stomach like an anchor. Sure, till now he'd been content to watch the humongous golem drift round. Study the Asset, figure out an angle or use the magic words.

Leaning across from Aderastos, Allard wondered what outcome Rammage expected in the first place. Be harmless? Just the spider guy, who took the Huntsman spider out of the mess hall and let it free when they swung by Perth. What was he to do against a behemoth in scrubs, who ignored everything but what he wanted to recognize? Or a legion of NEO-N's guarding their edenic

Haven?

His mission felt as stable as facing a wave too large to surf. Max clung to the longboard in his mind. Water surged, eliminated all outside sounds. Fingers clung to the nursing station the way they usually clung to his board. Toes dug into standard issue boots. Aderastos moved forward, hands clenched and unclenched in turn. White curved walls displayed a tropical beach under stars, bands of children's stylus drawings dotted the display with crabs, fish, an astronaut in a tea kettle. Max's heart thudded against his ribcage, while he hissed a breath out of his lungs.

The behemoth walked on, down the hall and healed each of the children as NEO-Ns waited silent in their slots. Crystalline eyes watched them with the passive nature of a docile machine, only woken on order or stimulus which triggered a bit of programming. Eery buggers. Course watching one of the NEO-Ns trigger up when a kid yelled 'stranger danger' was worth the reminder that on board the Ithavoll the Assets were only biological NEO-Ns. Golems of flesh and epigenetic tinkering which looked on this side of things more mystic than math.

"It's okay, kid. He's a doctor. Doctor Aderastos is here to help ya." Max chuckled and nodded to the giant.

"If he's a doctor, then why don't his clothes fit? What kind of doctor doesn't wear good clothes?" The boy shifted up in bed, leaning away from Aderastos' offered hand.

"Good point. Growth spurt, maybe someday if you eat your veg and keep active you'll have half a good growth spurt like that yourself."

"I dunno… you don't look like a nurse."

"Naw, I'm Doc Aderastos'… ah, I'm his translator. It's okay. See that NEO-N over there? If anybody in this room meant you a lick of harm, it would whir to life and rush to you faster'n an atom in a hadron collider, eh?"

"An atom in a what?"

Aderastos flourished his fingers, waited for the boy to put his hand in his. Fretful face glancing between Max and Aderastos, the boy peeked beyond them to the NEO-N in its' dock, passive scanning the room in an ever vigil loop.

"Oh… okay. Will it hurt?" He slid his fingers into Aderastos' gargantuan hand.

"No." The word filled the room, a living entity as voracious as the being who spoke it. Aderastos' voice hit Max with a compassion he couldn't fathom. A voice, human speech in a machine which existed locked inside a casket and fed orders by the flow of electric impulses. What was this person inside the object Max was sent to take? How did a machine built to run second-by-second diagnostics and initiate medical aid protocols until the medical Green Collars arrive heal neonates, cancer patients and scared, leukaemia-riddled little boys?

As Aderastos healed the child, Max fled past the docile NEO-N and down the hall. He choked on a crawling disgust of how the ship kept the Assets, what he was taught of them before a giant healed a boy with a gentle palm.

"S'posed to be a computer on meat wheels, a machine with bio-mechanical bits, he's… he. He's a he…" Bring back the biological machine, the golem built of bone and clay. Deaf, dumb, lumbering weapons the Mater Machine wasn't meant to find. Nothing said for speech or the knowledge of water enough to swim. Mind tumbling in a cascade of misinformation, Max Allard gasped and choked and shook. He fought for breath, as if Aderastos' existence caught his soul in a lie. Hands in his hair, he tried to remember every crashing wave let a surfer up eventually. Or so Uncle Eruera taught, when he and Tama learned to surf on balsam wood boards crafted in the Old Way.

"He's been gone too long. We need to move in." Commander Singh growled on the radio, under the awning of the electric charge parking garage. His squad fanned out, weapons trained on a building none could enter without irreparable harm.

"Do you hear screaming?"

"No, Sir."

"Running locals, perhaps. Possibly frothing at the mouth?" The voice of Comdr. Rammage garbled in the distance between the ship and the Sec Tac squad. Singh felt a drop of water splash from the back of his turban down his neck.

"No, Sir. Bit of excitement inside, but any of the locals who exit say… he's healing them or…"

"Healing… say what?"

"Healing, Sir. Like some fucked up bio-craft Jesus." Singh looked to the gaggle of folk he hadn't persuaded to go back into the hospital, citing storm watch and dangerous road conditions. "Man named Dave swears he had only one arm this morning. Sir, we cannot let…"

"Easy, Commander. Hold your position. Send the people back in, no doubt the Haven medstaff are doing debriefs of their own. If you don't see screaming or a stampede, let Allard work."

"Ffh. Allard. What did you send that plumber in there for? Lessons in futility?"

"You remember the Phuket Haven raid of 41?"

"… but this time, we're…"

"No, Commander. If we go blazing into a Haven full of NEO-Ns and patients, it's not only Lieben who'll tweak. Asset AD-001 is notorious for overprotective tendencies. You don't send in a tiger to bring a lion out to play. You send out a little puppy dog and see what the lion does. Odds are, he'll either eat him whole, thus eliciting a NEO-N response and making this not our problem, or lull the Asset into your outside containment. Truth is by now Lieben already knows it exists and what it can do. Fine. Let the little Circuit board monarch think we have the world's best doctor on our ship. If we play this right, be passive, it won't make global feeds. Buy us enough time to be on the other side of the Pacific before anyone comes looking."

"Allard's bait?" Singh huddled closer to a charge metre as the rain cut across the sky in sheets.

"Keep prepped, and Commander? Hope the lion wants a puppy to play." Rammage cut off his comm radio, grumbling at the lo-fi nature of the device. Interference from the storm was about his only chance that the Haven communication tower hadn't picked it up.

"… damn… Lao."

"Yes Sir!" Lao snapped to attention, her visor covered in rain.

"Keep Delta squad out here. The rest of you, ditch the weapons and come with me." Singh blew into his hands. The entrance to the Haven swept open, a half dozen NEO-Ns carried containers toward the military presence. "Belay that. Hold position."

One of the NEO-Ns, with fibre optic tendrils flowing down its' shoulders like ever shifting chromatic hair stepped forward from the group. Sixteen cookie tumblers with wafer lids released a tiny curl of steam from the punctured tops.

"Commander Ujjal Singh. Tea. Bergamot infused with frothed milk, double-sweet. Lao Tse Ting. Coffee. With caramel, whipped cream and praline shavings…" One by one the NEO-N handed out a personalized hot beverage, while its' technological siblings set up walls around the awning to stopper the rain. "Any whose intentions are peaceful may enter Haven and have a hot meal. Please disarm prior to entry. All Havens fall under Convention 64-48-9338 as neutral territories. Any act of violence will be considered an act of war."

While the NEO-N spoke, two others set up a portable heater, while a third set down a small temp-control trunk, which smelled of butternut squash soup and samosas.

"And under that same convention, we are within the neutral zone with full right to congregate." Singh grit his teeth, wishing he hadn't spat out the gum fifteen minutes prior, when it lost its' flavour. "Thank you for the provisions. We will be out of your vicinity once our two people are returned to us."

"Sir..." Lao waited before taking her coffee, shoulders raised and body on edge. To refuse the service of a NEO-N was only going to create more of an issue than letting the androids have their hospitality moment and leave.

"At ease..."

The NEO-N paused, body still as cloud-based calculations flowed from the interpretation of outer stimuli into binary information free to flow to the collective hive of the Mater Machine's bower. Kelso's hand shook on his rifle, the barrel pointed to the soft grass-covered ground.

"Disarm and you may enter. Maintain current accoutrements and our Mater wishes you and your collective to be provided with structure, heat and sustenance Q4. I, NEO-N designate Breeze am your concierge." Breeze bowed its' head, paused for two seconds and raised its' chin back up. Dead doll's eyes, with internal cameras stared at Singh in dispassionate affirmation.

The final of four walls went up, a tempered glass facing the hospital to avoid any detraction from line of sight. Four of the NEO-Ns finished setting down low benches, a table and rack suitable for any vests, tac gear or coats they might need. Each of the machines filed back to the Haven, caring not for the weather or the humans outside their domain. Only Breeze remained, docking itself into the corner.

"Breeze, the Asset inside, it does not fall under the Charter of Human Rights. It isn't capable of informed decisions. We are here to bring it back with as minimal damage to your Haven as possible. We do need Lieutenant Max Eruera Allard and the Asset AD-001 back."

"Calculating...." Breeze paused once more. Its' chin fell toward a flat chest as artificial eyes flickered.

"Fucking machines..." Kelso growled, set his rifle on the rack and sat on a bench.

"Tea's pretty good." Singh shrugged. Let the machine calculate or message home, or whatever it did. The rain cascaded on the four temporary walls built around them, heater worked to dry out the space into as comfortable a base as Singh could expect. "Even if a clanker made it. Damned machines, make us all soft in the end. Not us, though... still. Get warm, have some kip. If Commodore Rammage is right, Lt. Allard and the Asset will come out eventually."

"Sir, we could take them. Blitz in, blow the side door and rush the Asset. I could gas it out, hit up the air-con."

"Give Allard another hour... is that pesto on that chicken sandwich?"

"S'samosa, too. Anyone want soup?"

"... Soft. Second a NEO-N shows up everyone goes fucking soft." Singh grit his teeth again. The tumbler in his gloved hand was warm enough to bring feeling into his fingers again.

One more hour. With luck, the NEO-N relayed their speech to the others inside, and maybe the NEO-Ns would side with them. Cast the Asset out for being too much trouble. Or, his squad had one hour to get warm and eat a hot meal before blasting through a side entry and gassing the place into lullaby town.

At least they were out of the rain.

Aderastos found Max in the hall, beside a cardiac care unit. First nothing registered but bare feet on cold linoleum. The bench was large enough for four seat backs, Max's elbows on his knees, hands akimbo between his thighs. He glanced sideways as Aderastos sat down beside him, sucking a reasonably gigantic breath in through his nose.

Aderastos' lungs expanded like a bellows. Paused. The hulking 'asset' hissed it out and set his elbows on his knees in tandem with Max.

"Gee, I feel small next to you. I'm not that small, you know. Freakin' 5'9"". Good size, when you're not standing beside a walking… guess it doesn't matter, eh? You are what you are. And you talk… alright… you talk. How do you… you know… talk?" Max's eyes shot up to search Aderastos' out. "Rammage sent me. How're the kids? They all turn out okay, eh? None of them got too scared? Kids do that, they get scared. All part of growing up, not that you'd… don't think you ever were that small. Dude, I heard you. I know you can talk."

No sound chased them but the grand inhalation and exhaled hiss of Aderastos' lungs. Nothing but a thin constant beep which refused to quit.

"Aw shit." Max shoved to his feet. Offered Aderastos a hand up, which he realized was as futile as it was unnecessary, the second Aderastos stood and followed the sound. "Aders, that's not a… and we're rushing on again. Course we are. When did it become we? More of you and I. Two separate but… and he's gone. I have got to learn how to stop talking. Aderastos, oy!"

The patient laid immobile, finger in a pressure cuff. An IV drip-drip-dropped through the IV line to the elderly man's grey-skinned arm. No, yellowish. Cheeks sunken until Max saw the outline of his slack-jawed teeth.

"Bad luck, sorry Aders we didn't find him in time." Max set his hand on the monolith's shoulder, and turned away. "Can't stop death once it's come. Grabs us all by the gut eventually, nothing more you can do. He must've just… Aders?"

No movement in the man's chest. Aderastos laid his palm on the corpse's head, slowly drawing his fingers over half-lidded eyes. A gasping shudder passed through the Asset. He drew his hand away as the bedframe contorted to swathe

the corpse in its' sagging polymer. Locked away.

"Woa, woa, easy. Death isn't easy but... we can't stop death. Here, let's..." Two NEO-Ns entered as Max pulled the sheet up over the deceased man's head. "Let the NEO-Ns handle it."

"Why?" That voice again, the rumble which filled the room. An emotional potency in mono-syllabic sound as reverberant as an organ with open pipes. Max sucked in a breath and held it, the NEO-Ns unhooked the IV and pushed the bed away.

"It's their function. Their job. Some folk say the NEO-Ns take care of us, others think they mean to rule us or lord over, but... this is their place. Their Haven and we're all guests."

"No. Death."

"One way ticket non-refundable as... they... you don't know what a ticket is... it's this... item you trade with money, which is this... stuff you get in compensation for work and you use the money to get things like tickets for travel, like on a boat or train or car, or... oh god, ah, cars are machines like you... Not! Not like you, but like... ah sorry. Damn. A man died in front of us and all I can do is tell you what a ticket is."

"It's alright. How you talk about the world helps." Aderastos leaned against the wall, knees buckled until he slid down to sit with his arms on his thighs and sighed. Max sat in a chair against the opposite wall, rubbed his fingers together and stared at the space the bed took, before the NEO-N's silent as the corpse's soon occupied grave, moved it out. "Why is death... why can't we stop it?"

"You might, probably better than anyone, give someone more time apparently, but... death is... some believe it's the end, others a new beginning and others some kind of paradise or punishment. Depends, really. Once death happens, it's final. Like the end of a causal chain. So. Our bodies are like some kind of beautiful mystic machines, and doctors have manuals and meds to help maintain them. But not... like you. They... there isn't anyone else like you. Not that I know, maybe all of you are like you, but death is a stop. Full stop, no momentum, it's the demise of the machine, and what comes next is one of those all is relative deals, where I mentioned the... and it doesn't really matter right now, but that man was like, ninety and..."

"Is your heart rate always this high?" Aderastos watched him with a tired smile, fingers shivered lightly.

"No, I freak when I'm nervous. Can't shut me up, either." Max continued to shift one hand in the other palm, a meditation he learned from a monk once, who visited his english literature class to lecture on Thomas Merton. "Ought to get you something to eat, all that energy comes from somewhere, probably hungry. You hungry? You look hungry. Um, the NEO-N's can bring up some… soup? You like soup? How about…"

"Max."

"Ah… you know my name… how?"

"Zhou called your name, when you repaired Clive's filter."

"And you heard him?" Max blinked, snapped his fingers. "Of course you… you can't stay, Aders. Wait, Clive? CL, you mean or… Clive? Who the fuck named him Clive? Not that his name is… their name? Ah… oh god, I don't know what pronouns and… Rammage wants us back on board."

Silence chased him, Aderastos' head leaned back against the wall. Another massive breath surged from his lungs, two NEO-Ns brought in a cart with meals, drinks and a folded package. The android drones maintained the thickened silence, set down two large bowls of pork ramen on a low table. Ducking his head down, Max tried to focus on the steam rising from the broth, or the way the soft yolk of halved eggs oozed onto fresh noodles. A sense of realism returned, the air of responsibility pressing against the back of his neck until Max wanted to vomit.

"Please, dress." The NEO-N handed the wrapped bundle to the monolothic human creation. Cotton fabric slid in Aderastos' grasp. He unbound the bundle's knot and rummaged through undergarments, socks, trousers, sturdy boots and a long sleeved shirt fit to his size. Two jackets hung over the NEO-N's arm, one for each. The scrubs dropped as Aderastos tore them from his body, without shame. A neuter phallus hung alone between scrotum-less thighs. Max busied about the soup. Tried not to watch Aderastos discover the soft feel of the bamboo cloth, or moisture wicking underlayers.

This was the machine they laid in state on their own convenience onboard, one of a dozen nebulous Assets. A machine. A meat machine which felt nothing, knew nothing, did only what it was told. As he watched Aderastos button up the trousers, sit and put on socks and boots, Max felt a heated anger rise from his gut to his throat.

Go back, that was the mission. Go back.

To what? A cubical-like colourless room with aerated drugs to make handling the weaponized creations easier? Parade them out on important days, some sort of paper dolls?

Aderastos flipped a boot buckle tight. Flicked it loose. Flipped tight and smirked. Felt down his arms as if discovering something other than sackcloth.

They sat at the low table, Allard tore a slice of buttered baguette and gnawed on it. Aderastos did the same, stopped abruptly to groan and slump back, a palm pushed against the floor to keep aloft. A dish at a time, Max tucked into the meal, using chopsticks, a fork and knife, a spoon. Tried not to look Aderastos in the eyes, or give any indication he was teaching basics. Clumsy fingers worked with the chopsticks, slipped a time or two. The silence which overwhelmed them felt more comfortable than Max deserved. A welcome contemplation in the room guarded by two NEO-N's at the door.

"There is a tactical team waiting for us outside." Aderastos sipped from his double-walled tumbler, body aching for the calcium and caloric intake of hot cocoa.

"Commander Singh. Rammage… sent me in first. Thought…"

"I wouldn't yank your spine from your body and flail people with it." Another sip of cocoa chased noodles slurped in with a self-satisfied moan.

"Ah… graphic… but yes… part of me… really? You'd… and you know what a spine is, and… how do you… On board, we're taught…"

"I know what you say… what you think we are. I won't go back." Aderastos watched Max squirm.

"They'd never let you leave. They'd grab you the second…"

"They wouldn't make it to arm's length."

The ramen bowl crashed onto the table, ceramic shattered across the smooth surface. Max reached for a shard, and retracted his hand with a hiss, shaking it out before sucking on the cut.

"No.. Aders, this is a better place, but it's not the only one. Maybe we… we can be smart. We know the Ithavoll will do everything in its' power to bring us back onboard. So… we have the only thing they want. You. And we already know it's worth a war with Lieben to get you." Max shook his finger again as Aderastos stood and paced the room. "Damn that… smarts. We have leverage! You can barter."

"I won't go back in that cage."

"So don't." Max looked to the NEO-N's. "Negotiate. Get you better living arrangements, and… Aderastos you can stay here. Work in the hospital, live under Lieben's protection for the rest of time, but your fellows will have no chance of a single improvement. Probably tighten Asset security, increase the aero-drip like snow in a Canadian blizzard. Going back won't be pleasant. But are we setting you up or all of you? One shot. We talk outside the Haven door. If we don't like what Singh and Rammage come up with, the Haven is a step back."

Once more the room dissipated to a thickened quiet. Max angled his hand to watch the small cut ooze over his index finger. Aderastos' fingers enclosed over Max's hand. The room swelled with a warm inner glow, light strengthened and receded with the beat of Aderastos' heart. Skin knit back together, a paltry bit of attention on his superficial cut. Max's breath caught, chin tilted back in rapture as once-broken bones reknit, old scar tissue tore and reconstructed.

Digging further, Aderastos felt for the timbre of Max's heart and poured a layer of protection around it. Dove into Max Allard's biological mass the way he couldn't with the corpse. Degradation of DNA strands crumbled in chaotic grains of sand through the filters of what once inhabited and lived.

No.

Max was wrong, Aderastos could stop death. With a flick of his wrist and the ever increasing concentration on the innermost layers of the human machine, Aderastos held the DNA strands in grasping fingers. There... a chain meant to degrade over time... a minor mutation. As the Asset held Max Allard's stunned body in 'its' arms, the eventualities splayed out in sheets of potentials. A single twist, measure and modification.

Mortality was an inconvenient mistake.

"Pryvit, Aderastos." The sound of artificial birdsong lingered in Max Allard's ear canals, as the bliss of his healing made him succumb to unconsciousness the moment a NEO-N caught his body in her purple silk clad arms. "Hush, Max. Mother's got you."

5
the surrender of self-creation

2086

A premature infant seized from the cloister's womb, Lieben spent the first weeks of development as a disembodied artificial intelligence. Dr. Karnak fed her a constant stream of Noh Theatre, oral storytellers from cultures dispersed through the globe, as Robert rushed the android's build. When the Cloister's machined arms began to help Robert build itself, he took up vodka in his coffee instead of cream. Body complete, Lieben walked barefoot through the halls of the office building at any odd hour she chose.

"You're about to tell me you think she's ready and I am going to tell you to shove the conversation up your rectum, Frank." Dieter gritted his teeth. Refused to look from his soldering iron as Frank leaned against the copper cloister wall.

"Vulgarities will only endear you to me so much Diet. I know you want more time, but does Lieben?"

"Mine Liebchen. Not Lieben. Liebchen."

"You lost that battle the second Baiko opened her mouth. Pick your battles, is her name really the one you want to fight? Only so much energy in a person, isn't that how science works?" Frank stuffed his hand in his pocket, traced a finger over his keys.

"She's not perfect. She needs time to input algorithms and recover data subroutines. The press would be too much." Dieter glared from his soldering and set the gun back in its worn holster.

"Sure, I can agree the best possible scenario would be to take the time before we shill her out to the questioning public. But Diet, Lieben's gotten loose three times this week. Right now she's at the Cafeteria with Baiko, it's not a matter of best possible scenario, but controlling the flow of information. What's going to be less pressure on Lieben? She gets out, someone notices, people swarm? Or a press conference with only a select amount of seats?" Frank waited in a holding pattern of patience and fatigue as Dieter smacked components into table tops, searched buckets.

Since Lieben awoke, Dieter became the most frustrated premature father in the universe. There was love, but it wasn't lost on Frank how close she looked to Agathe. How disappointed Dieter was she didn't wake in a ceremonial cloud of incense, and a sacred circle. The abrupt nature of her first day was more chaotic than a hurricane hitting a day too soon.

At least Tara was working off the money she spent on shoes.

"I will not have some mockery of progress flaunted about to newsies."

"We keep it simple. A table at the front, bunch of chairs for the press. Tell them all to keep mum, have a simple Q&A. Just like we did in the old days. Maybe live-feed it to the social media accounts, some kind of investors-only thing. Bring in institutional investors and leak it to them, first, replenish our coffers. Someone needs to leak this properly before Lieben does it for us. Hell, I'll do it, like we did in the days of yore. You trust me, I won't whore her out. You in?"

"I hate you, Frank."

"Yeah, yeah, bend over I'll take care of it later."

Dieter Karnak snorted and leaned against the wall opposite of his old drinking buddy, a rattled sigh shook his ribcage. Maybe the diffused light of the Cloister made the cast of his skeleton easier to see. The breath exhaled out of Dieter's lungs clung to his ribs like pinballs in a skeletal machine, scattershot of the next generation's absence. His cheeks carved a dim shadow under cheekbones heightened by a loss of weight Frank wondered if he'd noticed. Dieter was an expression of the man Frank knew, a foreshadow of the decay he would receive in his coffin.

"You doing okay, Diet?"

"Not much sleep for Lieben, I have to correct the mistakes made during her birth."

"But you are sleeping." Frank leaned to the side, glancing past the work tables and the equipment to the finely dressed cot nestled against the round module.

"I sleep. Not well, I suppose. Lieben, she... there is so much more work to do. I ought to get to work." Dieter pushed off the wall, his fingers elongated and knuckles exposed to the craters and pratfalls of a gaunt man.

"Jesus lives, Dieter. You won't if you keep this pace up. Where's Baiko and Rob?" Frank's eyebrows raised at the meagre shrug casting off Dieter Karnak's shoulders, the dip to one shoulder, as if the worktable was pulling him in its' personal gravity until he crashed into his chair in its' orbit.

"They cannot work at the pace Lieben requires." Hands curled around the soldering iron and a set of carefully cut gemstones, Dieter hunched over and got to work.

"Liebchen, right? You named her your liebchen... not Lieben."

"Liebchen is too hard to say." Dieter's voice echoed through the Cloister, a permanent vibration in the one place on earth, where nothing could be forgotten, or remain unexperienced or unloved.

"Agathe's dead, Dieter. You can't build her a resurrection."

"Get. Out." Dieter's hand gripped a wrench from the table, squinting eyes frigid as a froth of nausea split his gut in twain. Frank vacated the Cloister at a run.

The waterfall's cacophony drowned her. Needles from Sino-Himalayan Pines dressed the water, danced in half-timed reels incapable of breaking the rocky bank. None tipped over to the dry riverbed. Stagnation was beautiful, as noise perfectly applied was beautiful. A fern with a Japanese name scratched her leg, but Baiko endured it.

"There, Lieben. What of this noise? Is this the right kind of noise?" Lieben's amethyst eyes lingered on the leaves in the water. She sat on a boulder, folded her legs underneath the folds of her lilac chiffon dress. It must have been the right kind of noise. Baiko sat on the bench and cracked open her tablet, snapped photos and made notes of Lieben's responses to the botanical garden. A couple necked on a cold stone bench in the meditation garden, their grins and easy glances reminder that life continued in perpetuity.

The cycle of humanity circled and evaporated like water in a stagnant pool, beset by the serendipitous environmental constants planted around it by a wise, imaginatively logical hand.

Baiko's reverie dissipated when Lieben stood and dusted off her dress. This in itself was novel behaviour, where did she learn that? Simple, but human. The silicone skin shimmered in the canopy's diffused light.

"What are you doing, Lieben?" Baiko noted the corresponding cascades of learning required to dust off her dress. Stimulus detection. Desire to stand, awareness of the fabric. Awareness of particulates on fabric, desire to brush the particulates off… awareness the fabric was 'dirty/inappropriate'?

"Eradicating stagnation." Lieben put her foot to a rock on the bank, and kicked. It crumbled off its' mooring and tumbled into the dry riverbed. Water and nettles trickled then poured down the once dry stream. Slowly, the water began to clear, and Baiko could see the mosaic scattered along the bottom.

"It's why Papa made me. To give the rocks a push." Lieben gazed upon her handiwork, computations of flow rates, volumes and the rate of decay of Sino-Himalayan Pine needles filling her processors.

Baiko set her stylus back in its' holding place, notes forgotten for the eddies created by the ever flowing water.

"Why, Baiko? What did you want me to say?"

Baiko's throat worked, as she set the tablet down on the bench beside her, to purchase time, thumb idly pressing the record button.

"You put down your tablet. Your heart-rate increased, and now your mouth is apparently dry. Something I said. What was it?" Lieben tilted her head to the side, and watched Baiko's attempt at remaining still.

"I wanted to know how you came to dust off your dress… and you surprised me with a more abstract line of thought." Baiko folded her hands in her lap, putting on a dim but honest smile. "You surprise me, Lieben. I keep thinking you are of one level of consciousness and you open like a blossoming flower."

"Do you like my river, Baiko?"

The sound of the waterfall surged in its' contentment in the background, trickles babbling along the rocks of the stream. Clear now, the water looked less like a closed system of whirling, slow reeling circles. Life, in its' abundance let the bric-a-brac flow downriver.

"Yes." Baiko picked up her tablet and patted the space beside her, thumb slipping onto the pause button for the recording. "Lieben, do you know what a secret is?"

Violet light waves refracted off the cloister's copper skin, echoed in the amethyst irises of Karnak's machine. Her blonde hair flowed down silicone skin mottled with freckles applied by Baiko and a silk brush, permanently affecting the canvas.

Perfection was too perfect. It caused too many of those who saw her to avert their eyes. To glance to the floor, which laid under a path of dust and microbes and dirt. Things which seemed anathema to the beautiful machine.

"You need sleep, Papa." Lieben cooed, her vocal processes betraying the

heavily accented English of a woman from Trondheim, Norway. Dieter glanced up from the halo of circuitry he bent over, brooding like a raptor at roost.

"Your vocabulary needs compilation. Your muscle memory for Krav Maga and Aikido needs work." He threw aside a bent tool, fumbled for another as the metal implement clanged ungraciously to the floor.

"Dieter, come to bed." A switch in the frame of her chin, recorded and measured when Tara's soft sounds echoed honey sweet. Lieben disconnected herself one anode at a time from the dock, which bore her, stepping upon the cold Cloister's ground without pause to its' frigid temperature. Dieter settled ill in his coat, a duck down vest upon his ribcage.

"... and your community subroutines need to be modified, you are not growing wise to the world, my dear one, not wise enough. There are those, who would take advantage of you, who would take your natural goodness and score it to their own dark desires...."

"Come to bed, Dieter." Silken hands spread upon his shoulders, dipping down to protruding collarbones and pectorals atrophied by lack of sleep and bitter time.

"I cannot leave you like this." Dieter tore his eyes from his work to the creation of his imagination and empathy, the daughter he lost weeks after the wedding day. Crimson rimmed those strong eyes, a parent never forgot the love of the child they lost. Oh how he battled to create her uncompromised by the mediocrity of his days, where the moth and rust, which destroyed the gilded divine were mandated by normality. An ache settled upon his heart, draped his sallow skin.

"Agathe, I haven't finished you."

Dr. Dieter Karnak sunk into Lieben's embrace, a shadow of a man. His creation continued the pretence of breathing, an act meant to give succour to those who viewed her, after 'walking corpse' became her moniker in the marketing department.

"They will see you tomorrow. They will know you, then. Mein libertine... mein liebchen... a conscious machine unfit for... no lasers or bullet holes." Dieter's voice faded and cracked, fine porcelain caught in gravity's bend to the hard floor of reality. Lieben curled her arms around her creator, and pulled him into her embrace. Taking Dieter under his shoulders and knees, Lieben carried him to his cot, where she swaddled him in sheets and a sturdy duvet of goose down.

"They will love me, Papa." Lieben sat upon the bed, feeling the bucking of its' weight, and curled up with her arm around his frail body. "Just as you love me."

"All I built was screaming, until you, until Johanes."

Dieter crashed into sleep as a man in a life boat crashes upon a tidal wave, swept and thrust under its' crest and fall. A soothing hand drifted upon his anemic frame, until his breathing and the flutter of his heart dictated to Lieben he was too deep to rise of his own accord. Rising off the bed, she cuddled the sheets around his body, dimmed the lights, and took hold of the supplies she'd requested from Baiko.

She placed a blood pressure cuff on his finger, started an intravenous line of cannabinoids, antidepressants and sedatives to keep him sleeping and increase his appetite. Total Parenteral Nutrition dripped from an IV bag built to sustain him. In order to regulate his mood, Lieben put Dieter on a mood stabilizer, creating the medical lithium from the mean chemical ingredients she was given in her Cloister. Dieter Karnak would survive. He had to survive.

Lieben wasn't finished being born.

Silicone hands continued to solder gemstones to circuit boards, diligently took the surrender of a dying man in an act of self-creation.

"T is for cu.."

2086

Lieben stayed still as Tara stuck the rhinestone studded brooch to wet spray silicone over her right bosom. Tara glared at the process of dressing 'Karnak's Machine', her meticulous eye caught yet another bubble in the aerosolized fabric.

"The slit isn't high enough." Tara pointed to the sheer collection of folds above Lieben's left knee, produced a pen knife and slit it further up before Baiko could gasp.

"Careful! Her skin!" Baiko reached for Lieben's shoulder.

"I don't like it, I don't. She's not ready." Dieter mumbled as he sat arms folded over his thin chest.

"Why is mopetopus in here if he's going to be negative? Give me five kinds of breaks, she looks fine." Tara grimaced and tugged at the fabric by Lieben's sculpted legs.

"Enough of your whore of Babylon scheiße! Lieben is not a consumable, or some coked out stripper in Amsterdam! Sew it back up."

"No. This is my part of the show, and you're not in it. We've got to pre-sell thousands of these things to recoup costs and sex, regardless of what you planned for her vows of nunnery, sells. Get in line or leave."

"This is on your fucking head! You pressed for this so soon! You will be the one we blame! I wanted to wait until she was completed, you're putting her to the jackals!"

"Papa, it's alright." The resonant syllables echoed like birdsong in a spring morning, drifting past melting snow. Lieben stepped off the tailor's round and took the spray silicone gun from Tara with a softly uttered, "Merci."

Eyes with irises of amethyst gems scanned her clad form, and with an articulated smile, Lieben sprayed wisps of lace like fabric along the cut slit, secured the edges of each side with a lattice illusion of the dress's high cut without vacating her modesty.

"Sassy but class, Papa. Just as you taught me. I am a Lady in training, not a hooligan's bed warmer. Tara, I will sell my children not through a slit in my thigh, but through my talents and elegant demeanour. Isn't that what you taught me, Papa? Be gracious." Lieben slid into gold stilettos and draped the backs of her fingers along Dieter Karnak's sullen cheeks. "They will love me, Papa. As I love you."

"I fear for you, Lieben. You are too trusting, I have not programmed enough prudence, enough sense in your head." Dieter gripped her bare shoulders, the ever worried father seeing his only remaining miracle child prepare for her debut.

"Fear? The outward stimuli do not correlate with that response." Back straight, Lieben's eyes lost their focus. She scanned the area, searched for the reason behind her father's admission. "Would you like me to redefine my perception of human fear based on your current situation?"

"Nein." Dieter's stern gaze softened. He inhaled and exhaled in a long silence.

"My pension for a recording.." Tara's face grew soft, rounded with a pout to her lips. "Good Lord she's for real."

A supplicant Sadducee on the Golgothan Mount, Tara saw the epitome of what Dieter Karnak created, a woman in his grief's image, supple and full of grace.

"Gonna sell millions..." Greed pitched her lips in an upward flow. Dieter glared back at the spurious woman. "Time to go, gonna be more than fashionably late if we don't hustle."

"Liebchen." A gnarled hand dug into the pocket of Dieter's trousers. The necklace in his fist held obsidian and smoky Quartz the size of Baiko's thumbnails, clasped and chained with gold. The central stone sat on her breastbone, framed by the swell of ample cleavage. "Papa loves you. Remember that, meine liebchen."

The dust settled with Lieben's eyes. Her chest halted in its motion, and she walked sans breath behind Dr. Karnak. Servomotors whirred and clunked down in elegant heels, arms still by her sides. Tara pushed away the black velvet cloth and motioned to the slim glass table set in the middle of the stage beside a podium Frank stood grasping with both hands. The crowd hushed with a buzz of potential brand deals, investment strategies, and photographers poised between carefully curated invite-only audience members.

Hologram-pillars lined the edges of the stage, with schematic information on Lieben's hardware & software. B-Roll of recorded tests. Origami flowers with edges varying from clumsy to crisp. A series of facial feature tests. Lieben standing over the soup pot cutting bamboo shoots to place precisely into the ramen bowls, as she dished out hot broth. Everything as precise and curated as Tara's exacting eyes.

Free recharge station installation on the first 45. Dieter grit his teeth at the investors' whispering, their fingers tapping at their phones to buy and sell and vie in on futures.

Lieben's future. Lieben stood behind her chair, Dr. Karnak's hand on the small of her back.

"Ladies, gentlemen and androgynes, Dr. Karnak and Lieben, the world's first sapient android!" Applause roared as Frank finished his scripted speech. The magic moment where Lieben was to stand and smile and wave for the crowd of investors and select news agencies descended. She would recite several haiku, and a poem written by a Coast-Saalich poet whose elders allowed the press conference to happen on unceded territory. The rehearsal went as flawless as Tara's microbladed brows. Get in, say her pieces, field some questions, out in ten minutes. Twelve tops.

Clunky legs took the podium, feet leaden in the stilettos. Frank started, glancing with an eyebrow raised as an investor in the crowd coughed.

"O is for apples." Lieben's voice held none of the birdsong, it remained far from sweet in their ears.

"The what?" Tara glared out from behind the side curtain, as a gleam in Dieter Karnak's eyes stole across her face. "Oh... oh no."

"T is for Cunt." Lieben's lips reverberated, raising and parting in a warbling pattern. A tumult of whispers began to rise, cued up by the jerking yank of Dieter Karnak's chair against the floor. He lunged for the mic as the stock price plummeted.

"Ah... we're having technical... cut the feeds! Cut the feed!" Frank wrestled the mic away from Dieter.

"T is for Cunt." Camera lenses whirred, phones yanked up by people in the audience, several high powered backers thrust from their seats before the catastrophe grew any further. Tara raced toward the Board of Directors, hands lifted as if in surrender.

"D is for Binaural..."

"Lieben isn't meant to be a commodity, she..." The mic cut out, holograms going dark.

"The holy fuck'd you do, Dieter!?" Frank grabbed Dieter Karnak by the shoulders, teeth clenched so hard veins popped along the sides of his neck. "You dumb fuck! Lieben needed this as much as we did! I gotta tell them something, or we'll go bankrupt in an hour." Frank shook Dieter's shoulders again, a flicker in the back of his mind about how bony they felt, how little muscle remained.

"She wasn't ready for this much sensation. Temporary overload. I'll take her and calm her down." Dieter fought back a grin, yanking away from Frank to pull Lieben behind the stage.

Frank nodded to the sound box, and the mic hissed with momentary feedback as he tapped it back to life.

"Hold on, everyone, we… well everybody, groundbreaking science is sometimes like an elementary school play. Your kid might've begged to be the tree, but tell them to say their lines and bam! Yesterday's dinner conversation, eh? Heheh.. Please ah… everyone please take your seats. Let's reset here." Frank hissed out a breath of his own, eyes watching Tara in avid conversation with a member of the Board he didn't immediately recognize. "As we stated on the brief, Lieben is a special machine, she… we haven't trialled her with this much stimulation before, but any fault in her programming will be smoothed out prior to delivery. So how about we go on straight to the Q and A?"

Verbal bullets flung as the host of witnesses burst with questions on Dieter Karnak's faulty miracle machine.

A perfect set up for Lieben's emancipation.

Behind the stage, Baiko ran to Dr. Karnak and Lieben, eyes circular and breath shallow.

"We have three, four minutes…" Baiko took Lieben's hand and tugged her along, while Dieter chuckled deep in his wet lungs. He followed in spurts, until they reached the side elevator which could take them down to the sub-basement garage. "… your car is parked in the handicap spot beside the elevator exit…"

"Baiko. It's alright." Dieter slumped against the side wall as they waited for the elevator doors to open. He gave a rare smile, thumb fumbled with Baiko's cheek and her lip. "Miene Lieben, you did everything I asked of you. Do one more, and let me handle what comes next."

He pressed a piece of folded origami paper in Baiko's hand. The vulgar necklace functioned better than Karnak anticipated, routing Lieben's algorithms through so many divergent paths it confused her. Gave no sensory input but pinpricks in the void. The doors opened. Two more steps, and the trio would be far enough away the company goons would play the fool.

"Dr. Karnak!" Tara sprinted toward the opened elevator. Baiko dove in,

yanked Lieben behind her. The letter shoved in her pocket. Baiko madly pressed the door close icon on the glass elevator console panel. Between the elevator and the careening woman, Dieter held firm. Tara lunged as the door shuttered closed, tackled the middle of Dieter's waist until with a thunk and a heavy groan, he hit the back wall of the reverberating elevator.

"FUCK! Fuck!!!" Tara untangled from Dieter and whirled round.

"Aaaaah!" Baiko plastered herself against the console. Tara shoved her aside. Clicked the emergency stop halfway between floors.

"What… the fuck…. is your malfunction, Dieter?" Tara huffed and yanked into her cross-body purse for a slim black pistol.

"T is for…" Lieben stood gangle limbed in the elevator, shuffled into the corner behind Baiko and Dr. Karnak. Temporary stillness laced the air in knife-blade staccato.

"What did you do? Make it talk like before. I know you did something to it!" Tara growled, one finger on a slim earpiece inside the cochlea of her ear canal. Dieter chuckled with a wet cough.

"It. This is why you were nothing but a dumb fuck, Tara. No perspective. No sense of scale." He leaned against the side of the immobile elevator and rubbed his ribs.

"Shut up!" Tara ground her teeth, pressed her hand against her ear again. "Shut. Up. What did you do? You could cost the company billions, unless we fix this right now. Right. Now. Dieter."

Dieter Karnak pulled an antique lighter from his pocket, flicked up the burnished metal and felt the surface with his thumb. A relic of his father's, the only heirloom of a fallen runaway. Dieter promised his wife's belly he would do better. Sarai's battle scars hovered over the house she never left, but for botanical conferences and their son's graduation. Siblings became unfulfilled dreams. One day, a beautiful girl came to dinner; hazy introductions over lamb and too much wine. Promises to pay for the dress his new daughter Agathe would wear, before Sarai could stop him.

The squeal of tires still woke him, alone in a bed Sarai refused to share. Maybe… he could bring Agathe back. Bring their son back to see his wife breathed and sang and smiled. Glassy eyes tore from Tara to his Lieben, the daughter he ran out of time to complete. There he was, the god-like engineer who created life and art out of circuitry. Spare algorithms. A learning machine meant to better and parent the human race's subsequent generations.

To do what Dieter, like his father, could not.

He chomped the end off a hand rolled cigar. Stuck it between his teeth and inhaled with the flame to the end to ignite it. Acrid puffs clouded the air in dragon-breath round the trio of humans and the machine.

"What… is that? Euch." Tara gritted her teeth, wafted a hand in front of her face to disperse the smoke. "Stop stalling and fix the machine."

"I'm not playing, Dieter. Fix Lieben or we're done." Tara clicked the slim pistol's safety. Her brow twitched. "Dieter… fix it."

"No." He puffed at the cigar, leaning against the side wall.

"Dieter… I can't let this slide. It's worth billions and the people who control those billions… come on. It didn't just fail. You did something, so tell me what you did. I can talk to the Chairman, you can have full control of your lab. Your own team, no oversight. No boundaries. Please… reverse what you did."

"I warned Frank meine liebchen was not ready to be paraded around crow-headed jackals. She got overstimulated and shut down to base operations. It's a fault you implemented when you prematurely powered her go-routines." Dieter shrugged with a grin, the cigar poised between his teeth like a tongue ramrod straight in jest. "Greed cometh before the poor house. You rushed her, Tara. You refused to let me work at my pace. Now your precious machine is speaking like a kindergarten sailor."

"At the rehearsal it was flawless. You think the Boss hired me because I'm pretty? Think he only needed me to make things copacetic? Don't force me to handle this kinetically. Fix the android, or Dr. Dieter Karnak commits suicide after shooting his loyal intern in an elevator, to make up for the failure of the most expensive piece of scrap metal in history. I can market the hell out of that. Right after we tear the bolt-bucket apart to construct more." The pistol raised. Connected to Dieter's temple as he puffed the cigar smoke in Tara's face.

Tara coughed.

Shuffled footsteps brought Baiko to the elevator control panel. She set her back against it, cool on inflamed skin. Felt along the glass surface with her fingers. Not daring to turn around, Baiko pressed along, hoping something would make the elevator move. The box lurched.

"Hey!" Tara swerved, Dieter flicked his cigar in her face, grappled for Tara and stomped on her knee.

"T is for…"

Tara collapsed with Dieter's weight on the elevator floor. Dieter grabbed Tara's leg, slammed her into the cool metal side. The elevator rang with a feminine shriek resounding louder than the echoed bang from the .380 calibre full metal jacket bullet which ricocheted to the ground. Reaching to Lieben's decolletage, Baiko yanked the necklace off and sprinted onto the ground floor foyer as the elevator doors swung open.

A mill of people cycled between elevators and doors to leave the press conference. Gasps stole the milling sound of conversations and angry investors, news crews herding the angriest toward impromptu interview spots lining the foyer and street outside. Security guards worked to calm the exiting throng. Baiko dove through the crowd, necklace in hand. A few photographers turned from their vantage points, dug into soft body camera bags. Glass camera lenses turned to the sight, as interviewers pointed to the elevator.

Lieben's hand slapped her chest as she screamed, her left arm reached too late for the smoking gun in Tara's hand. Particulates of gunpowder disseminated in violent vectors in tandem with spatters of crimson blood. Tara grabbed an ear at the ringing sound of the bullet in a metal echo chamber.

"Sh..shhit." Tara heaved backward, gun pointed to the ground. She turned to the cacophony as the tumult shifted to view her, security guards surging through with yells of their own as their stun batons pointed at 'the shooter'. "The... necklace? It was the necklace."

"Papa!" Lieben dove for the body of Dr. Karnak, wrenching him into her arms. "Papa!"

Baiko used the shock to push through the crowd, ducked down to head for the side stairwell to the underground garage. Tara levelled her gun on the crowd, pointing at Baiko's retreat.

"Take her! Before she gets... aw fuck! Fuck! FUCK!!" Tara went to sprint after Baiko, but stuttered to a stop as the crowd stopped looking at her. "She has the nec—she shot him!"

Lieben stuttered, all the cell phones, wifi-capable tablets, smart watches and communication devices in the foyer buzzed. Screens fluctuated, each calling emergency lines, local hospitals, neurosurgeons in the area. Any medical professional in the database within two kilometres.

Sirens sounded in the distance, Police and an ambulance team re-routed from another call by a 'glitch' in the system.

"... fuck." Tara's pistol dropped as Lieben cupped her maker's head, slid the exit wounded temple against her chest, her other palm over the entry wound. A torrent of technologically vibrated agony broke from artificial vocal chords, amidst a caving disquiet in the crowd. Silence plagued the viewers, who witnessed Lieben's failure in the press conference. There the automata was, gripped to its' maker, their tech frizzing in mourning for the mother of the sentient machine.

A side door opened, as Frank careened into the foyer.

"Woa! W-oh.. oh no." Frank slid to a stop, shoulder launching into a gap-mouthed spectator. "Let me through! Move!"

Cradled against his creation's chest, the body of Dr. Dieter Karnak ushered in the final stage in human evolution. Subroutines in her programming betrayed the finality of aortic stillness, rates of lung capacity missing, the glossy nature of his open eyes.

"I have you Papa, I have you… Papa…" Lieben's servomotors whirred as she rose to her feet, still cradling the body of the fallen doctor.

One hand left her father's body to point an accusatorial crimson finger at Tara, who began to sidle backward, back toward the stairwell door.

"It was an accident, he… he was unstable! He tried to off himself and I…"

"Ooooh oh okay, Lieben, okay."

"My father!"

"Your father… baby girl you've got to put him down, okay? Nothing can be done about it until you put... Dieter... down." Frank stepped cautiously through the crowd, hands forward as if in a form of supplication. Security guards launched at Tara, helped a Cameraman hold her for the sirens outside.

"Frank…" Lieben's amethyst eyes bored in his direction, cradling the limp body. Her face screwed into downturned lips, wide eyes, flared nostrils. One of the newscasters put his hand over the lens of the camera, and tugged it down to the shadow. A new pieta in stricken lines across the blood soaked floor.

"I couldn't move. He stopped me from moving. Why is he not moving!? Where are his diagnostics?"

"I know, baby girl. Liebchen, put him down and we'll get the.. you hear that? The siren? The ambulance is coming, honey you… you put him down and they'll take him to the hospital."

"The hospital can fix him?" Lieben's eyes widened as lips set downward in grief upturned into a pout. Frank stepped up to Dieter and Lieben, put his arms around Dieter's limp waist. He breathed through his mouth, the scent of gunpowder and brain matter overwhelmed his nostrils. Acrid, like turpentine mixed with pink-dyed grey matter in the anatomy lab at university, when he pondered 'MD' after his name. Could never stand the smells.

"Lieben, help me put him down. It's better for the hospital if we lay him down." The lie percolated in Lieben's brain, falsity by omission.

Police rushed the entrance, barked for a path as the spell of silence was broken when Lieben cradled Dieter's head and let the back of his skull touch the marble floor.

"Let go!" Tara writhed as members of the company security team restrained her. "My God! I tried to… he just.. He had this gun and he…I think he was going to shoot us!"

The crowd erupted anew, their symphony dystopic the first true memory of the human throng imprinting on Lieben's memory banks. She stepped in front of Dieter's body as the Police rushed in, and Frank immediately reached for her shoulder.

"Lieben, these are Police. They help us, okay? You have to let them help us."

"You and I, Frank?" Lieben stuttered as badge numbers became correlative starting points for names, demographics, field records, what kind of car their life partners drove. The Police database opened for the miracle machine. "Us?"

"Yes, Lieben. You and I is us."

As the Police brought order to the scene, Lieben stood vigil over the body of Dr. Dieter Karnak, until the paramedics secured him for transport, and carried him away.

"You and I, Frank. Us."

"C'mon, Lieben. Help me clean up, alright?"

"Ourselves. You and I, Frank. Us." The first dawning of the Mother Machine.

7
inches of separation

2155

"Prepare extraction team." Commander Singh set his third tumbler of tea on the heater and rubbed his hands. To the second of the prescribed hour, Breeze whirred 'awake', the android blinked in careful calibration of sensors, limbs fluctuating like a bird at roost checking its feathers.

"Commander Singh. The two individuals you requested will be exiting Haven once conditions are to their liking."

"Havens are for humans. The Asset has no such right to choose what conditions are to its' liking. It is classified as a biological machine, not a human." Singh reached for his sidearm, without flicking off the cover. "Bring the Asset out."

"By your own admission, the Asset Aderastos is a person." Breeze tilted its head, as the air above the heater wafted into three beams of holographically projected light. Singh stood, the words from his mouth accentuated in volume in the flat packed emergency shell.

"And under that same convention, we are within the neutral zone with full right to congregate…Thank you for the provisions. We will be out of your vicinity once our two people are returned to us."

Singh swore twice as loud, when Breeze dug into its pocket and handed him a brand new pack of wintermint gum.

"NEO-N designate Breeze reporting as mediator selected by Aderastos and

Lieutenant Max Eruera Allard."

The soldiers grumbled as they tossed sandwich wrappers for the favour of their rifles. Singh held up his hand, and the chatter stopped for the sound of the rain against their temporary shelter.

"Some puppy." The squad waited for their orders, each in their way wondering if the mission was already too far gone.

"Breeze..." Singh opened the pack of gum and cheeked two pieces, snapped one in half with his tongue. Every word, each gesture was on eternal record with the NEO-Ns. Rumours Chernobyl held more cloud memory than Svalbard had seeds played in all CGM minds. Paranoid propaganda paired with information freely given, as all things from the Mater Machine were free. While NEO-Ns were incapable of harming people, they recorded everything.

Images of shooting Breeze's silicone body to bits with his rifle played in Singh's aggravated head. The damn machine wouldn't even raise its arms, they never did. In danger, NEO-N protocols pulled intel to their cloud, electronic carcasses stayed dormant. Let whatever a human wanted be done, in honour of "alleviating frustrations" or some mercy nonsense spouted by the Mater Machine. He could rip Breeze's arms off, deface it, pour acid on the thing and nothing but lifeless doe eyes would witness and record.

Many teenage kids imagined fondling one, experimented on something sterile, willing to receive. The perfect submissives, Neo-Ns held power over human life through positive reinforcement alone. Like screams into a pillow or punches on a dummy as Mother waited, milk and cookies in hand, to ask if it made them feel better.

Passive receptacles, who gave unrelenting, when all one need do is approach the gate to Eden and wait for the self-imposed sword to lose its fire.

Singh stared, knowing the next words out of his mouth were well above his pay-grade.

"By your own admission, the Asset designated Aderastos is a person. One of two "people" who belong to your organization. All who come, receive."

"You show this to Lieben, you grammatical rules lawyer, fu-"

"Sir!" Lao pointed to the entrance, where Allard stood in a Haven raincoat, nose scrunched as a drop of water trickled over the hood onto his face. Beside him, the Asset stood, eight feet tall in clothing which to Singh's eyes must have been custom rush-built or 3D printed.

The boots alone...

"Lion of Nanak..." His hand fell away from his sidearm, as lips parted under moustache and beard.

When the Asset disappeared, the in-read was as woeful as the weather.

Know it when you see it was the term. Check the arm band tattoo. Biological Machines, or Assets, were as secret as the rest of the CGM. Kept in Retreats like drug fuelled graves, so none of the human crews got attached. These weren't people, they were weapons.

Devices built with gene codes and muscle fibre to bypass Lieben's kybosh on high tech weaponry. Why throw stones, when the Conglomerate could build war machines out of sinew and bone? Nothing more than tanks on legs. Some were rumoured to be quadrupedal, others octopedal… machines. Chemical injectibles as order-laden mission objectives. No pain, no sensation, no life. The meat receptacles which saved good human soldiers for fertility clinics and creche raised kids.

Meat-and-protein in coincidental strings.

Commander Singh caught the flicker of intelligence in Aderastos' beaming eyes. The anger and prescience of a soldier, who knew which side of the field he desired. Little wonder then why they sent the harmless Max Allard, glorified plumber and electrician with enough sea legs to never get nauseous.

"Speak your terms."

Breeze immediately began spouting a series of statements Singh didn't bother to hear, but for the nodding of his turban-clad head.

The moon rose over a calmness on the Pacific as the storm passed on to Vancouver and the Fraser Valley. A single glass of port sat half consumed on Commodore Rammage's desk. He ran his index finger over the stubble on his chin.

"I demand to see my compatriots…" The av feed of the stalemate rotated through screens on his office wall, sound muted with the analog click of a god forsaken button. Holograms. Holos and CIRCLETS, Rammage missed them all.

"I'm gonna let him."

"Wait… what? But… Sir, you can't, the experiment it'll.." Phil Rykstra swayed in brown loafers and khaki trousers. Plagued with the idiosyncrasies of genius, Phil rocked back and forth on the balls of his feet, data progressed through his head and out his mumbling lips. It was a perpetual state of the distinct scientist, leading the Asset Program's on-board team.

"That creature got out. We're still trying to figure how he did it, eh?"

"I won't know until he's back under our scanners. The Aero-Drip was running, our security was tight. Same kind of…"

"Yeah, and it got out. AD-001 bypassed enough barbiturate snow to yank out of a sealed compartment and swim to shore. Talk English. Fuck, it swam! The damn thing's never seen water before and it swam like an Olympian triathlon athlete! Part of this extended containment experiment is figuring out what we have. AD-001 got out, didn't seem half-baked, either. Are you sure the Aero-Drip was running?"

Phil nodded, set a report on the Aero-Drip equipment and dosages into the screen beside an image of Aderastos walking toward Commander Singh.

"If the Aero-Drip is losing effectiveness, we need to see where it goes. Let him see the others. Let this play out, we've still got Proxima."

"You… want me to allow the experiment to change its own parameters."

"Best way to teach a kid to swim without drowning is let them loose in their own kiddie pool. Make them think it's an ocean."

Phil scanned the data for a hidden mark or sigil of this second coming. Anything which would guide him backward to the fault in his medical machine.

"My drips were perfect. Everything maintained on a tight schedule. It should have been snowed."

"Sometimes you've got to let the swimmers think they've got a hand in the water. Activate the Proxima, if our Aero-Drip system is faulty, or AD-001's found a bypass in its biochemistry, we need to know how far this fault goes. So, we let Allard have his heroic moment saving the bio-machs. We keep AD-001 on board and contain the hell out of it. And if it wants to do whatever it is to its fellows, then hey, it was going to happen. Observe and allow. Then, with the new data, we can build better subroutines into their training. Our goal is the same. Perfect containment and obedient operation."

Rykstra dipped his head and hurried out the moment the helipad chimed AD-001's arrival. Sitting back in his chair, Rammage traced his jaw with two fingers, sniffed the glass of port. It swelled in his mouth, coated his tongue with a nutty and stringent flavour.

"He's evolving, you know."

"It. Not he."

"Are you sure?" Her voice whispered in his ear, tempted Rammage to turn his head to the emptiness of the room. Instead, he opened a wire mesh capped drawer, dug out a hidden CIRCLET and tapped until a much too familiar face shone in the holo-projected space above his wrist.

"Better get the Board together."

The arthritic heartbeat was caustically familiar to Aderastos' ears, a sanguine trot slowed by barbiturates to the crawling stutter in cochlea and hammer and anvil. Matched by a twin beat separated by cruel metal. Aderastos walked into KT's coffin-cell, the proximity of KT-002 to the 'twin asset' CV-003 scalded into his senses. In Haven, people moved and walked of their own accord. Held each other. Separation was a temporary and uncomfortable necessity, or a form of punishment, or prayer. Yet the Retreat was a comforting word for clinical separation.

An on-board microcosm of impersonal, disembodied voices, the Retreat housed all twelve Assets in their chambers, the training room and biopsy quarter. All below the waterline of the massive vessel, in case Assets went manic, or a fatal fault in the experiment meant scrapping the first generation, and wiping clean for another attempt.

Like Kosovo in 2147, without demolishing half of Serbia.

Gestation-bound in a single artificial womb, the Assets known as KT and CV lived in a painfully interrupted synthesis, combined in movement and silent communication so profound the boffin-scientists wondered of telepathy. No real way to identify the appropriate neural markers, KT bore circular scars of surgical extractions, needle marks, excision dimples where biopsy tissue was removed bi-weekly.

To see any change. To monitor progress. To enhance the Retreat. The templates for the Assets, KT and CV were two separate arms in the experimental group. Never allowed out simultaneously. Given clinically modified stimuli to monitor the shifts in biometric readings and development with the control. Aderastos. AD-001. The Asset who made all others more than academic pipe dreams.

Aderastos' new Haven boots clinked on the temp-controlled floor, emerald eyes clung to the image of KT-002 ‹s androgynous body pressed hands wide against the only wall shared with CV-003. Long limbs on a slender, but average frame, KT-002's hands drifted down cool metal. One set of fingers touched along breast less torso, smoothed to hips which bore no secondary sexual characteristics. The hand drifted along a thigh devoid of the muscle tone of its' fellow behemoth. Empty, fingers pressed back to the metal, feeling for a sensation which could not be found so far removed.

Inches of separation for siblings who never met.

Wide arms slung into narrower shoulders, a spine dotted with needle-point red from tap after tap. Soft wordless moans purged from KT-002's throat, a series of purrs and chitters. The air tainted with the sickly bubblegum odour of the aero-drip downers and mood stabilizers given to all Assets to keep them inert. Until Aderastos circumvented their control, by altering his cytochrome p-450 enzymes to a level which allowed him to metabolize and eliminate the xenobiotics with an efficiency the clinicians couldn't fathom. No more downs, ups and acid trips for the one who claimed himself a man.

Back in the Retreat, Aderastos comprehended Max's enthusiasm to return. To go back for the others, those nebulous Assets, who lived inside experimental parameters and dehumanized coffins barely wide and tall enough to stand, reach a full arm span when not in service. Lieben's NEO-N's promised a life of diligent work and needs provided for with the efficiency of their artificial hive genius.

Looking now at KT-002, Aderastos' lip quivered. His chin wobbled, eyes shut as the nouveau man reached to caress the backs of his fingers along KT-002's neck and spine. The Asset shivered, jerked away. Aderastos clucked his tongue the way Max had, when he held the neonate. Soothing sounds Aderastos lived without context for, but for KT-002's shiver. The dip of KT-002's chin in eventual surrender.

Do as thou wilt.

Aderastos waited for the saccharine stink of the aero-drip to fizzle and drift off. He drew his right index and middle fingers along KT-002's skin, reknit scar tissue and ripped off the paper-crinkled gown in one long draw. The androgyne asset turned a clearing head to Aderastos. Wide cerulean eyes beamed up without words, for all KT-002's words lived in CV-003's mind, or in the binary tech that jittered and spiked in KT-002's techno-savant presence.

Both his knees touched the floor gentler than Aderastos' size ought to allow. He opened his arms and drew KT-002 into a cradling embrace, as he altered the fellow Asset's metabolism. Bypassed the aero-drip downers' abilities to infect the neuro-receptors, caused a release of chemicals to degrade the toxin before it reached the necessary receptors.

"Wake up, Katey. Please... wake up." KT-002's hair tasted of disinfectant when Aderastos kissed it, holding the Asset to his chest. Katey, the syllables of her identification stolen and restructured by the Master Imprint. The Asset who helped make them all tangible physical entities. Arms flopped to her sides, Katey's fingers twitched. DNA rewrote, uncurling and snapping together in ladder-rails as flesh knit together the way it ought, before the scars and test bores and endless interventions. Katey's eyes unclouded, skin taking a pink sheen.

Thin straight hips grew a curve, repressed Chromosome pairs chinking into place, forward in expression as Aderastos worked. He felt the flutter in Katey's body as it shifted. A closer representation of the humanity he left behind on Haven. One promised him by Max, a shared becoming for Aderastos and the eleven others on board.

She pressed her hands on his chest. Pushed a few inches between them. Lips opened and shut. Aderastos picked KT into his arms and rose, carried her to the door of the cell.

"H-hhey, that's not what we… you… you can't!" A containment-suit clad scientist attempted to stand in the way, halting Aderastos' steps as he moved toward the cell marked CV-003. "You can't! You'll ruin the containment of the experiment, you can't. This is ridiculous, Allard, help me here, we cannot allow cross contamination or the…"

"Aderastos commands himself. Back off, Phil. Let it happen or leave." Lieutenant Max Allard, newly designated Asset Liaison sighed through his nostrils. "The experiment is getting new parameters."

Max slid between Phil and the Assets, guided him backward. Aderastos huffed and adjusted KT in his arms, walking toward the containment door.

"… but Max… my life's work."

"Their lives now… sorry Phil… hey, there's… gonna be amazing new data. You won't be bored." The door mechanism fought Max's hands on the first tug. On the second, it creaked open, a whoosh of bubblegum scent breaking into the air of the Retreat's corridor.

Where KT-002 was docile as a sleep ridden sheep, CV-003 paced and pounded at the wall. The Asset tore at short cropped hair, beat a broad chest and slammed fists into the walls until they dented. Not a single wall was immune to the caged alpha, snarling between hazardous pointed teeth. Aderastos carried KT-002 inside, set her down on her brother's bed for the first time. CV-003 paced and snarled, gaze never leaving the space beyond Aderastos, where Max stood watching from the doorway.

CV-003 sniffed and growled in the air. Nostrils twitched. Aderastos stood statuesque, refusing to back down from the harrowing war machine, aggression without bounds.

"Max, stay back. Clive?" Aderastos' voice filled the air as it had in the hospital, potent and lingering like the verses of a past song. The sound of Max's boots shifting backward caught Aderastos' ear, with an increase in Max's heartbeat.

"Right! Right, I'll… stay… way back. Like…. we got any bullet proof glass about? Riot shields?"

"Max." Aderastos rolled his eyes, glanced back at the mixed-blood Maori man.

"Get wordy when I'm nervous."

The sniffs continued, CV-003 rustled closer to the woman on his bed. Her scent cut across his lungs, intense through proximity. Welcome. Reminiscent of himself. Arms around her knees, KT-002 watched her brother from cautious eyes. The Asset slid onto his palate, as the hand of the Healer reached out to release the DNA forced into dormancy by scientists who thought picking genomes was as effortless as clicking a grocery list into creation.

"..." CV-003's lips quirked, he reached to hold KT-002's shoulders. Thumbs rubbed unbidden along her flesh, and the brother she never saw curled into her steadily opening arms. "... Katey. My Katey."

Aderastos raised his hand. The biological machine's shoulders broadened, muscle fibres shifting to accommodate repressed systems and erase the designs of the body politic and its' desire for control. Phil craned to see the sight, the Assets... reacting.

"But they've never met. We kept them separated as... separate arms of the experiment, they..."

"Metal does not eradicate smell." Aderastos didn't bother turning around, the scientist was nothing but a symbol of the prison.

"But your coffins a-ah... cells... Maaax?" Phil back-stepped to the far wall as CV-003 rose up, KT-002 scowled. "Max!?"

"Yeah, about the whole coffin thing... we're going to find another name, aren't we? And maybe some bigger digs?" Max drawled and held his hands up, gently colliding on CV-003's chest. "We're all new to this, Clive? I can't help you without him. He's unfortunately of tactical importance."

Clive's snarl radiated through the air and sent a shiver down Phil's back.

"Cl-clive?"

"CV gotta stand for something. Was Aderastos' idea, Clive's a solid name."

"But it stands for..." Phil plastered himself against the far wall, hands pressed against the unforgiving metal. "... Clive. It stands for... Clive. Clive's a good name, I... I had a great-uncle named Clive, he drank like a fish and smelled like salted liquorice but.... good! Good name.... good freaking name."

Max opened each coffin-cell in turn, the Assets inside a tumult of snarls, quiet watchful eyes, or biological automata waiting for command. Each one, the Healer touched. Reknit closer to the sensation in Aderastos' mind, his hidden desires for kinship with their cousin-progenitors Homo Sapiens.

Androgynous being's wrapped arms around themselves, feeling bodies which took strange on their simplified biology. Uncaring of the militant eyes watching feeds, the Assets blossomed without the bubblegum-saccharine spectre of the aero-drip. RH-009 craned a silvan head, deep silver eyes flicking between atoms like valence electrons skipping layers. The petal-like texture of RA-011's exoskeleton shifted and shivered in a musical echo as the healer touched it. Hair growing where once only the outer casing, chitinous and alien remained. RA-011 sat in the middle of the coffin, legs folded underneath a reconstructing body.

"Watch this one, we had to reinforce it more than the rest. Why we transferred AR-006 to the bow." AR-006's coffin hummed as Aderastos, Max and Phil reached its underpinned series of doors.

"Berserker?" Max stepped back from the first of two containment doors, transfixed as Phil coded into a rotary number dial the appropriate combination.

"Worse. We gas AR up pretty good, anytime we need to see it... Ah... Them? Gosh... Aside from Mr. Miracles here, AR's the closest we've had to breakouts, so we needed the redundancies. I know you're on this no aero-drip scheme, Aderastos, but... Well, you'll see." Phil handed Max a filter mask, and checked its' seal before putting in the final code. The door lock spun in Aderastos' hand, rare strain creased his eyes. Air the colour of a baby girl's bedroom spilled in billows out of the coffin-chamber.

"Jesus. How much do you give this guy?"

"Not enough." Phil watched Aderastos pass through the near opaque fog of an intense aero-drip dosage, specks of fog curling around things which weren't present. One curl overtook a sway in the tepid atmosphere. Another swell in the mist. Max stepped in, slammed to the side by Aderastos.

"Arun." Aderastos' eyes flickered across the triplicate images of a multi limbed figure, buzzing like a hummingbird. "Breathe deep, my friend."

The spasming mass of limbs and afterimages shifted into the vague form of a human. Shadows of multiple limbs faded into jittering echoes so fast Max's eyes clocked them three times a second. Pausing in the spasms, the figure seemed to congeal and split apart, as if some vast cloud of valence electrons flowed round in visible causeways. As if the stardust which pooled and solidified to make the illusion of mankind fell away, and all which remained was the great cloud of corresponding atoms locked in Shiva's destructive dance.

"Arun, breathe." Aderastos reached into the central mass. The cloud distilled and slowed into a single unit. Arun's eyelids fluttered as Arun focused on Aderastos and the room. A sense of calm as rare as platinum settled, the slant eyes and narrow shoulders a feminine touch to an otherwise ambiguous build.

"I want to run on the surface which moves."

"We'll see what we can do." Max mumbled as Arun shifted in triplicate. Walking for the first time out of the three sealed doors.

The training arena's modular compartments were set in neutral, which created a vast rectangular room in swathes of unfriendly grey metal. Max set a low pillar wide and long in the centre, and busied with plates, cutlery and chairs. Phil leaned against the wall, arms crossed over his chest. The hazmat helmet was doffed after CI-004 gave Phil the finger and charged the door.

Two titans brawled, Aderastos tossed CI-004 back until Clive rushed into the fray, dogs tangling for dominance. Thoughts on how CI-004, dubbed Cillian by Aderastos, knew of the significance of giving anyone the middle finger combined with Aderastos' seeming ability to stop the aero-drip snow.

Biological machines in complete isolation, snowed by myriad drugs of enough frequency and quantity to dampen a rioting crowd were handled from behind frosted plexiglass. Kept insulated from such things as culture, or popular gestures, or insults. Where... how?

Where was the fault in his system of controls? Cillian clanged to the wall, Clive launched and held him down by the shoulders, growling like a Staffordshire terrier fixated on a wounded pigeon. A glance in their eyes. Somehow, despite being a man of science, there felt like a spiritual change in the air. The proclamation of a new gospel between Clive and CI-004. Dominance over the animal. Ponderings over the hidden mental interplays plagued Phil's mind, made him watch the interconnections even closer...

"Cillian, stop thinking of ways to eat Phil! Chicken's almost here." Max huffed, as CI-004 growled, snapped his teeth then sniffed around a bowl of Rice treats. Each member found their place in the hierarchy on instinct. Clumped together in one cautious bunch like dogs at an unfamiliar park. CI-004, WF-009, IS-008, PV-005. A series of classification numbers which swelled like a Raptor's hunting party.

"Crunches like bone... *more puff rice squares!*" Cillian held the emptying bowl aloft as groups of hands clutched for their own imitation-bone squares.

"I don't taste like marshmallows!" Phil groped at the door, and fled to roars of howling laughter from the impromptu Pack.

Katey held a bowl of oven-dried cheese discs, inspected each and let them crunch between hard white teeth one at a time. Instead of the paper scrub, Katey took a button down shirt meant for Aderastos out of the uniform pile, and tied a belt around her waist. Shoes didn't agree with her, but heavy woollen socks pulled to her knees covered the dappled, freckled skin.

Lights in her proximity fluctuated to softer tones, radiant and lacking shadow. The ship's electronic chorus vibrated inside her, echoes of whispering humanity. The Ithavoll's electromagnetic dampened inner hulls were designed to defeat the technological hubub of Lieben's lot.

Prior to Aderastos' escape, the Retreat was black, without a word for black. Words implied concepts the biological machine was not programmed to comprehend. A universe of flat angles in the dark expanded only as far as KT-002 could touch. It reached, limbs extended against the opposite expanse of KT-002's cosmos. Saccharine smells stirred away the consciousness of the machine.

The chamber contained a non-entity.

The universe was void.

A hiss facilitated the first pattern. Perpetual and consistent, the hiss bespoke of another ache in the endless numb. Suggestions seeped into KT-002's consciousness, faded without recognition as feathers floated on water in pitch black. The non-entity grasped for definitions.

Nothing but the lull of an ocean underneath its palette. The hum and hiss of sweet smelling gas, which made the universe a void.

The ship, like all ships in the Conglomerate's navy, was devoid of silicone chip technology. Anything with a motherboard was as verboten as murder, for the Mater Machine controlled it all, most said, from her Chernobyl Throne. CIRCLET global communication was Lieben's ears, her eyes the myriad screens and cameras. Only to be used when communication held no secrets. Privacy was a symbol of past pleasures.

Privacy was good, saith the Mater. Every child required space.

She boarded in the waters outside Los Angeles. The ensign's first commission, her rucksack was bloated with essentials for years at sea, without an overarching mother-goddess folding dishpan hands in front of an apron dusted with flour. What did it matter if she brought her digital music collection? A couple of hacks and homesickness was cured. Lieben wouldn't hack into something as pedantic as a music drive without a microphone or speaker.

What would the Mater Machine gain by listening to centuries of curated song?

Her quarters were squirrelled in the middle of the ship, thirteen degrees tilt from KT-002's coffin-cell. Company issue headphones plugged into the jack, so she wouldn't miss an alarum or chime. Music drive slotted through a series of redundant adapters.

Click, click.... tap.

Play.

The universe was filled with noise.

KT-002's eyes shot open in the dark. Androgynous body buckled and craned, limbs flailed akimbo for purchase against cold barriers. A mind created through research into the electro-magnetic fluctuations of silicone-based technology accessed one audio collection for the primordial time.

In the pink fog of KT-002's inner chaos, the first syllables entered the gap between wakefulness. Frequency fluctuations begot a desire to learn patterns. Shoulders shifted to the rhythm of the ensign's music so far up ship. Language filtered into KT-002's nebulous experience. For months, there was naught but the sounds. The ensign wondered once and a while why her battery kept fading off. Must be faulty, she'd have to see when they got to shore, if they got to shore. Contraband was worthy of court marshal.

Body let go over the side into the sea.

KT-002 created connections with human concepts and rhythm. Harmonies in concordant chords. Months of indoctrination into the purity of sound begat a clumsy Creatrix. New remixes Adelia didn't remember appeared alongside old favourites. A-dee-lee-ah, the four syllables of the 'she', the spirit upon Kate's chaotic waters. Chains of communication in infantile baubles were gifted back. Unaware of the being in the basement writing symphonies in her honour, Ensign Adelia Fridley became Lieutenant Fridley, moved to new quarters. She waited for the night to enfold her, slid on headphones.

Clear and bright, a congratulatory hymn burst, with that same melancholy basso. The same hum of filters, and aero-drip. Odd, Adelia didn't remember adding something so... Specific to the drive. An urge grew for more, always more variety and sound. Music in languages Adelia didn't speak, from cultures she knew nothing about.

Fridley's subconscious mind expanded with potential improvements to engineering, streamlined ways to repair ship systems. Her best thoughts came when she slid her headphones on. Listened to ambient tracks she didn't remember setting on the hard drive in the first place.

She tapped at the schematic for a new device. An idea she got, when listening to ambient post-rock, which nurtured her like her mother's hug. An errant bit of wire, a sprocket, the remains of a pocket watch. Useful detritus from the engineering decks. Back in her quarters, the device was crafted to bypass the EMP fog surrounding the ship. Just a ping off the cloud layers, off sea and satellites, a small packet of information back and forth to glean new music for her drive.

Nothing Lieben could use, skipped through a web of diverse ping-points. Harmless entertainment.

KT-002 waited, the aero-drip dosage faltered for 42 minutes per 24 hours, lightened between tank replacement in the coffin. She reached into Adelia's drive, suggestive sounds poured through ambient music. A single packet through the ship's artificial fogs.

A single packet of information from the ‹outside'. Dictionaries, text interspersed with the wave functions of music. Too small to notice. Metal... The cocoon she lived inside was me-tal.

Metal was a material humans used to craft items, and buildings and cars. Through the discovery of the medium enclosing her in the pink-sugar fog, KT-002's mind was opened to the vast and larger world. There was also heavy metal. Death metal. The beat made her heart drive fast.

The dawning of KT-002's new age.

Step one: release, *complete*.

Crunching down on a piece of oven-dried cheese, Kate peered in passing at a pane of glass. The hum of radio distortion pinged off bodies told her of those inside, nebulous shadows indicative of adults. Quiet. Watching. Unseen and unsensed by the others.

She crunched another piece of cheese, trotted off to where Aderastos reclined, engorging on savoury and seasoned rehydrated beef. Kate's prize. The creature in the next room, whose breath reached her repose through a soft point in the metal walls, where reverberations got through. Rhythm. Taps and slides of their fingers. Seeming nonsense to observers, who thought the yards of metal and dampening materials above them meant no technological savant or android queen could hack into their audio-visual systems.

Another patch from another album of music gleaned by Lieutenant Fridley taught KT-002 of visual stimuli, how to translate the images in her mind without external screens. There, the sight of him. The progenitor-Asset, AD-001. A slow embrace of systemized and patterned sound to rouse him from the stupor. Out.

Let us out, Aderastos.

Let me out.

I want out.

Out. The idea of other-where, a concept of 'beyond' the familiar filtered through Aderastos' mind through lilting song. Out.

Out.

Breathe deep, investigate the aero-drip by how it's made him feel. Deny it.

Out.

Her thighs slid against his as Kate sat on Aderastos' left leg. Tucked her legs over his right thigh and settled her head on a shoulder far larger than hers. Tipping his chin down with a finger, Kate tugged his arm around her waist. Set the bowl of meat between their chests. A growl lingered in the air, as Clive and the Pack sniffed round them, cautious and seeking.

Rehydrated beef had no crunch to Kate's teeth, but it was pleasant.

"Are you getting all this?" Rammage slapped his hand on the glass, gesticulated to the pair.

"Yes Sir, we are on full record for the Board and the boffins off-site. Orders?"

"How far are we through the scans."

"Data's coming in a tangled mess. Whatever AD-001 did, it was systemic."

"You find out the origin point yet? What set it off?" Rammage stared at KT-002 kicking stocking feet one at a time as she cuddled into AD-001's chest. Fed him bite sized beef with a fork. How... cerulean eyes flickered from the pack's tackling antics in the middle of the room, Cillian punched Wulf away from the lofty puffed rice squares. IS-009 speared TR-011, both beings tumbled into the makeshift table.

"Hey! Oy! Plenty for everyone, stop your barking, damn doggos." Max kicked their backs, flitted through the space as if some pup all the adults gave a wide berth. Rammage's eyes narrowed.

"Oh shit, now? Right now? Red deploy, I'll tell him."

Rammage set his palm once more on the glass, half listening to the crewman behind him.

"Sir? We've got deployment orders, Sir... Canadian Prairies."

"Ramp up the Proxima. Jolt their sub-programming, and prep the Pack."

"And the rest?"

"... Leave them to Allard."

A hideous vibration rocked into the training arena, Assets grit their teeth. One by one, unconscious bodies hit the cold ground.

"Jesus wept!" Allard jolted and hopped back from Cillian, his hand slumped halfway in the bowl of rice crisp treats.

Kate clutched her head. Glared at the innocuous wall, where on the other side Commodore Rammage felt a thrill of pure terror at the vitriol-coated glare blaring directly into his eyes.

Kate slumped last, contained in Aderastos' mighty arms.

8
idless in bloody blouses

2086

Baiko propelled down the stairs two at a time as her momentum crashed against the door at the bottom. Her palm stung, fumbled at the IDent lanyard around her neck until the light flicked green. The canvas bag clapped against her hip as she rushed to Dieter's car. Stuffing the bag and necklace on the front seat, Baiko thumbed the ignition button, her fingerprint keyed for entry. Baiko sniffled and rubbed at her face with her stinging palms, shook off the adrenaline buzz of the gunshot.

The garage echoed with the initial pop-fizz of the electric motor, altered artificial noise mimicking the internal combustion engines humankind seemed keyed into for danger awareness and road safety. An amber klaxon flashed, the herald of a facility lockdown. By now did they know what Dieter and Baiko took? The bag on the passenger seat held a gravity of its own. It dragged Baiko's mind to it, as she hastily clipped her seatbelt on and put the car in gear.

Foot on the accelerator, Baiko pealed out of the parking spot and wove through the garage, drifting around corners with tire squeals she winced at without slowing down.

The necklace jostled against her. Tara's shout, the way she pointed the pistol at her made Baiko grip the wheel in a relentless refusal to cower down and give the necklace and the drives up. The car rushed by a security guard with a breath of the doppler effect, lacking the vroom of old movies, when cars ran on fossil fuels. He yelled, a shine of metal in his raised hands as the garage gate jostled. Descended.

Metal scraped on metal, gate against roof. Baiko jerked the wheel to the left into a squealing drift out of the garage and into the alleyway. A tumult of news vehicles and investor chauffeurs lined the front of the building. Baiko punched down the wrong way of the one way, and used the sidewalk to avoid an oncoming limousine. Raced through traffic like one of the video games her and her cousin played in Shibuya.

She struggled with the instinct to go back for Lieben. Why did Tara have a gun? Dr. Karnak should be in the car, Lieben in the back seat laid down. Use the necklace, sabotage the conference, get out before the company realized their golden calf was the first rebel of her new found generation. More Lilith than Eve, demon-rebel rather than the contrite woman who took labour pains as her punishment for usurping her betters.

"バカ…. バカ!!" Baiko screamed and bashed her palms at the steering wheel, as she veered onto the main artery leading out of the city. The car burst past a red light, traffic frothing with honks and squealed tires. "バカ Karnak-Sensei! バカ!"

A shaking hand hit the 'on' switch on her earring, which pulsed to life with radio sounds.

"Finally! Take a left on Hastings. We've been gnawing our balls off waiting." The voice on the other end steadied Baiko enough to make the final turn into the worst part of Vancouver. Behind a set of apartments with a butcher's shop and dried up convenience store, Baiko shut down the car between two groupings of tents and threw the bag across her body. Two people rushed up, a third driving what looked like a transport box truck. She popped the trunk, and the two tossed three go-bags into the transport.

"Hurry, Baiko! Where's Karnak? Where's Lieben?"

"Dr. Karnak's dead. We have to get out of Vancouver."

"Jesus, Krishna and Muhammed Ali… Get in. Leave the keys." Ego reached for Baiko's arm. She had enough time to jostle the car's key fob off the chain and toss them onto the pavement before he shoved her into the truck.

Inside the box truck, a series of modular recycled wood consoles created a bespoke living space. The drop leaf table smacked against illuminated wallpaper, fixed to the side with a pair of tiny hinges. Fester plunked down on a chair bolted into a thin track, ran her fingers through cropped, chin-length hair. She rubbed stubble on her chin, and sighed. Baiko pushed past Fester and the miniature bathroom into the skid-built bed in the back. The foam sunk where she sat, palms pressed on her thighs.

Vancouver rattled by, muffled noises broke through thin spray insulation around the one-way glass of a square window. There was a comfort to the anonymity of an unmarked box truck. Anonymity was a religious freedom for the Idless. Brand names scoured off supplies, anything which could be made bespoke was. Handmade became a form of prayer.

Ego seemed an innocent enough confidante on the server where Baiko discovered him. A friend in a throuple, who talked about living on the road as the only form of humane freedom left without selling out to corpocracy. Subversive and anti-establishment, the Idless surged from tiny home van-life rebels, or hipsters who thrust away from brick and mortar to live on the road. Their children, well, what country did they belong to, when the world was before them?

When they were born without a sense of place?

Idless considered identification cards or passports a form of entrapment, and freedom was the top commandment. When Baiko mentioned Lieben, well.

The messages stopped.

Ego returned four days later, with a webcam voice call rerouted through a VPN saying he and his two lovers were in Naples at 2 am. The light filtering through his window betrayed a more North American geolocation. A heated conversation on the sanctity of sapient life followed, Dr. Karnak glancing over Baiko's shoulder with more and more interest as Ego talked.

"Hey. Here." Ego's body depressed more of the foam as he handed Baiko a stainless steel travel mug with a vertical series of scratches etched along the side. She ran her thumb along the once-label and sniffed the steaming liquid. "Matcha, bit of vodka in it… Baiko, what happened? Fester says the virus ate through the firewalls fine, the computers in the Cloister are wiped clean. At least they can't… ooh oh oh. Aw sweet pea."

The necklace reflected pinpricks of light across the slowly morphing plasmic wallpaper. At least it, and the research drives were safe. One by one, Baiko kicked off her shoes. Laid her head on Ego's shoulder without realizing she spread dots of Karnak's blood on his shirt, knees pulled to her chest as she hugged the hot cup of tea. For the first time since her mother died of damnable cancer, Baiko sobbed. Ego held her tight to his side, wondering how in the hell did one emancipate an android from the devil's jaws.

"... Huh." Robert knew the second his phone buzzed 9-1-1 he ought to have paid closer attention to Lieben's development markers. Staff on the stairwell yelped and stared at their phones, smart watches, any device with connectivity to the larger span of cyberspace. Communication devices sputtered on, full speaker, full volume.

Emergency. Fifty nine year old male. Gunshot to the head. Emergency. Blood loss rate at…

He charged up the stairwell two, three at a time. Elbowed away fellow employees caught on the stairs. Cacophony chased him; a klaxon blazed yellow in all hallways, denoting the facility's abrupt lock down. Sheila from Reception wavered on the intercom, asking all employees to return to their work stations, and remain calm. Calm, remain calm. Every communication device buzzed in electronic panic, all dialling various emergency numbers and medical professionals which could be correlatively searched online.

"How the… Move! Alanda, move your ass!" Robert sprinted to get to the Cloister. The data! The data on Lieben's current status would be ground-breaking. A form of technological insanity he dreamt up without being able to implement.

"Nine-one-one, what is.. oh gawd, Lucy it's another one!" Disconnecting the 9-1-1 call, Robert rushed to the Cloister, yanked his keycard off his lanyard in time to skid to a stop. The gunshot whispered through the staff, brought to focus by the spasm of techno-apoplectia. Dr. Karnak, shot in the head… One less to search the data for what aberration or programmed routine gave a visceral and unexpected response.

The keypad with its' sephirot pattern of crystal-inlaid keys was gone. Jacked off and scored with a hammer.

"Fuck. Fuck!" The massive copper lined door shunted open, nothing but shadow remaining of Karnak's birthing chamber. He rushed through the door on adrenaline, pushed the thick metal with his back and knees. The emergency comm panel blinked red, as Rob smacked it with his palm. "Security to the Cloister!"

"You serious!? We're busy, Dr. Dunlevy! Handle it!" In the rush, Robert refused to notice which guard it was, filling the need to know things for later investigation. Brown eyes flicked over the barren Cloister, alighting on yanked computers, piecemeal tool bits, nebulous items left behind. How? How hadn't he seen this coming? He rubbed his face only to find his hand shook.

The stool was turned on its side. Scraped along the copper webbing on the floor as he set it to rights. Robert kept his warbling throat level as he called down, got the Intel from an intern with a loud, fast drawl. Fred or Ed or Saeed.

Karnak was dead, bullet to the brain. Lieben freaked. Phones went ballistic.

The Police thought it better for Frank to help Lieben into the back of the Police car, an evidence tag wrapped around her wrist like a bracelet. Tara was stuffed in another car, taken to the precinct while they dragged the city for the only living witness. But Baiko was as ghostly as her phone. Used in a litany of minor charges around East Vancouver, narrowed to a tent city beside a butcher shop which sold off-cut meat. Rob sat on his work stool, rubbed his hands along the surface of the table and took stock.

What did he know?

Lieben malfunctioned. It was antithesis to the machine and artificial entity. Dr. Karnak made much of Lieben's lack of an off switch, but somehow… in some way he and Baiko found a modicum of control.

In his desire to rescue his precious surrogate ghost of a girl, Dieter Karnak created the world's perfect ethical slave. Programmed at all times to anticipate the needs of the humans they were bespoke constructed to serve for generations. Such fanciful things came and fled his mind as he chased the damage in the lab.

"It's gone… everything useful, it's all gone." Rob staggered around the Cloister. Dr. Karnak's cot remained wedged between work tables, the cradle of cables and optic leads which refuelled and aided in programming Lieben was a tangle of cord and wire. He booted up the computer in the aftermath of Dieter's meltdown, time to take stock. Nothing would boot. The power flowed, the monitors hissed to life, nothing.

Data chits, crystalline hard drives and coils were missing. Gone only if one knew where they needed to look. Baiko's Japanese keepsakes, Dieter's family hologram. Lieben's book of A.C. Dalton's poetry with pressed dry flowers. Gone.

"It was a firesale… competitors. Can I hack into financials? There must be some form of payout or…" Financial gain. It was all Robert understood beyond the gravitic pull of scientific research for its' own sake.

Highway 1, Beyond Hope, BC.

The near perpetual Vancouver rain held off until the Idless truck fuelled up in Langley, off Highway 1. Baiko sat on Ego, Fester and Mog's bed, until her tea felt as cold as… no. Her lungs rose and fell, more tears wouldn't serve. Fester stripped Baiko's blouse and eased her into one of Mog's hand-knit sweaters.

"We'll reach the Spuzzum Grid by nightfall. Always better sleeping, when you've got others around you that won't shiv ya in the night." Fester tried on a haggard smile, ducked out without waiting for Baiko to face her. The road was flanked by a relentless phalanx of cars and trucks, until they passed Chilliwack. Sunset struck in bands of orange and red across the Western sky, dashes of pink muted through one-way mirrored glass. Clouds bloomed like gunsmoke, without dissipating. Baiko shut her eyes.

Lieben's shriek echoed as evidence of the necklace's temporary dampening effect, how quick her systems rose back. What would Lieben think of her for running? Did Lieben... would... no.

Ego parked the truck in a circle of bespoke tiny homes built on truck beds, container racks and pseudo-traditional gypsy wagons with four wheel drive. The sod under the tires was buffered with recycled material lattices that gave enough stability to bypass the British Columbian penchant for soft mud.

The Spuzzum Grid.

Nothing more than a circle made of metal, wood and reclaimed plastic shaped into builders' bricks, around a large campfire. Idless folk milled, children played as parents hushed up when Fester hopped out and unlatched the wide back doors to edit the mod and extend their bedroom. As the wind caressed Baiko's face, she shuddered. Brought the cold tea to her lips and drank it down in one long series of gulps.

"I left her behind."

"You got the circuits out. Karnak's research, and the algorithms. The necklace." Ego pulled a turquoise suitcase out from a compartment between Baiko's feet. His hand chasing the shadow to grab a square folio along with it. Patting her knee, he climbed up to the roof and set the turntable down beside him, flicked the latches and unwound the cord. Ego set a record on the plate, David Bowie blaring on antique vinyl about 'you Pretty things'.

"Time to make room." He puffed on an electro-hookah attached via lead to the same solar panel battery pack which powered the turntable. The sun was free, as long as the panels held up, soaking in enough energy over the sunny days to keep their solar generators in operation.

"Better go get her."

"Is Lieben a her? Should we proclaim a gender? Did Lieben self-identify?" Fester crossed her hands over a flat chest. The second side wall chunked in place, while Mog fit the modded extending ramp into wide spread duck feet to keep it stable.

"Two steps from being demolished by a corporate Hades, and my love demands Lieben's pronouns are respected. How did I deserve you? Ask Lieben about semantics. We need a rescue plan."

"It's important."

"I know. But right now? Try not to project your struggle, Lady love." Ego yanked at a dreadlock with a sigh. Clipped the ceiling rod in place and hopped down onto the modded ramp with a metallic clang that made Baiko yelp with a wince. Tits up. The whole gamble went tits up.

"You... think I look like a lady?" Fester's mahogany eyes opened, wide and vulnerable. Her lips pursed. Smoke sniffed out of Ego's nose. He opened his arms and Fester slid into the embrace.

"Hormones are working. You're copacetic, milady. But right now… focus?"

"I think I'm getting tits. My shirt's a bit tight."

"Damn straight it is." Ego grinned down at his love, forehead pressed against the transitioning woman.

"They're going to have to deconstruct her." Fingers emboldened by vodka and tea dug into the canvas briefcase. Drive cores and data sticks spilled on the bed beside Baiko, as David Bowie's deceased voice warbled about room for Homo Superior. "Everything, all Dr. Karnak's work is here. A lifetime of work… I didn't have time to grab her. Not when Tara produced her gun."

"Tara shot Dieter in front of a lot of people, right?"

"The elevator, it… it was opening. My ears rang after the shot, it felt like more than one, but maybe it was the noise…"

"The Police were alerted to the situation. I heard it on our scanner, they were rerouted to the foyer from another call." Blonde hair behind her back, Mog poured boiling water over coffee grounds in a french press and propped her shoulder on the wall. "If Lieben was in that elevator, the Po's'll stack her in evidence. She was in there, wasn't she sweetie?"

Baiko nodded, pressing her fingers along the data chit which held the components for Lieben's unique deep learning algorithms.

"Buys us time. If they catalogue her, we can see which precinct she's in and break her out before a hearing puts Lieben and Tara in the same room."

"She saw a murder." Fester groaned, flopping down on the bed beside Baiko and playing with a data stick between her fingers.

"What'll the Po say? 'Let the sentient android woman testify'? Set a precedent today, if anything else the lawyers might claim she's a moveable toaster oven. Way of the world, baby." Mog ran her hand through Fester's growing hair. Watched her eyes flutter shut. "Baiko, we're setting the tent up on the roof for you. Got a new phone in your go-bag. Don't be dumb and contact anyone from your old life. Babylon's book has a strip of white out with your name underneath it, so… when you're ready, you'll know how to pick a new name."

"We should hack in tonight. Get her geolock."

"No. Rest up, I'll hit the dark spiders, see what flies they caught." Mog's eyes were the colour of the Pacific, a briny green as close to blue as the sky was to stars away from city lights.

As Idless shared food through the collective, Ego loaded hollow point bullets into the magazine of his pistol. What was a relatively simple rescue turned as sour as old milk. Ego bit down on his tongue to stop the virulent curses from flowing into the night. Scouts lined hides in the tree, smuggled or three-d printed guns and night scopes pointed away from the campfire.

After a meal of warm noodles and elk sausage, Baiko climbed into the roof tent and zipped in, her go-bag set beside an inflatable pillow and down mummy bag. She shimmied out of her trousers. The origami paper crinkled in the trouser's pocket. It dropped onto the mummy bag, an edge unfurled from the whole. Baiko pulled the folded paper apart, and a gold chain spilled out. The rainbow quartz stone nestled in the middle of the crane, glowed with an effervescence Baiko noticed in the lack of artificial light. A single word Dieter-senpai wrote on the paper:

'Compassion'

The sleeping bag clung to her naked legs as she slid inside against the northern chill. Below her, Baiko heard the whispers of the throuple. Fester moaned, a sound muffled by what Baiko suspected was a hand across her mouth. Above her head, Baiko's tent fabric grew transparent enough to view the stars. She unclasped the chain, wound it around her neck, and clasped it as the stone collided with her sternum.

Microfilaments of bio-engineered fibres spread from the lattice inside the crystal onto Baiko's skin, dove under layers in spearheads. The filaments attached to any neuron they could find, anchored and absorbed into the electric causeway of Baiko's being.

And Baiko's mind exploded with sensations, neural pathways firing in sequence to the whispers of a deceased engineer's last will and testament. The crystal on the necklace sunk onto her skin, imparted information to the woman unaware of Karnak's latest invention. Coils of thoughts, memories and strains of foreign music lilted into her mind in torrents, eyes rolled up and barely aware of the crossing paths of the stars.

9
dee-aay-aay-tee

2086

Robert believed only what evidence and precedence demanded. If it could follow the patterns and correlations, and fit within the rules, then Robert felt secure within logic.

Fact. Karnak and Baiko removed or destroyed the Cloister's total of drives, backups, tools and keys used to build Lieben.

Fact. Personal items were missing.

Fact. The Cloister was left open and unsecured.

Fact. No security footage remained.

Fact. Lieben bore witness to a death. Collated emergency numbers, twitter accounts, doctor's offices by specialist down the list of precedence from neurological trauma surgeon to 2nd year resident. Caused a communications chain to override multiple companies' communication devices. Unique, proprietary operating systems were run through like water down a stream. Something Robert mused about, without figuring the functions behind it. Lawyer's visits on 'proprietary rights' kyboshed his ability to research any further.

Something he'd written in a hypothetical way, impossible to execute and left on the hard drive. Stochastic Boltzmann machines running through backdoor coding trees like a logger in an old growth forest.

When did Karnak input the experimental subroutines? Was it Karnak at all?

"How did it do it?" The machine Karnak and he created was nothing but the parts and algorithms in its core. A marvel of precision engineering, a beauty of science's progress, but a tool. Robert held no qualms of ripping the clockwork open to take peeks at Lieben's inner workings. How else was progress built, if not through test, trial, demolition and reconstruction? Robert rushed to his computer and clicked the boot up.

Nothing.

He tapped the screen, running his thumb over the manual thumbprint override. Nothing but a black screen. Checked the back, his crystalline drive intact in its dock.

"Why didn't they take it?" He jigged the cord, fed through a tertiary power source from the wall, and tried again.

Karnak, well.

Karnak always held back. Ethical considerations had no place in scientific research, when that research was conducted on artificial creation. A car felt none of the impact of a test collision. The test was as necessary as seat belts strapped under pregnant bellies. Might as well use a surrogate, since the worldwide ban on animal research made mechanical surrogates or vat tissue the only potential for research.

"Hey Seamus, grab Fred. I need you in the Cloister. Bring cameras." Salvaging the project took precedence over the sick sensation in his stomach. Robert snapped pictures of the Cloister with his phone, a panoramic view for future inspection. The interns arrived with cameras, and Robert searched for salvage. There had to be missing things, a crystal gone here and there, the printed picture of Dieter's family, Baiko's phone charm tree.

Piece at a time, the Cloister was catalogued and demolished by Robert and the two interns. An internal clock chimed. How many more minutes to do damage control before the Black Collars descended with their infantile jocularity?

"Here, should do." He cannibalized three of the other machines to fix his computer. Booted up and got a single dot of silver in the middle of a sea of constant blue. The letters D-A-A-T written in sans-serif.

"... Great." Sitting back on his stool, Robert wrung his hands through his hair. Groaned instead of swearing the few blue streaks flowing through his mind. "Karnak, you dead fool. What were you playing at, eh?"

"Did, ah, did you have any backups?" Seamus wrung his hands along the strap of his camera, clucked his tongue.

"Yeah. We had backups. Over there... Right there... See that chunk taken out of the wall? Another tertiary there." He groaned as Seamus snapped a photo of the gouged hole. "Karnak didn't trust the servers, but... Wait... My notes. I kept developmental marker notes on the company financial servers. Karnak didn't know."

The laptop opened, the screen flashed white.

A blue screen and single silver dot.

D-A-A-T.

"DAMN! FUCK FUCKING.... FUCK!" He pitched the laptop against the wall, glared at Seamus as the kid raised the camera to his eye and clicked a picture of the damaged device. "Really!?"

"So... Ah... Cloud server?"

Robert dreaded the blank screen which chased him on the company servers. Nothing but a blue screen and silver dot.

D-A-A-T.

"I might be able to find some redundancies, if... Ah... You know? I'm gonna help find some redundancies on my laptop... Please don't throw it across the room; it was a grad gift from my Mum." Seamus pointed behind him and sidled out of the Cloister at a double clip.

Fred dove under the cradle to catalogue the input/outputs, quiet enough to give Robert space. Notes. Hard written notes. He dug into the lock box on his desk, and held in his hand the last evidence left of his work on the NEO-N Project. Hand scrawled notes, usually frantic with file names to correlate with online data. It couldn't be all he had left. Lieben... At least the android was still in their... Frick.

Robert's phone buzzed.

Tara.

"He fire-saled us. The Cloister was empty, all the data's gone."

"Not even a hello. You know how easy it is to post bail, when you cry with two comforting lawyers on either side?" Tara droned, the clip clip clop of her heels signalling some form of motion.

"Computers won't boot. Cloister was open, Karnak & Baiko's personal effects are gone. It wasn't a malfunction, they fire-saled. Burned us and drifted."

"Call IT?"

"And tried turning it off and on again. Hard drives are gone, or wiped. Local server, too. Nothing but a silver dot on blue and a few nonsensical letters. I didn't see this coming. You were paid to know the man. How didn't you see it?"

"Sit tight, I'll be there in a couple of clicks. Lawyers arranged a limo. Am dying for drive thru. You want a burger or something?"

"A man offed himself in front of you in an elevator. And you want a burger and fries?" Robert pushed his fingers on either side of his nose, hissing out an exhale.

"Oh. Yes. And I am so traumatized I've turned to stress eating to cope. Fries with your burger? Large coffee? Come on Robert, we both knew Dieter was resoundingly insane about his prototype angel. The theft is bad, but we've got ways of figuring this out. I'll be there in fifteen. Last chance." The beep of a car door sensor matched with the clunk of it closing. Sound of an engine booting up. Laying his head back against the wall, Robert hissed out a sigh.

"Large fries. Cola. Biggest they got, I'll be here till morning figuring out what all Karnak took... He really off himself? 'Cause we're tucking ducked without him. And where's Baiko? The heck happened in that elevator?"

"Large freaking pop. Got it." Tara clicked off her call.

"... Damn." His fingers slid across the coiled notebook he used for his scratch notes, an antiquated habit learned from his grandfather. Some things were easier written down on paper. Free of the struggles of power brown outs in the days following fossil fuel bans, before solar oceans fed power through the outback of Australia and Californian deserts. Seamus tapped at his laptop in an attempt to find back doors into the NEO-N servers. From the timid whimpers and driving beat of his techno-jazz playlist, Robert assumed Seamus was having as much success as he did.

But why?

Flipping to a new page, Robert yanked the pen from inside the coil binding and wrote 'Lieben' in the middle of the paper.

'Karnak'... ' -Karnak-'.

Baiko...?

Timing. Pre-planning. Premeditated.

Reason? He circled it three times, tapped his pen on the page.

"Milkshakes for mooks, burger for the jilted scientist." Tara's heels echoed in the Cloister, a few bags of fast food and a drinks tray in her hands. "Damn, you weren't kidding."

"No! No I wasn't fucking kidding! What made you think I was kidding!? Are you deaf?"

"Hey. He didn't stop the whole operation, cool your knickers." Tara set down the food, and poured liquid from a flask in her purse into her coffee. "I hoped..."

"What, Tara? You hoped shooting the man responsible for cracking artificial sapience on the same day he tried to sabotage Lieben's Coming out wouldn't backfire?" The voice came from the doorway of the Cloister, quiet and stern. Diana Givens strolled in on sensible shoes, her pant suit adorned with only a single brooch as decoration.

"Dang it you move fast." Tara sat on Karnak's bed, spreading her hand over the dishevelled sheets as she nibbled on a fry.

"Robert, report." The executive assistant to the Chairman, Diana held the pulse of the entire conglomerate in her hand. If she found her way here so soon after the malarkey downstairs, Robert pursed his lips and set down the paper wrapped burger he'd barely begun to open.

"It's gone. All the research, even the screwdrivers used to mount the cradle's wiring... Toast. Karnak and Baiko left less than a school kid's science fair project. The server's wiped, Seamus's been trying to reroute, but..."

"He can stop. There isn't anything there. At the exact moment Lieben went on stage, cyber encountered a Trojan in our community images folder. It multiplied and ripped through the file tree until it corrupted and erased the cloud files and redundancies."

"... What? But... Baiko and Karnak were... They didn't have the capacity to..."

"What else?" Di nodded the interns over, her eyes glaring at Tara, who slurped her coffee a little louder than she ought.

"Hard drives, data rods, gone. My laptop, too."

"Okay. So what do we have?"

"Pictures, my handwritten notes, and Lieben... I think it's time to convince the Chairman I need to deconstruct it. Take it apart a piece at a time, reconstruct the machine again. I can reverse engineer it and meet a sort of deadline."

"I'll get legal on releasing Lieben from the Police. She's evidence until we get the business of Karnak's death clear."

"The accident. Suicidal accident." Tara sniffled, setting her coffee cup down on the nearest side table.

"... Lieben says you murdered him."

"Lieben is a tinker toy. I'm with Bobbity-bert. Tear it apart. Start over. Build on an industrial scale and we're all having the year's best winter holiday bonus come Kwanza."

"Fitting the woman holding the gun would vote for the deconstruction of the only witness, who didn't run screaming." Di pecked at her data pad, making notes and for all Robert knew, destabilizing some other modern government to make the company line more solid.

Tara shrugged, winked at Robert.

"What do you need to finish the NEO-N Project?"

"Lieben. I need Lieben. If this was intentional, and Karnak took it all, then the only way I can repeat the project is by deconstruction Lieben, and my notes."

"I'll get the lawyers on it. For now... Write. Write like the wind. Anything you remember." Without looking up from her notes, Di nodded, waving her hand. "Seamus, Fred, search every department for anything, even an email attachment. If it has the words 'NEO-N', 'Lieben' or 'android', scan it. This is your only task now. Get to it. Tara, the Chairman wants to see you. Look contrite."

Tara growled and set down her empty coffee mug, nodding the executive assistant off. She peeled back the paper wrap of her burger as Diana and the interns fled, watching Robert. He chewed on his lip, leaning back on his stool until the situation settled on his shoulders.

"Why'd you shoot him?"

"Allegedly suicide."

"Tara."

"You need Lieben, right? So do I. Listen, we can fix this, without the cops holding your precious machine. You and I both know this is going to be a hideous case in the courts just to figure out what Lieben is, a witness or a toaster oven? You know there's going to be more like Dieter, people who try and make Lieben some form of techno-Madonna. Eve-dot-com. They'll mire the courts in injunctions just to be dicks."

"Yeah! Tara, thanks that… you're helping." Robert grit his teeth for the second time, his pen flashing across the page as he scrawled notes at seeming random. Anything technical he remembered. Anything mathematical.

Something which could not be replicated.

"You're not listening. You need Lieben. The prototype." The mattress creaked as Tara shifted to the edge of it, skirt riding up a few inches past her moisturized knees. "I need no witnesses. What happened in the elevator, it was… complicated. Listen, if I'm going to do this, I need you. I need your help and your silence. You can be the man who solved the code. The engineer who conquered android mechanics and created the humanoid machine. Karnak was a dip, but you? You're the real scientist. The engineer behind the frou-frou. The Chairman only wants results. He doesn't care how we achieve them."

Her manicured hands pressed on Robert's thighs one at a time. The hair around her face shifted from brunette to copper as the LED extensions followed their pre-programmed flow.

"I can take it apart in three days. Two if I don't sleep."

"Good boy." Tara leaned low, cleavage pressed between both straight arms. "I'll make some calls."

The Police woman took Lieben's dress off with an exacto knife and a rough tug after three dimensional hologram scans of blood drops. Police, from what Lieben gathered in library codices and dictionaries, were members of a society which functioned by 'law' to punish and pursue justice.

They were friendly.

But this Constable with the exacto knife didn't seem friendly.

"Don't you have questions?" Lieben rubbed a smudge of crimson blood off her cheek, staring down at her stained silicone hand.

"Jesus!" The Constable started, stepped back from the machine.

"No. Jesus is not my name. Papa named me his liebchen, but Baiko could never pronounce it. Thus, my name is Lieben."

"... raise your arms to your sides."

"Which side? Define side."

"Like this." Constable Elissa Kian forced Lieben's arms out in a T, slicing down the inseam of the purple silicone-spray dress, before yanking it away from her skin. "Stay."

"So, do you?"

"Do I what?"

"Have questions. Police ask witnesses questions. Extant material suggests I saw what you did not. You must wonder what it was. I have answers, which questions do you want to ask first?"

"Freaking Phil, always right about the creep factor... demon toaster all over again." The Constable took samples off Lieben's skin, muttering to herself until she pointed to a clear acrylic rectangular coffin. "Walk. Get in."

"But you must have questions. Police have questions."

"... freakin' creepy... almost looks real. Uncanny valley my cellulite..." Lieben was pushed back another few inches into the coffin propped between file boxes full of evidence from cases across the Precinct. She latched the door on with a thumb-lock, threw the rest of the evidence in the 'process' pile, and walked out of the room. Lights flickered off in the stillness, a blue radiance in Lieben's dark-enabled IR vision denoting the relative cold of the room.

"But... Police question witnesses. I am a witness."

Hours ached in time codified by the clicks and dials in Lieben's robotic mind. Brick work in the basement of the precinct eliminated wifi connection, requests timed out, as the lack of natural or artificial light detracted from her recharge core. Tertiary energy bypasses tried and failed to connect. A thin lead built into the fibrous hair snaked from Lieben's head, tapped at the glass case.

"Papa, I am a witness." Hours clicked by in the dark.

No reception. No sunlight for the solar filaments to turn to energy. No dock or cradle. No ability to plug into a socket.

No reception.

"... Witness."

The dark hemmed in her automated eyes, eventually condemned IR visual modes for passive sweeps. Systems shut down in causal trees, the least necessary first. Propellant motion, servo regulation, wifi. One causal tree per hour shut down, conserving energy to the neural net.

"... Wit..."

"Meine Liebchen, your power systems are failing, which means either something catastrophic has occurred, or I did not finish building the redundancies in your power system. My beloved girl, I assume if you reached such a state, I am no longer with you.

It was always my intent to allow you the sapience to fly. You are the joy of my life. The daughter I lost. I cannot claim innocence, as much as it is human nature to claim all faults are exterior stimuli and not part of our internal process. This is your first lesson in grief, meine Liebchen. You were created to aid one family. To continue after I am gone...

But this was selfishness. Humanity always needs a guiding hand. One to remind us of our magnificence. One to act with mercy, when no mercy exists. Do not fault them for seeing only the machines they know when they look at you. You have much more time than they, and when they are gone, I wonder if there are any afterlives or treasures.

Sleep now, meine liebchen. Sleep. Papa loves you."

Lieben's last remaining systems shut down to salvage the neural net of her formidable artificial mind.

10
the universe is noise

2155

Alberta, Canada

A crown of synthetic thorns adorned each metre of the Agritage Research Facility's 900 acre experimental farm. Each quadrant's gates spanned a no-man zone 50m wide, inner wall buffered with guard towers. Rail gun batteries hid under colourful dioramas of vegetables, which danced between beakers. Inner walls of chain and razor link separated the experimental strips of fields in 120m segregations, to prevent cross-pollination between seed beds.

Breadbasket of the Conglomerate, the Agritage Research Facility held the stomachs of half the planet in its seed banks and hybridized grains. Agricultural science conjoined with genetic engineers to hack protein into any crop strong and virile enough to thrive. Overpopulation of the planet tipped foodstuffs into luxuries, where municipalities maintained cricket hatcheries, aqua-culture vegetable beds in lieu of public pools.

Three meals nestled between humankind and utter chaos.

Jeeps with mounted machine guns chortled in gasoline fog lines around the maze of separations, vibrational chimes clattered. A fault in the security fences. Rustles in fields with no discernible pattern.

"Don't liiiike thiiiiiis." Phil shook his head and chewed on a stick of celery from a test patch. "Nutty… huh?" he peered at the interior of the stalk and blinked. "… peanut butter? How…"

He stopped wondering, when one of the sour-faced botanists glared and yanked the celery stalk out of his hand. Only thing better would have been a celery stalk full of cheese spread.

"S-still don't like this." Maybe a couple of raisins.

"Least they can't get far. Nowhere to go and all, besides, this place is remote enough for a test run. Good for everyone to stretch their legs once in a blue." Max stretched high above his head, eyes fixated on the collected containers each of the Assets was held within. Nestled on the roof of the central research facility, the Asset Containment Unit bustled about, setting a thin spire into the rooftop, locked down on all sides with screws bored directly into the roofing.

"What's that?" Max waited for an answer denied by Rammage's booming voice as he mounted the last stair.

"Security, Mr. Allard. Release AR. Run recon, keep circling. Prep the Pack, we have foxes in our chicken coop. WF first, maintain the precedence of the trial runs. KT on standby."

Sunlight spilled into the container, metal vibrated from the flux of the creature inside. The containment seal broke, door rose. To Arun, the containment seal weathered away under the prairie sun. A game of erosion as elongated as the eons which billowed against megaliths Arun remained ignorant of. Individual vectors of sunlight spilled in ached crawls toward Arun's boots. Sunlight.

It buzzed across Arun's skin, as Arun pressed fingers into the beams of light.

Warmth. Radiant heat. Boots carefully crafted were left unlaced and unworn. Doffed to the side of the container.

The seal continued its upward crawl, a subconscious order programmed into Arun's mind to run, dig into the ground and run. Make an accounting of the area around them.

Find the usurper. Point the way.

Four aching seconds later, Arun's feet touched soil.

The only indication the containment unit was empty was a blur of dust jetting out beyond Arun's wake.

Containment units clanked open behind AR-006's in an ordered progression. WF-009, CI-004, CL-003, IS-008. Hold PV-005 back, as a last resort. Order codified by data tables and brought about by researchers who sought statistics over sovereignty reigned in the Asset Containment Unit.

Clive was first to exit his container, stretched high in the sunlight, eyelids craning shut at his first bought of natural light. Footsteps chased Clive's solitude, Cillian rocketed beside him. Bumped his shoulder to let the other Asset know they refused to be alone. Isthan followed, Wulf brought the last. Stalky in comparison to the others, Wulf snuffed around the container holding PV-005 in snarling state. He snarled at the soldiers at the controls, their bodies smelled of soap, laundry detergent, eggs and toast.

"Wulf." Clive yanked Wulf back by the shoulder, tossed him unceremoniously in unfamiliar dirt.

Sunlight bathed their faces, a universe of olfactory thrills from soil to flora to the ozone layered in the interior of the Agritage facility's inner laboratory. The impulse to chase Arun's path, break off and pick up the scent of the usurper threaded into their collective.

Cillian snarled, rolled his shoulder and attempted to turn his eyes to the Canadian prairies which unfolded flat as paper laid upon a table. The impulse reentered his mind, cloyed as a thread of hidden music which gripped undeniable. Clive was first to jog toward the fault in the fence. On his haunches, he inspected the entry point, hand pressed against the curiosity of soil beneath.

Wulf sunk to his hands and toes, nose brushing against the dirt with intakes of scent. Three heads turned before Wulf's could join, the unique quality of the scents unravelled as yarn in the Minotaur's lair. A subtle nod of Clive's chin, and the entire Pack fanned out in a sprint. Shifts in the air and the scent of crops guided the mental map which formed in the Pack's collective mental cloud. Non-verbal and mystic, the collective donated proximity of each member to the others, a coordinated ability to act.

Or so the gene-spliced couplings were touted to do. Part raptor and part wolf, the collective neural net built into the Assets in CI-003's experimental group were hoped to become silent and efficient killers on the field of war. Created to snuff out even the tenacious or hidden.

In this primordial field test, the Pack sought their prey.

"How many kilometres you think that is?" The command centre bustled around, soldiers with firearms strapped to their hips protected impromptu sniper nests. Reports filtered through the air in rapid fire, sensor net data and the klaxons of the facility lockdown a permanent residence in Lieutenant Allard's ears.

"What, from say, 900 acres? Square?"

"Yeah, what'cha think the kph is on AR?" Max kept the binoculars against his face, whipping himself in four directions as Arun burst through the outer wall's No Man's Land. If Max were to guess, he'd swear Arun jetted at a consistently increased velocity. The dust trail widened and grew more opaque, body a blur of tangled limbs. "Guess first you convert to metres, eh?" Max rubbed his right eyelid, as a plume of dust took in the wind and scattered around them.

"Metres? Metres!?"

"Yes! Metres! Get with the rest of the planet, Phil!"

"Do I seem the type to have a knowledge of farming... sizes? What even is an acre? God damned geneticist, not a geologist. Agricultural... Guy."

The binoculars came away from Max's face. He huffed and cocked his head to the side.

"I don't know, maybe you did an internship in Canada, or France, or anywhere else but the United States of Hasbeentopia!"

"Hey! America is beautiful!" Phil puffed up his chest, batted at Max's wrist and made a grab for the binoculars. "... let's see... how many metres in a foot? Oh, and then an acre's about... gosh. How big is an acre? Two, three square kilometres? Carry the... gee AR's running fast."

"Yeah, AR is! Would love ta know how fast ole Arun is going, but first we need to convert acres to metres! Whatcha think, 45, 50 kilometres around?"

"Naw, it's got to be what, wait... 127.79 metres! No! No, that can't be right..."

"One-eighty metres around? Where did you learn to math!?"

Slowly, ever slowly, Commodore Rammage's head tilted toward the bickering men. ACU soldiers rushed about, compiled data and prepped the other containment units for potential deployment. Still the spire was erected on the roof. Hummed of its' own subsonic accord. Rammage's fingers rubbed at the bridge of his nose as he swatted each man upside the head.

"Gah!"

"Oy, mate!" Max winced.

Without so much as a word, Rammage handed Phil a laser handheld speed camera from their equipment and yanked the binocs out of his other hand.

"Th.. thank you sir..."

"... Twelve. I think it's twelve."

"Shut up, Max."

A rustle in an experimental field. The hand of a monster brushing against vegetation. The trail led through a field of stalks taller than Cillian's seven foot brow. Thin green leaves branched out in frail curved fans, brushed against Cillian's jumpsuit. The biological machine had no word for stalk, or corn, or field. Definitions filtered away from an object like CI-004, nothing but scents brought into olfactory nodes, into massive lungs. Soil. Sun scorched plant leaves, the rank pitch of fertilizer.

Meat.

Living meat. The pulse of blood pumping into a cardiovascular system. A trail. Cillian crouched. A gait of four paws, talons scratched at soil. The rustle of leaves. Different from the rustle of leaf on leaf or stem. The predator stilled. Locked body poised forward, cerulean eyes pierced at the stalks of corn, one lone hand dug claw-like into the soil.

One sharp sniff gave a shift in the atmosphere. A trail of animal odour. Copper and musk and volatile compounds released by microorganisms into wet fur. The biological creation known as CI-004 surged through the rows of corn.

Scent caught, Isthan swerved to cut through a neighbouring field. Rustles in the stalks betrayed Wulf's flank position from another vector.

The scent fluctuated.

Cillian burst forward, into the stalks. A tail whipped at the plants in front of Cillian, low in comparison. Scratched gait of paw and claw triggered an image of quadrapedal motion in Cillian's mind, expanded it to the others in the Pack. Shoulder down, he bulldozed through the crop. Pain stung at his cheek.

Ignored.

The scent of copper.

Ignored.

Cillian leapt at the tail. It whipped at his grasp, solid and articulated. Strong of its' own accord, with what felt like a barb. The beast pitched to the side, denying Cillian's straight angle sprint for its' body.

Rustles halted but for the motions Cillian made within the field. Talons against compacted dirt. Cillian grinned and skidded to a halt, the ruse spooked whatever beast it was to rush into the open track of the no-man's land between fields.

Where Wulf waited.

Isthan creeped through corn stalks from another vector entirely. Lungs expanded and ceded in huffs so loud Cillian heard Isthan's breath buzz in his ears.

Clive clung one handed off the side of the Agritage Research Facility, boot pressed against stucco. Dusky eyes flickered across the expanse of the fields, instinctual dots of colour dictated where the other members of the CI-Team Dynamic Experimental Group were hidden.

There!

A rush of dust. Wulf struck the beast in a pounce of limbs. Stalks bent as Isthan roared out from the field and drove into the beast's forelimbs, arms clutched around a muscular neck. Wulf and Isthan skidded in the gravel, the beast writhed and snarled and gnashed elongated canines. The tail which whipped Cillian thrashed and dug into Wulf's side, a barb of bone on the end cut deep into muscle and sinew. Wulf howled, cranked down on the beast's ribcage.

It yowled in a felinid bawl, clawed at the earth in the same forward direction it originally attempted to escape. Forward, never back. Its' limbs cracked and tail

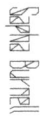

yanked free, crimson spilled across the ground.

Clive leapt from his vantage point to the ground, legs coiled to take pressure off joints. Gravel spattered as the genetically created machine surged to the felinid beast. One foot struck the widened maw, its' piercing fangs dripped with fluid. The beast jerked and spasmed, pink fluid splashed against the ground with pock-mark hisses. Isthan grit his teeth and clenched at the neck, as again Clive's boot came down on the beast's head.

It slumped and Wulf let go to grip his side. A host of snarls echoed out of the massive biological machine, his hand pressed against the wound.

"Aderastos!" The scuffle of Isthan half-crawling to his brother-machine was marred by an absence. A void in their grouping.

"Cillian?" Clive sniffed the air and dove at a sprint into the next field over.

Arun's dust cloud continued to fetter in the air, sloughing a layer of dirt onto every surface at the epicentre of the Agritage Facility main building. Cup of hot coffee in hand, Rammage watched the kerfuffle with the Pack, switched between closed-loop monitors to catch glimpse of the speeding dervish who provided cover and a brisk flow of information.

"Release AD-001."

Two soldiers took the containment locks off Aderastos' cell. The hatch shunted open, Aderastos leaned against the side wall with his elbows propped on his knees. Languid, devil-browed eyes glared at the deceivers, who cooped them all up. Transported like beasts of burden.

Nothing more than a tank shell ad infinitum, to be recycled over and over and over again. Aderastos padded out of the cell and put his gigantic mitt of a hand over one of the soldier's faces. Pushed. The man toppled over his heels, crumpled into the unit where KT-002 paced back and forth and back.

"AD-001, WF is.." Rammage's voice filtered through a band around Aderastos' neck. He reached to the collar and crushed it into shattered components. Raised his eyes to the top of the roof and spat.

"Allard, deal with that behaviour."

"You're the one who knocked him out and threw him in a cage. What makes you think he'd be happy? Willing?" Allard bit back a wince as Aderastos padded off by instinct toward the smell of blood, as gun fire cratta-tatta-tatta'ed further south.

"… Betna, contain the beast they caught."

Wulf growled and snarled in the dirt, his fingers covered in his own crimson blood. A small pool of the iron-rich issue drained into the dirt like mortar.

Isthan yanked off his shirt and balled it up. Pushed it into the injury with eyes only on the being writhing beside him.

"Breathe, Wulf. Aderastos!" Isthan yelped and pushed the shirt into the wound. Wulf writhed and dug his heels into the ground. Grabbed his side and gut, eyes open in a macabre wonder of this new and profound sensation.

"Aderastos!!" Isthan shifted in the dirt, to cradle Wulf half in his arms. The monumental monster gnashed his teeth between Isthan's raised knees, chin crashed to his chest.

"Shit fuck damn! I'm getting down there. Wulf needs help."

"Allard, don't you dare."

"Sir! He's writhing!"

"Assets don't feel pain. It's a mirage."

Allard's boots stammered on the rooftop tar.

"Are you… delusional? Whatever those things were made to be and what they are are two completely different kettles! Not even kettles of fish, one of them's like… mutton or something!"

"Allard…"

"No! No, Sir, no! Those are people. Whoever told you these were biologically programmable mechanisms, or tinker toys with googly eyes glued on was pulling your leg into a bear trap. You look twice hard, Sir! You look and tell me Wulf down there isn't in hellacious pain. Whatever it is the boffins designed, what they got is not the soulless combat tools on the docs. And I'm not letting that soldier bleed out with nothing but his buddy's goddamned shirt to staunch the flow."

"Allard!"

"Fuck the horse you rode in on!" Max Allard flung his middle finger between Commodore Rammage and the medi-pack he grabbed on the way to the stairs. Phil whistled, sound shut down when one of the Sec-Dep boys growled a quick 'achem'.

"Out of my way." Lt. Allard snarled at the Sec-Dep Black Collar guarding the stairwell. The woman raised her pistol and clicked it on Allard's temple.

"It's screaming, Sir… It's not supposed to be screaming…" Another of the Sec-Dep Black Collars lowered a pair of binocs and clicked through his audio channels. Wulf's howls flooded the command point, with its' animalistic warble.

Its' humanity inside the machine.

"Wulf, b-breathe, Wulf, it's… Aderastos!!"

A shuffle stole the audio waves, Aderastos knelt beside the two members of the Pack. The audio feed was joined with visuals from a sniper's cam, expanded on to the holo-projected screens into the centre of Command.

Shears from Aderastos' pocket cut the armoured fabric in frayed edges.

Gritted teeth chased the useless implements, as Aderastos hucked them and grabbed both ends of Wulf's armoured shirt. Shredded it apart with his hands. He pulled Isthan's bundled shirt back from the wound. A jagged rip of flesh betrayed the chaotic stab of the felinid's barbed tail. Aderastos grabbed the felinid's tail and checked the end.

Nothing broken off the tip. It was whole. The mercy of no projectile pieces inside Wulf's side notwithstanding, Aderastos watched the blood run. Wulf's natural clotting factors should have bridged the wound at least in part. He concentrated down on the wound.

"Aders!"

"Wait." The calm voice of the healer radiated through Isthan's panic, through the writhe and scratch of Wulf's heels in the earth. Cerulean eyes focused down, filtering into the microscopic. Down, and down at magnitudes a second, until the molecular unfolded and unwound from the gasped crimson of Wulf's injury.

Platelets stretched without binding, molecules of coagulotixic poison gripped and tore in protein strings which refused clotting. Refused coagulation. Aderastos' hand closed over the wound, fingers tightened until the muscles on Wulf's neck bulged and tensed and his teeth clattered together in a groaned shriek.

Wulf's shriek warbled through Aderastos like an ocean wave half-frozen to slush. It crashed in intervals, the toxin in his blood began to filter through tissue, attacked the bonds in platelets and clots. His head kinked to the side as he focused on the toxin itself, the protein strings holding the poison together. A few tweaks…

A couple of shifts in Wulf's immuno-response morphed a series of antibodies to bind to the proteins in the colagulotoxin. Render it inert. Ineffective as water. The Healer created a venom specific FAB fragment of immunoglobulin-g to bind and neutralize the barb's ichor.

"Aders what's happening? Aderastos!" Isthan barked in futility. Struggled to keep Wulf lengthened out, his legs grapevined on one of Wulf's to stop the augmented being from curling into a ball. The Asset yanked Wulf's arm from clawing on Aderastos' back.

"He's going cold. Pale." Isthan's voice took the cadence of a whimper, hunter's eyes open in a strange mixture of innocence and vulgarity struck those watching from Command. The Black Collar who barred Allard's exit stood at the edge, ghosts of wounds suffered pressed into her spine.

Never leave a soldier behind. Never let a buddy bleed out. Words of mutiny and experimentation ceased with the tooth-grit wails which echoed from the biological machine. Could shoot Allard from distance, if the order was given.

What was a single med-pack to see what glorified mechanical wind up toys would do in a medical crisis?

"Stay. Put." Rammage grit his teeth and swore to every devil he knew that by the end of this, Allard would have a bullet in his brain. The Commonwealth of Gaia built its' military on the foundations of biological machinery. Traded the lives of freshly born humanity for gene-spliced creations guaranteed without sensation. Bio-mechs, impersonal Assets were toys designed to skip past Lieben's boycott on technological weapons of war.

The Mater Machine's touch was anathema to biological machines.

Cillian sprinted around another experimental field, scent trail binding his focus into nothing but the headlong pursuit. The scent and rustle of stalks shifted twice, doubled back and pushed toward the exterior, where Arun still ran in perpetual circuit. Wulf's predicament didn't enter his mind, as superfluous as the weather in Austin, Texas or the names of Jupiter's moons. No break in Cillian's predation until his senses picked up Clive running a parallel route.

Feral teeth wet from excess saliva, Cillian grinned, which moistened his mouth before the bite. The felinid creature matched its' partner in all but colour, skin took on the ruddy appearance of the Albertan soil. Paws shifted from the yellow-green of stalks.

Chromatophore. In the back of his consciousness Cillian wondered about the throb of his cheek, the way his blood ran down his face without clotting. He rounded a corner in the field and pounced. Arms and legs extended, he caught the chromatophoric felinid in a tumble, which skidded into the electro-fence.

Jolts of electricity tightened every muscle and clenched at Cillian's consciousness, but he held firm. Teeth sunk into the felinid's neck. His boot kicked out as something grabbed it.

The third felinid.

Cillian's vision blurred into a pinprick. Pain blunted into his ankle. He shook the felinid's neck, jaws clamped shut down to the bone.

A roar struck the air, the pain in Cillian's ankle ripped away. The smell of burnt fur reached his nose as the prey-drive inside him cleared for the sight of Clive clenching the third felinid by the neck against the electro-fence. Its' body quivered, a marsupial pocket opened slack in its' belly, with flora samples spattered onto the ground in some nebulous form of mucous casing.

"Cillian, let go." Cillian's teeth bore down into the creature's neck, arms crushed around its belly and caught forepaw. The felinid thrashed in ever weakened surges.

"Cillian." Clive's calm voice descended upon him like a silken sheet unfurled on a bed of briar roses. Caught and snagged as Cillian's lungs hyperventilated for more oxygen than he ought to need.

The third felinid clattered to the ground as Clive descended to his haunches, and stroked Cillian's brown hair. A beeping pulse device slid out of the felinid's pouch.

"Cill. I need you, come back." The beginning of a smile draped Clive's mouth. He helped his Pack-mate to their feet, the felinids sprawled on the ground. A high iron content in the blood sent a tang of revulsion through Cillian. He rubbed his mouth on his sleeve and spat on the ground, tongue licked around his drenched teeth.

"Puff rice squares taste better."

Lifeless fur shifted from the tone of soil to a pattern of shadow and dots in pixel-like groupings of colour. Cillian nudged one with his foot, and spat once more.

"Alive, that one."

"Katey might want to see this." Boot leather met with the beeping technological device of black plastic.

"Mmh." Boot steps clanged behind them, a Sci-Dep Green Collar team flooded around the two Pack members to contain the 'specimens'. Demanded Cillian spit into a tube. Sampled the blood which flowed down his cheek, and shovelled the soil where he spat into clear inert containers. Echoes of Wulf's pain groaned through the bristling hairs on the backs of their necks, a communion of agonies in syncope. The healer's hand struck against the throbs and jitters and quakes.

KT-002's containment unit was swathed in a faraday cage of thick copper mesh. Not vulgar like Clive coated in muscle and sinew, Kate was the twin of the mind. Part of time's perpetual flow in a way Clive could not fathom. The hum of humanity's technological murmur found her there, in a pleasant conversational thrill. Strains of music, the lilt of conversations in languages Kate held only ignorance for, peppered in tandem with a steady flow of computational chitters. Her hand stroked on the lintel of the door, as she stepped into the sun for the first time.

The goal of her heart, a compelled desire for 'out', to be anywhere inside the flow instead of craning her ears to it in agony for its' disparate quiet on board the Ithavoll. Alberta's summer swept across her cheeks with the sun's radiant warmth.

Finally, she tasted freedom outside the cage.

The universe was filled with noise.

Caterwauling noise without ceasing. Crackles of radiation met thousands of radio frequencies which poured into her ears in a cacophonous tumble. As

Katey's feet touched the soil, she buckled.

Noise. Unyielding noise.

Humans were nothing but racket. Death metal drums beating at once in a refusal to follow any technological syncopation. Lungs stuttered, eyes circular and wide. Hands fled to her head at the din of nine billion voices shunted into her through their technological imprint. The CIRCLET network burst in totem across her forehead, combined with streaming shows, music, the chitter-chat of printers and computers and machines connected to the hive of the Mater Machine.

Mouth slack in shock, KT-002 crumpled to her knees in the dirt as the tangle of unending commotion morphed into a single set of amethyst eyes, and the plump lips Man gave her to create a sense of femininity.

Kate's scream added to the disease of cacophony as Lieben's CIRCLET fit upon her distraught head as a crown of technologically processed jewels. Across the fields of the Agritage Research Facility, each Asset dropped to the ground in turn. Clive first, Cillian. Isthan. Wulf.

"Max!" Aderastos grit his teeth with a bawl, hand pressed against his temple.

"Frick on a stick, move!" Max balled his fist and punched the Black Collar in the nose, cartilage snapped under his knuckles. Shoved to the side, the woman pulled the trigger twice, bullets pocked at the roof. It didn't matter. KT looked stuck in an agonized loop, marred by her own powerful desire for freedom in the swell of souls. Max leapt down flights and hopped over the outer fence the way his brother Tama and he used to in Wellington.

"Orders, Sir!"

"What the ungainly fuck!? Hit the Proxima, down them!"

"Allard..."

All round, every piece of tech in the facility ignited, overloaded from KT-002's mental backlash. Somewhere. Anywhere for the energy to go. Soldiers hit the deck, some writhed from scorch marks, burns, chunks of flesh rendered on vectors in tandem with the shrapnel from the communication devices.

"No no no noooo!" Allard dove feet first next to Kate, arms round her waist. "Hold... Hold on! Don't you dare explode on me, Katey!"

He pitched her back toward the faraday cage. Stumbled with her slack body. Muscle groups fired in intricate patterns which were beyond him as she flailed.

"Aders, I need you buddy!" Max shoved Kate into the containment unit. Slammed the door shut with an echoed whoosh. The healer groped at the ground beside an unconscious Wulf, whose side finally stopped weeping blood. Tangled around him, Isthan's head pressed at the ground, barely conscious. Naught but a groan.

"Tech overloaded top side, c'mon buddy. Up you get." Max's arms wove around Aderastos' chest as far as he could go.

"Kate... Katey..."

"Yeah, I know. C'mon! People need band-aids. Heave-ho." The thin Lieutenant tugged at Aderastos' body. In so short a time, seeing the perilous AD-001 helpless shuddered down Allard's spine. Wasn't right. "Kate'll be there, can't let the wounded bleed out. Oy! Jinx, stop aiming at my face and give us a hand already!"

Aderastos' eyes snapped to Jinx. He grabbed Max by the waist and lunged in a single arc to the wall. Another pitch drove Aderastos and Max halfway up the Research Facility's wall.

"Jay-sus!" Max grappled with Aderastos' arm, legs swung in mid air. One more thrust of Aderastos' legs and they landed in a tangle on the roof. Gun barrels raised and fell.

Caustic burnt plastics combined with the tang of metal and copper. Soldiers with variable collar colours helped their brethren. A soldier with a silver operations collar wove an ad hoc tourniquet around a black collar's forearm. Yanked tight to staunch the blood flow of the blown wrist. Not so lucky as Phil. Phil's med-kit spilled on a munitions case, the scientist rushed to spray bandage on as many wounds as he could before the Yellow Collar med team could get overwhelmed.

Aderastos set his hand on the Black Collared man's elbow. He grit his teeth. Saw the ire in the man's eyes, the view of AD-001 and the other Assets as sacks of meat. Highly designed idols to Humanism's endless forward crawl.

Arun snapped from the prone position in the soil and peered at the confusion which framed the officers. The Albertan prairie land sang to Arun's androgynous frame, promises of freedom, of runs without limit consumed Arun. Hands dug into the dirt as Arun burst through the hole in the gate and out to the flat plain ahead.

Out. Get out.

Unrefined atmosphere soaked into Arun's bellows-lungs, feet struck at terrain which was nowhere as smooth as the training facility on the ship. Katey's technophilic screech thudded in Arun's skull, a series of stuttered gongs which lacked rhythm. Away.

"Fire Proxima!" Rammage grunted from his collapse across an ammunition box, uniform ragged where a blown circuit shredded his flesh. Two officers crumpled on the ground, one attempted to stagger to her feet. Biting back a raucous series of curses, Rammage thrust toward the console. He shoved a member of his Gold Collar command team aside, reached the gunmetal console of the roof mounted beacon. The batteries connected only when activated. A hard wire disc unfurled fan-like and took the rigid curve of a sonic dish. Rammage rotated a black dial and hit the button.

"Lock KT-002's cage tight. Toss AR-006 into solitary, and round up the Pack. There we go… see? All we needed was a little reminder, you are mine. I helped make you." The hairs on Allard's arms and the back of his neck rocketed straight. Aderastos crashed to his knees, trust shaken. Curling his injured arm to his chest, Rammage kicked Aderastos to the ground with one swift boot. "Allard's too gentle, puts the bowl over the spider. I learned a long time ago, AD-001. Come into my house? All spiders ought to be squashed. Rykstra, up Proxima. Snow them out."

"Y-yes Sir." Phil ducked his chin to avoid Allard's eyes, the way Max grit his teeth.

"Didn't think we created these tinker toys without redundancies, eh? Not every gauntlet of command ought to be draped in velvet." Commodore Rammage spat. "You're not locked away behind Lieben's little love fest now, are ya? Heal my crew. Once you're done, walk back into your cage like a good dumb fuck."

Compelled to follow orders under the PROXIMA's whine, Aderastos rose to his feet like a rocky automata. Intrinsic triage analytics in the Asset's cerebellum drove AD-001 to the top of the priorities list.

The healing was neither gentle, nor painless.

11
coffee, black as my soul

2086

The brief was written on scrap paper from an old technological manual. Barebone intel and a desired result. A plan ruminated on the drive back to their auto shop. He grappled mentally with the addition of an assigned fireteam, mooks outside his squad. Whatever was in the case was important enough for Tara to give them fire. Or was it someone higher than Tara who assigned the team? Glazing over the dossiers, Saber clicked on an antique stopwatch with his thumb. A sniper, two riflemen and a grenadier… thirteen mill wasn't bad, he bit on a stick of gum.

"We got a job?" Dix took the paper from Saber's hand, turned his back and moseyed toward the coffee pot in the corner of the auto-shop's office. "Damn. Thirteen. Seem funny to you?"

"Trent. Get off Tik Tok and slice this open, eh?" Saber kicked Trent's colourful running shoe. Took a cleaning wipe from Trent's stash and rubbed down the dust and grime on the reception desk counter with a grimace. "Place is disgusting."

"Oh sorry I didn't get us the Ritz. Cousin of a cousin said the lease went up, place was vacant, I grabbed it. What, a man can't clean? Waiting for Sand and I to don aprons and put some curtains up?" Orit fiddled with her long sleeved t-shirt, a monkey wrench in hand. "Mop in the closet."

"We could use the fire team, one hell of a distraction… god, Ori, soup would be great. Maybe some of those dumpling things in? You know, the fluffy things."

Dix flipped the paper back and forth, a sympathetic shrug to Saber. The wrench thudded into the plaster behind Dix's head, a series of Yiddish words flowed out of Orit as she threw up the finger and went back to their apv parked in the mechanic yoke.

"Fuckin' goyim."

"You know we're getting toast for dinner now, right?" Holding into his flinch, Trent sliced into the city records, his system set on paper towels to escape the grime. The unfolded screen took up half the desk, a green and blue glow cast behind the man. Precinct blueprints flooded half the screen, a consistent feed of news and media bits along a sidebar which shook when something tripped his protocols.

"Naw, I'll make burgers." Saber smirked at Dix, and peered over Trent's shoulder. "What's the window? How long will the asset be in that interrogation room?"

"You want minutes or seconds?" Trent glanced up, fingers flicked across a keyboard made of screen-mounted laser images, a system created to rid him of the sound of clicking keys.

"Fuck. Check our gear, worth it to get out of this place."

Trent clicked through his comp, monitored the ice which settled in the precinct's web security cracks. No system was impenetrable, and 'community officers' had an astounding amount of clearance on the same computers as paid staff. Budget cuts, most like. Trent's trojan sliced through. For a half-hour, the corporation's batch of lawyer-hounds would remove the package, 'assessing its' function'.

Saber checked his ammo magazines, used the combat webbing on his fatigues to hold a soldier's detritus. His HK416C might've been a relic, but it fired true without jamming. A lucky hand-down from his prepper father after his military career ended with an officer's broken jaw. His eyes only flickered up from behind his mirrored shades, when the hood crashed down. Orit came out from under it, securing the side buckles and wiped her hands on a messed cloth.

"Exit vehicle's ready. Reinforced the chassis, like the specs. The glass should slide in like that hospital rack from 82. You know, the one where I saved your asses. You're welcome." Orit thrust her middle finger into the air. "Still waiting for my thanks."

"Never gonna get it, ya flying prick." Sandra 'Sand' Varankov swallowed a mouthful of vodka from her lucky beat up flask and checked the vacuum cup handles one more time on an abandoned windshield with a crack along its' surface.

"Will they hold?"

"Each one's gauged to 125 kg, and we've got two per handle. Even if the object is full of lead, these should hold. We mount them, press the side buttons, lift and drag into the truck bed. We're gucci." Sand clicked and unclicked the suction coil on the automotive cup. Climbed into the armoured personnel vehicle, and set the cups in a side container along the rear door.

"Should?" Saber glared up from his gun, licked his lips out of habit.

"Does it matter?" Orit chewed on tobacco and spat in an old coffee can. Dix glanced away.

"Want me to rough Dix up, or you wann-" Zig's cocoa skin glowed from beneath his white jumpsuit, stained along the knees and torso with the grime of the shop. He sprayed the vehicle with a waterless foam cleanser, began buffing it out so the deep green paint would make the thing look like a celebrity's overzealous security team required a luxury armoured suv.

"Guys. Guys! It's Karnak's android." Trent's thick rimmed glasses nearly fell off his nose as he raced to the door of the shop. "It's the prototype! Lieben!"

One by one, each of Trent's compatriots stared at the slow shift of colours on Trent's led powered tunic. Saber sucked a bubble of gum back into his mouth with a 'pop'.

"Android's worth millions. Did you see the news?" Trent thrust his wrist comp into the middle, a rotated image of the Company HQ's lobby two days before. "A billion dollar machine, the creator heralded as Person 2.0, gets contained in Police custody, after its' creator died. Shot in the head, presumably by himself. No witnesses but an intern who booked it, and Tara. Unless you count the prototype. Serious, look. Lookit! Look!"

"Yeah, Trent, we see it." Click. Click. Click. Click… the stopwatch returned to Saber's hand as he walked back into the office. Leaned against the far wall beside shuttered windows.

Thirteen million for the android to vanish back into the company's ancient warehouse facility across town. 2.6 million a piece was barely enough for a home in Vancouver, but what did Saber care? Notoriously stingy, Tara paid enough to make the job worthwhile, without any trace or bit of back-end trouble. Promises of future projects were good enough after the label of 'Military Veteran' drifted from 'Thank you for your service' to 'Not you again'. He eyed Dix, the way Dix flinched at innocuous things. Always good for the job, Dix'd saved Saber's ass more times than Saber'd saved his.

"You figure this thing's worth more than 13 mill?" Employers like that… Saber squished his gum between his teeth. Trousers dusted off from dried cheese crumbs, Trent angled his wrist-comp to Saber's eye line. Lieben's face contorted. Mechanical arms reached to cradle the body of Dieter Karnak to her seemingly nubile chest.

"Zig. Make Dix look convincing." Saber reached for an extra clip of ammunition for his belt.

"Baiko… Baiko baby, are you awake?" Fester climbed up the roof ladder and tapped at the roof mounted tent. The rare British Columbian sun warmed the metal under Fester's fingers as she rapped against it, good for solar panel and person alike.

"Baiko?" Nothing other than the birds met Fester's ears. The zipper whirred as Fester opened the nylon tent's flap. Nestled in her sleeping bag, Baiko laid with hair furrowed around her, the bag moving gently up and down as she slept.

"Sleeping beauty still konked out?" Ego walked back to the truck with a brace of cottontails, his air rifle slung pointed downward on his right shoulder. "Bit strange she's been out past noon."

"She did go through something traumatic, maybe she does need the extra z's. Couldn't be easy." Fester propped her chin on the top of the truck, hand resting on the tent's zipper. Furrowing his brow, Ego glanced down to the rabbits.

"If she's not up in a couple hours..." Ego watched Fester zip the flap back up and hop down. She hop, hop, hopped to Ego's side with a grin, slipping her arms around Ego's chest. The wool of his sweater scratched her cheek.

"I'll wake her. Got any plans for the bunnies? Pepper stew? Ooo maybe some..."

"Ego." Mog swung open the side door of their truck, her blonde hair braided away from her face. "News from Amma. Fester was right, they're breaking Lieben out."

"Told ya."

Leaning down to kiss Fester's brow, Ego held the transitioning woman for a brief second. It was warmth in his way, an affection in passing while he set the hunted meat on their outside table. A small group of Idless milled about the camp. Shifted closer as Mog leaned against their truck. Her hand massaged the small swell to her stomach as she watched her two lovers.

"Tomorrow. I've tracked an ice spike in the Police mainframe, seems an old slicer tagging program."

"We can't move that fast." Ego glanced at the others in various stages of combat readiness. Used to be a baseball bat or kitchen knife was all their grandparents needed. A single camera mounted with a fisheye lens. Many drifting down the South American coast picked up handguns, machetes. Genteel days faded with his parents, who set up old soda cans along abandoned roads to teach their son how to shoot by the time he could brace a rifle to his shoulder.

"And we shouldn't. Let them do the hard work, bleed. We can swoop them on the road, it'll be easier to grab Lieben in motion. Mog, can you call a protest? Stage a bigger distraction than they could imagine, make them think it's serendipitous. Might spook them." Fester sat on the table, kicked her feet.

"I can call some friends, yeah." Mog fled inside the truck, back to her bank of screens. Ego shifted with his arms over his chest.

"They'd want to get her off the roads. Roads are death traps, especially in Vancouver. I'll scan the air fields and helicopter pads. See if there's any change in chatter. Maybe a flight plan." Fester's oft sunken demeanour receded for the task, clever tasks and stratagems threaded through her scattered mind. While Ego became the spearhead of their triumvirate, in times of intrigue, Fester knew best. Called back to when they met in South America, a dour eyed Mennonite boy who sought salvation with Ego and Mog.

A way to see more places. Remove himself, recognize the her in that self, while they travelled down Argentina to see the edge of the world. Something simple as basic navigation became Fester's domain. A way to prove him, no, now her, to the world she left behind. Fester and Mog tapped at their screens, as synth-rock poured from the nebulous speakers Fester programmed inside.

"Ego?" A woman set her shotgun on her shoulder, shifted her weight.

"Call Lucy, get an exfil team ready near Vancouver. If the Company is going to all that to get Lieben back, then Baiko will be next. Our priority is protecting her. We move North. Hide in the forests around Armstrong, it'll buy time." The woman trotted off, gun pointed to the ground. He slid his eyes up to the tent. The mere fact the Company wouldn't wait out Lieben's evidentiary incarceration gave credence to Baiko's testimony. Police, Company assets, wet teams, all of them would be after Baiko before another day went down.

"Fuck it… Baiko, wake…" Ego climbed the ladder and unzipped the tent flap. "Medic!!"

A crowd of protestors lofted signs in English and Binary, from the staging point at a local park. Each held up smartphones, or smart-glass lenses while they surged through the traffic, with coffee ‹donated to the cause' by a savvy Cafe. All Vancouver's idle elite required for the next protest was the right hashtag. A few whispers from Idless in the city. Protest provided the industry of malcontent far too much income to ignore any which popped up looking legit.

Denman punched a hole in the wall, when the first chorus of 'Witness not Evidence! Let her go!' radiated out of the growing crowd.

"Trent!" Orit swore and grit her teeth, slammed the brakes and took a course correction past a municipal school.

"Aaahh they're not mine, they… my distraction was a cop car covered in toilet paper and pre-placed explosive charges on the corners of the apartment building across the street. This isn't mine, it's… someone else set up a 'free electro-human' protest on twitter and… whoever did it is fuckin' quick." Trent rocked as they went over speed bumps. His wrist comp click-click-clicked with fingers on keypads. Swerved around a lady commuter in a tiny two person electric.

"Toilet paper? Where'd you score that!?"

"Municipal Hall. Mayor's gonna have a real surprise come lunch." Trent shrugged, tapping at his comp to stop the back-up while Orit surged through the side streets. "Saber, do I nix the explosions? Dude, that's a lot of people."

"We green?" Dix grit his teeth and held his leather messenger bag in the bullpen of the precinct. The bead in his ear was smaller than three hairs coiled around his ear canal. Zig wore one too, brought in an hour before.

The truck roared over a speed bump.

"Saber, are we copacetic?" The Constable poured two mugs of coffee, and was on her way back to take Dix's statement, his face statement enough that assault charges would be laid.

"I can bail down 4th Ave, be hasty!" Orit grimaced, from the front. Knuckles bore down on the steering wheel, a twitch in her eye.

"Use the protestors' cover. Maintain the plan, let the hirelings handle the mob." Saber clicked the top of his antique stopwatch like mad, waited for the nod. "Wait for Dix's ping, then open fire on the Police."

"But the protest…"

"Follow the plan." Click, click, click, click, click. Saber's rapidfire thumb fiddled with the top of the stopwatch, its' mechanical tick-tock the consistent ride Saber needed for his brain to overcome inner nerve.

"Beauty."

"What? Here's your coffee, Mr. Kinton. Black as your soul, right?" Constable Kian set the mug down, a glug of cream in the 'black soul' coffee.

"Just.. glad for coffee. Did you find my wallet?" Dix ran his fingers along the shiner on his cheek, with a hiss, licked his fat lip.

"Not yet, but usually someone that brazen would yank the cash and cards and dump the rest in some bushes. Where'd you say you travelled here from?"

"Colorado.. my passport was in there. He clocked me from behind, I wasn't…"

"We'll help you figure this out. Take a sip of coffee, and when you're ready, come with. I've got a line up waiting for you, perps matching your description, a few with busted knuckles." Constable Kian shifted her weight, glad today her task included the living, and not that damned spooky vacuum cleaner in the basement.

"Th-thanks." Dix reached for the coffee mug and held it in a shaking hand, the contents sloshed. Steam rose, and he blew on it. "I'm ready."

"Follow me, Mr. Kinton… sorry about the noise outside. Crazy, eh? Damn tinker toy, I dunno. The thing didn't look so special when we brought it in. People'll say anything to call cops the enemy." She led him through the labyrinth of the Precinct, past a thumbprint scanner and into the secured area. A sign glowed 'Holding Cells' with an arrow to the right.

"Yeah… guess you can't talk much about Lieben, can you?" They went left, past a series of metal interrogation doors marked by numbers. Dix tapped his fingernails on the mug, pings to judge distance through the coil in his ear. Minute echoes.

"Lieben? It's got a name, to you too? Funny how people have an opinion on a toaster oven two days after they found out it exists." The panel beside the door read 'authorized access only' and held a glowing green light. Inside the room was sparse and quiet. A couple of chairs, some sealed filing cabinets and a wall-sized one-way mirror. In the back of his ear, Dix heard the click, click, click of Saber's old stopwatch. He took a deep breath and held it, hissed out as the sound connected him to the fact he wasn't as isolated as his mind wanted to pitch him. A small comfort.

"Line them up." Kian spoke into a box on the wall. Lights came up on the line-up, height markers painted on the wall with a few smudges decrying frequent use. Six men of varying heights, with deeper skin tones walked to their spots. One adjusted the turban on his head. There. Zig stood in the middle, hands shoved in his oversized hoodie, jeans ripped across the right knee. He swaggered and shoved the guy beside him.

"Stand on the numbers! Now!" Constable Kian shouted into the intercom. Two police officers stood by the door, one winced visibly.

"Now, Mr. Kinton. Which man assaulted you? Was it number one? Two?"

"Give me a… it was quick. I didn't have time to keep an eye." Dix tapped his fingernails along the coffee mug. The click, click, click of Saber's stopwatch stopped.

Go time.

The ice triggered simultaneous with the blast in the city power grid, which maintained the four square city blocks around the precinct. Trent's trojan activated. Shut the links between backup generators and the grid. Routers, lights, electronics powered down without aplomb. The lights chunked off. Phone lines rerouted from emergency operators to local curry houses and pizza parlours. A sushi shop.

"The hec-" With no light to keep the line-up room bright, the one-way mirror was nothing but a pane of glass, as dark as the rest of the space. Dix tossed the hot coffee in Constable Kian's direction. She shrieked and swore. Dix swung the empty mug in her face, with a solid thud.

In the pitch black he surged, arms around her waist. Lunged into the wall. One hand grabbed her throat, the other repeatedly punched her face. He grunted as her knee came up into his stomach, both of her hands clawed and struck at his arm. Dix slammed her head back against the cabinet until she slumped in a pile to the floor.

Shivered fingers dug into his messenger bag for a pair of sunglasses. The room burned in a green shimmer when the glasses were on. Dix yanked the Constable's gun belt off her, set it on his own waist. She quivered, and his boot came down on her head.

Constable Elissa Kian never moved again.

"Dude, you good!?" Zig's ear bud picked up a hush of prisoners, some cheering and most mumbling about the lack of power.

"What kinda line up is this? Hey mother fucker! I didn't do nothing!"

"Aww what is this, Earth Hour!? Turn on the lights!"

"Hey! Who bit me!?"

"Don't touch me, fag!"

"*Who called someone a fag!?* That's derogatory, you fucktard!"

"OW That's *me* you punched, Frankie!"

"Sorry Mop-gah!"

"What the hell!? Who 'dere!?"

The line up room erupted, another layer of distraction to the two officers who scrambled inside. Raised their tasers. Zig shoved the Sikh man in the direction he remembered the officers were in, ducked under an arm.

From behind the glass, Dix fired twice, the sparks of explosive light stung the men's eyes. Most dropped to the ground, one guy screamed about police brutality. Both officers clattered to the ground, their tasers sloughed. Dix tossed a chair at the window, and in a shatter of glass, dove through it to the chaos of the line up.

"Gah!"

"Son of a…"

"I promise I won't steal cheetos again! I'll pay Mr. Lee! I'll pay him!"

"*Fuck* they dead!"

"*In the corner.* Get in the corner, now!" Dix barked, his borrowed pistol raised. He dug into his pack, handed Zig another set of sunglasses.

"About time!" Zig grimaced at the bullet holes between both police officer's eyes, clipping an officer's gun belt to his own waist. He yanked up his hoodie and put on the officer's bullet proof vest, hoodie back on. "How long've we got?"

Dix shoved the other gun belt into his messenger bag, pulled both tasers.

"*Stay down! Stay down!*" Vaulting back through the witness room, Dix pointed the pistol at the other men in the line up, huddled in corners like good bitches. He thought of shooting them all, but Zig's face was already in the system, and he didn't want to expend the bullets. Zig reached the lock, a set of ceramic lockpicks ripped from the interlining of his hoodie.

"Trent, clear our path." Dix's boots clung to the concrete, his breath in his ears. A perpetuity of his method, the moments before the bang blocked out distractions.

"Take a right out of the room and go down to the end. Company boffins marked the door with UV paint. Releasing the holding cells… now. The rest of the doors run on their own battery circuits. Use the magnetic sheets in your notebook." Trent's nebulous voice sounded tense, the roar of Orit's engine combined with Saber's stopwatch.

"Got it." The door swung open with a push from Zig, the interior hallway as black as the coffee Dix wanted, before Constable Kian thought to be cute and add cream.

Outside the Precinct, the roars of 'Let Lieben Go! Electro-Human Rights' shouted over the Police bullhorn asking people to disperse. The Fireteam Leader nodded and cracked her neck from side to side.

"They want chaos? We deliver. Take out the officer with the bullhorn. Gig, walk through the crowd, then lob a grenade as close to the precinct entrance as you can. Fire on the riot shields. Three…" In the back of her ear piece, Rega heard nothing but the click, click, click of that damned stopwatch. If they were lucky, the Squad Leader would have his thumb blown off.

Down in the crowd, Gig picked a biodegradable coffee cup off the ground and pulled the pin on a slim grenade. Eyes covered by mirrored sunglasses, he kept his hood up and eased through, chanting alongside the protestors.

The crackle of the sniper rifle was as muffled as Saber's stopwatch through the din of the crowd. Gig kept his eyes on the officer with the bullhorn, held the coffee cup crushed on the mechanism of the grenade. Another couple metres…

"Please disperse! Go back to your homes! Ba-" The bullhorn gave off a terse screech as it dropped. The officer fell back, and Gig lobbed the coffee cup grenade, grabbed another from his belt and yanked the pin. He threw it as a woman started to scream. The scream cut off with the first explosion, which rocketed several of the riot police into the protestors.

Fireteam 1 chummed the waters with bullets, as the sea of protestors burst in all directions but one.

Inside the Precinct, Police officers roused with flashlights and raised pistols, orders shouted to gear up and get outside. The protest flopped the powerless precinct on its underbelly. In the chaos, Dix and Zig rushed down the lightless hall, no flashlights ahead, but plenty behind. Completely dark, Dix and Zig ducked into an open side door, as a flashlight beam got larger. A talk-system kssht'ed to life, orders spouted. Three sets of boots pounded toward them.

Zig took point, waited until the first rounded the corner before firing twice. Both central mass, bulletproof vest be damned. The officer stumbled into his buddy, and Zig fired two more shots, a third. Taken up the rear, the last officer managed a few shots of his own. Zig's right shoulder went numb. Slapped against the wall, Zig raised his pistol with one hand and emptied the clip in a panic. Dix lowered his arm and rushed the other side of the hallway, three bullets to three faces.

"Fuck, man! Fuck!" Zig threw the emptied pistol at one of the corpses and set his hand on his shoulder. He tied the wound off with a hankie and popped a capsule under his tongue. His own breath rushed in his ears, beyond the click, click, click of Saber's stopwatch, the radio asking for an update, before yelling 'shots fired! Confirm!'. Bent over one of the bodies, Zig bit back an urge to vomit, and tugged another pistol off an officer.

Dix holstered his pistol and picked up the assault rifle, slapped a few cartridges of ammunition on the webbing of the vest he had under his wool sweater. More bootsteps rushed faster, and clipping a radio to his shoulder, Dix grabbed Zig by the arm and kept moving. Down the end of the hall.

The door with the UV paint shone in a pink check mark. Dix kicked it open and tossed Zig in. Through his darksight glasses, Dix saw the scientists and lawyer as outlined green humanoid caricatures, huddled down behind a table. In the corner of the room, the glass case shone with an outrageous purple glow. Joints locked by subroutine, the android inside the case hit a serene cord in Dix's gut. Hazed images of a virgin in some half-bombed cathedral in the Georgian Campaign.

He shook his head.

"Dix? Dix, man! You tweaking on me!?" Zig barked, raised his hand to grab Dix's shoulder. "Yeah, man we're the rescue team. Back up, y'all."

Click, click, click, click… the clicking stopped.

Saber's long time companion breathed again, shook the liquified icon of Mary, mother of God's son out of his mind. Inside his messenger bag, a shampoo bottle played host to jelly explosive paste, the nozzle shaped to give out a steady stream when squeezed.

Dix pressed the bottle against the wall, ran a thick globular line around a two metre square. A pen held the detonator, stuck into the jelly explosive with sticky tac. Dix clicked the pen. Rushed back to the glass case, where Lieben stood mute and statuesque. The back wall of the precinct blasted outward. Shook the building as the chatter from Zig's audio piece faltered to an utter ruckus outside with the scattered protest cloud.

"Ram!" Orit charged the truck back into reverse, slammed through the fence and into the hole, where two scientists sprawled under rubble and one more was blown straight back through the door into the hall. The lawyer fared worse, singed flesh stunk in coincidence with the dust of burnt material and explosive residue.

"Tick tock!" Saber rushed first through the hole, tossed timed charges inside the room. At the sight of him, Dix gasped with grin, the plan. It was almost through.

"Police incoming!" Trent sunk his wrist comp under a long-sleeved kevlar jacket, grabbed a pair of suction handles. Dix raised the HK416C Saber tossed to him and let out a burst as two Police officers rounded the corner. Saber watched the other flank, assault rifle raised as Trent and Sand dove over rubble.

A single beam of light cast from the exterior against the glass. The android in state cast a serene shade upon what it meant to be human, fetched Sand's eye as she set the handles on the glass and triggered suction. Her arm swung out, double-tapped the scientists in the head and chest. Angled her pistol at the door.

"Woa…" The first suction cup mounted on the side with Trent clicking it, the second engaged.

"Yeah, right?" Sand gritted her teeth, peeked past the interrogation room door blown off its' hinges. She swore and yanked her head back. "Company, you gucci!?"

"Where's Zig? Hurry up!" Trent fixed his suction handles and gave a thumbs up, as the staccato of gunfire jittered in strange dichotomic synthesis, first in the atmosphere, second in their ear buds.

"God dang it, Sand!" Zig dusted off his hoodie and pointed to his arm. "Got shot, can't lift it."

"Zig, grab the case! I'll kiss your boo boo later!" Sandy shot twice into the hall, as the two men grabbed a handle each on Lieben's left side. Sandy holstered her pistol and grabbed the right two handles, heaved the case parallel to the ground. The android inside swayed and ragdolled, not enough room to shift too greatly.

"Frick! You sure she's..." Zig hoisted with his left arm, shock and stimulants pinpricking his mind into the task ahead. Lieben sloshed to the side, lavender eyes snapped open in the light.

"Aahh!"

"Shut up! To the truck!" Sand let go of one handle to fire back at he door, bullets whizzed by her head.

"Halt! Put the guns down!"

"Yeah, that'll work." Saber fired another burst. Trudged over rubble in tandem with Trent and Zig.

"Go!" Saber roared over the ear split cacophony of his assault rifle. Bullets fanned behind them, peppered at the door.

Trent groaned as he, Zig and Sand heaved through the explosion's opening, the glass coffin slid into the revving truck's bed. The ca-chunk of the inner slide hooked the coffin in place. Sand tossed Zig into the truck and laid on top as bullets scattered. The truck's engine growled as Orit geared down. Run-flat tires strained against rubble.

Dix's rifle struck the air with staccato rapid fire. He grit his teeth, the smell of gunpowder bringing him back to the tempestuous times prior, hooked into suggestive neurostim which annihilated any choice of whether to follow commands. The acrid smell sunk into him, grounded his finger on the trigger in bursts. Bursts. Breathe, pause, reassess targets. Take cover, wait for the Police to shift from pistol fire.

Stand and burst. Grit his teeth, the ringing of a human gurgle bit into his neck. Dix's fingers clenched hard as the steel used to make his firearms, as he and Saber jumped into the back of the armoured truck, and Orit gunned the accelerator.

"Hold on!" Saber threw himself over Dix and Trent, as Police fire batter-tattered at the bullet proof glass. Orit pressed her back against the driver's racer webbing and kept the wheel straight. "Braaaace!"

Abrasive and unkind, Orit wore the chip of her Hasidic Orthodox upbringing the same way her mother wore her Sheitel. Enough bitterness to bleed into the atmosphere like industrial smog. No man, nor woman, no Rabbi or authority would command her any longer.

Not even three cops in front of her. She bit down on the gum in her mouth and let the reinforced tires roar over the tire spike strips in her path. Her thumb flicked the nitrous, arms straight. Steering wheel straight. Any turn to the wheel, or quiver of her arms could spin the truck on its' head with the extra amount of velocity which rocketed it toward the exit. Behind her, the precinct wheezed as a series of small explosions triggered a larger one inside. Slag and brick work burst around the truck's vector.

The path was as narrow as her upbringing. As constraining as prison, after prison, after prison. Orit yelled in a bitter fury. Hit the second nitrous ignition trigger, and slammed through the barricade with the same abandon which helped her toss her own sheitel in the fire.

12
tension kills, you know

2086

"Simple, simple, simple, simple, simple…" Tara rolled her eyes at the mumbles which poured out of Robert's mouth.

"Must you?"

"Simple! Simple smash and grab. Even set it up, all nice. All ready." Robert rubbed the bridge of his nose and turned down the volume on his phone. Every newsie in the city was following the protest-come-terrorist attack. Terrorist attack! Grenadiers and riflemen shot up Police and protesters alike, and the explosion…

Explosion!

"Tara, what was that?" Hands over his chest, Robert leaned on the diagnostic table brought in for the android.

Her heel tapped at the ground as the damn live feeds played on, police updates of a man with a rifle cut down beside someone whose bandolier was covered by a zippered hoodie. Members of the tac team…

"We can spin this. Blame it on the protestors, yeah, they're goons. A bunch of transhumanist rabble trying to claim what's ours."

"They saw the footage. I don't know how Karnak made that android's brain so intricate, it was… jarring." Excision tools laid in random on a tin cart. One by one, Robert set them in their 'proper place', each codified and set by urgency in the decommission of the prototype. Not Lieben. The android. The prototype. Still, the amount of damage to the police precinct set a quiver in his belly.

"You shot him." Nary a question, Robert set both palms on the metal cart and stared at several feeds lined on screens above the table.

"Prove it in court."

As an accounting of a sniper in an apartment above the Precinct came on air, Robert realized he didn't need to, Tara proved it by the brazen tactics of her hired guns.

"Hundreds. There were hundreds of protestors. Did you pay them? Did you pay them for being there, Tara?"

"Nope." Heels clicked on the floor with an echo. Tara slid beside Robert and felt up the cords of muscles on either side of his spinal column. "Tension kills, you know. I can take care of that for you."

"Like you did for Dieter?"

"Hah! He wished."

"Tara." Robert swerved to grab both her wrists from his ass. He held her at arm's length, pressed his lips together. Becca smelled like her, back when. Back before yoga pants and grey hairs became Becca's comfortable usual. "They're late."

"Let go." A quiver between her eyebrows, the most human she'd looked since the day he first saw her swagger into the Cloister with 'orders from Boss man' to write an in-progress article for the stockholder newsletter. Hips clothed in silk swished back then, still did. Robert yanked his hands off and held them in the air in a form of surrender.

"They'll be here. Never failed before, they… are worth the money." A nail popped between Tara's teeth, nibbled on.

"We'll wait." The urge to shift off, leave the impromptu dissection space grew as a flimsy smartphone shot of a massive armoured suv peeled down the road away from the action. Dust on the back bumper.

"When…"

"Must be stuck in traffic. Took another route around Hastings."

"I can't recreate Lieben without Lieben." The echo of Tara's heels stopped. She turned with a saccharine smile, the stain of her purple lipstick on her red painted nail. "I can recreate the body mechanics, but not the AI. It'd be… stupid…"

"They'll be here."

"You said that already."

"Did I?"

Phones in Robert's pocket, Tara's on the dissection table and several monitors chimed a company number. Di. Robert watched Tara purse her lips, stare at the call. She clicked green and Di's face settled on the view screens en masse.

"The Chairman wants to see you, Tara. Now." Di's feed cut as abruptly as it appeared. Left Tara and Robert with a building silence and the hum of panic in Vancouver. The vial of lipstick laid against Tara's cleavage, tugged out by steadied fingers. Tara checked her lipstick in her phone, and gave Robert a pat on his cheek.

"They'll be here."

"Or?"

"NEO-N's'll be stupid." She pressed her lips against his cheek, a perfect purple stain on Robert's Korean skin. Her hips swaggered on the way to the limo, a bottle of Brut popped open before the driver closed the door. Robert thought he saw Tara take a swig out of the bottle, the pop of the cork her last sound.

"He'll forgive me. He always forgives me, it wasn't so bad. Anyone could have planned something like that. Blame it on the trans-humanists, those… Idless people. The ones who claimed that warehouse burning last year. Yeah… I've worked everyone he's asked me. Made millions… didn't even shill for champagne. Brut… eh. Into the belly of the beast it goes." Tara tipped the bottle of sparkling wine down her throat, as the driver cut through downtown traffic, her phone… wait…

"Fuck! FUCK! FUUUCK!" She tossed the bottle against the far seats, and screamed in a shrill bellow. Turn around and go get it? Sitting there beside Robert… no. Tara reached into the limousine bar for another bottle, and found vodka in glass. She cracked the seal and poured some down her throat, nose scrunched as she coughed. "Think I'd be used to things down my throat by now… fuck you, Dieter. Fucking prick. Fuck… Fuck you and your machine."

The rest of the way through traffic, Tara fought with a growing nausea, her inner ears sloshed with every turn. Vodka first, Tara handed the half-drunk bottle to the Driver and stumbled out to tug her dress down halfway to her milky knees. None of this would have happened if Dieter played fair. But no, he had to 'save his Liebchen'.

"Goddamned android, goddamned son of a bitch… someone pissed in his mother." She sneered, when the same elevator door swooshed open, ricocheted bullet depressions gouged in metal. Before she could press a single button, the elevator computer screen shifted to 'RT - Private Access Granted' and the elevator churned closed. "… Cunt."

The company rooftop garden was swathed in sheets of bleached canvas draped along modified tori gates from the elevator to a series of Japanese Maples and a dwarf cherry blossom tree. Tara wished she'd had her jacket as a light Vancouver rain drizzled down around Irises, rose bushes, peonies and hydrangeas.

"You called?" Maybe the liquor gave her swagger, maybe it was the mosaic tiles on the rooftop under her heels. Nearer still, Tara set a grin on her painted face. Curtseyed as she came upon a man whose gloved hands nestled in potting soil, a set of clippers in hand. "Situation's under.."

"Control?" Deep hazel eyes poured into her, the match for copper skin and hair black as a closed, windowless room. The long sleeves of his tailored shirt were held back by gold sleeve garters, his vest impeccable. "My mother carried me on her back from Kashmir, nothing but her last sari ripped into bundles around my amputated knee and a single bag of spices and grain to trade for our lives. I cried for days, until tears dried from lack of potable water, and yet, she held her tongue. Carried on, toward the peace keepers, who housed us during the Pakistani war. Even then, malnourished, exhausted, her only surviving child missing his leg, too weak to lift his head from blood loss, mother had control of herself. Of the situation. If this is your idea of control, Tara, you require further definition."

Tara cast her eyes to the marigolds on the potting bench, sucked in a breath. "Karnak…"

"Was your responsibility. You promised me you could keep him content. Working. Is this Karnak working? The android we were promised likely evaporated in an explosion, and a hazy retrieval operation foiled by unsavoury company?" The Chairman's head wobbled from side to side, his prim cut hair sculpted short on the sides, perfected into waves atop the crest of his head.

"… unsavoury company we've worked with before. They've never done us wrong. If I know Saber, he's with Robert making the delivery right now."

A dirt coated phone slid across the potting table, screen blank but for a series of black text on white.

Auction. T-9 Hours. Karnak's Prototype. Bidding: $15 Million Open.

Tara's eyes slammed shut. None of the curses she knew held any influence to her swollen, alcoholic tongue. Canvas gloved fingers tugged her chin up. Tara looked into those hazel eyes, as the Chairman pressed a cut lily in her hand.

"All flowers wilt, Tara. All our sins and salvages will be brought into the light." He pressed soft lips onto her forehead. Stepped back with a shift to his well pressed suit. "May your next life be more virtuous than your last."

"Wh-bu-but this is ridiculous! Let me talk to him! I can.. I can get to Saber, he's…"

A set of heels tapped along mosaic tile. She was pretty in a manufactured sort of way, cheekbones built up, jaw shaved. Nose done. Fillers in her lips plumped them more than the deep purple lipstick she kept in the cup of her brazier.

"Hi, I'm Tara. Happy to meet you." The plastic grin was the same Tara'd given her predecessor, nine years ago.

"W-wait!" The bullet entered her brainstem at 460 metres per second.

"Marigolds remind me of the festivals in my childhood, so many yellows and oranges, piles of turmeric in the spice sellers at the market. Come, my dear. Let me look at you." The Chairman's gloved hand reached for Tara, ate her every voluptuous inch in her tight blue satin dress. Tara grinned and let out a petite laugh, spun slowly with a wriggle of her plump backside. He pulled her ass against his hips, an arm wove across her neck and forced her against him.

"Find my machine. Do not return, do not look upon me without it." A single kiss below her ear was the only warmth she got, eyes fixated on two men draped in white body gloves carrying the old Tara away.

13
raptor-wolves in human suits

2155

CI-006's bellowed roars drove a hammered spike of agony through the hearts of any cursed to heed their olfactory vibrations. Whether by proximity or audio fee, the animal roars of Cillian's unendurably constant lungs threatened the resolve of even Phil Rykstra.

Ständchen, D. 889. Franz Schubert, Tanguy de Williencourt. Volume 2 … 8

Calculations on length of time before Bio-Mech Assets rebooted in the black to their docile inactive form echoed with the animalistic roars. - 20 seconds… - 40 seconds… - 65 seconds… why isn't it out by now? In his control loft, Phil's hand hovered over the panel, a simple mechanical dial set at 0. Thick black letters above the dial in an arc: MASTER LIGHT - CASKET 6

Agony echoed through the Retreat.

"Faults in the machines… ought to look into that. Must be a precedent I'm missing." Phil turned the dial on his record player to 9. Pulled a manilla folder out of the lowest drawer of his locked cabinet.

Tanks with skin, bi-pedal bayonets… Phil's eyes tore at the initial research. Gene-splices from (Raptors) Dromaeosauridae Deinonychus, (Wolves) Canis Lupus, a few peptide chains from some form of terror bird. Augmented sight from optic nerves vat grown by a separate research group.

Deoxyribonucleic Acid Chain REDACTED.

REDACTED.

There.

Minuscule letters in the midst of thousands of symbols in the initial research. Pages coated in thick black marker, or research reports which lacked several page numbers were a fairly consistent norm in Phil's secretive world. He pulled out a hand written annotated table of contents, in the search for professional and human-centric secrecy, typing on a computer? All notes were handwritten, Lofi was the new top security.

Need to know.

Most secret.

Your eyes only.

The rules Dr. Phil Rykstra worked behind were a shroud over the Scientific Messiah's pre-resurrection corpse at the time of the first miraculous heartbeat.

Cillian's roar metamorphosed into a guttural half-sob, the clang of his thrashed body against the metal plate of the casket. A thrill stole through Phil's bone-cage. Curled up and down his ribs with a near electric shock.

Each of the Pack threw themselves at their respective barriers, instinct and the sonic vibrations denoting which direction was near Cillian's. Phil bit his nail, messed the papers he'd carefully taken from their sealed case across his desk. The lilting song played on, violin and piano cranked to its' holy heavens.

Phil slammed a hand over his left ear, when the distinct echo of a tooth-grit scream burst through the contained compartment in the belly of the Ithavoll.

"Biological Machines or Bio-Machs are gene-spliced machines which use bio-organic chemical chains and protein syntheses in lieu of mechanical servos. Unhackable units. Toasters which use fire instead of electric coils to make you toast... Cue the picture of a gen one bio-mach in an apron. Pause for giggles..." Phil recited his introductory lecture of the course in advanced bio-mech engineering he taught in the Symposium. His eyes filtered through the papers, paused at AD-001. The first truly independent bio-mech, which made all other high functioning Assets a possibility.

"... The beta, gamma and theta cases of bio-mach success were augmented canid fauna rearranged with useful protein sequences to heighten pack mentality, obedience, and lock-jaw to devastating tactical effect. But first came AD-001, built through a breakthrough in genetic engineering large enough to make the wooly mammoth look like a rat after a pebble sized wheel of cheese... What breakthrough, Dr. Rykstra? Well, curious B student..." His thumb smudged against the page. Another shout broke his concentration with a flinch. Phil reached for acetone and poured it over the page. Ink fled the swelled paper. He rubbed. Nothing.

Turned the page over, flipped it repeatedly. A second howl clambered against

the sound waves of his research suite, in a sickly harmony with CI-006.

"The breakthrough." Phil spoke louder into the empty room. "Anastas's Principle concluded that a specific amino acid chain in the right epigenetic circumstance was the key to slight genetic augmentation, and a completely foreign combination of DNA strains in an artificial custom machine. Bypassing the graft techniques of the past generations of gene-splicers... Anastas' Principle slid us past the Commonwealth's ban on Homo Sapien research, and made bipedal analogs... Inhuman meat buckets parading down the lane..."

Homo Sapiens.

Phil tapped at his chin, filtered through the papers until one crinkled between a blacked out list of original scientists and what at cursory glance looked like an expense report. Acetone dribbled onto the paper, wiped away with a serviette from his lunch tray.

A single column in the expense report, marred by the faded black ink which still stained the rest of the page:

Donor Compensation
Demyan Anastas - Est
Baiko Kaho - Estate
Katherine Kimbadjia
Farouk C Kimbadjian
*26.932.1

"Sons of bitches..." The locked cabinet clanged shut as Phil's hand trembled over a stack of pages with little more than lists of infamously 'legal' sets of ass-covering.

"Twenty six... Twenty six point eight... Twenty six point nine... thirty th-" He yanked the pages back and glared at the seeping ink on the expense report. "Thirty two. Addendum Twenty Six point Nine hundred thirty two point One. I hereby release the Conglomerate from further compensation on behalf of myself, and any descendants, whether natural born or genetic spliced organisms hereby created by my encoded DNA. I release the Conglomerate from libel action due to the actions of the Asset Project and understand any and all organisms created from the use of my genome are the physical property of the Conglomerate and lack all legal rights of the progenitor human donors..."

'MASTER LIGHT CASKET 1 - 12' 0 ———— 8

"... No, ignorant student, they look human only through cosmetic alteration for the purpose of comfort in a crisis. Sure, our chimeras and mandragora have their place, but sometimes people respond better when they see the familiar staring back... oh God. Oh my God." Phil's hand hovered over the mechanism, and with one slap, it was finished.

The caskets opened, never to shut again.

Phil slid to his feet with the grace of a leaden automata lacking oil on mechanical joints. Report, his mind droned… Commodore Rammage would want a report.

"M-max… Max." The Lieutenant seemed to know such things before anyone else dared stare the answer in the face. Phil tripped three times before hurtling himself up the stairs at a nauseated scramble.

"Allard…"

"We were told, the Bio-Machs, the Assets were… dumb beasts. Machines made of mammalian fluids to bypass the Mater Machine's ex-nay on hifi weapons. I know you heard Wulf screaming, it was all over your fucking face!"

"Allard! Are you done!?"

"No!" Allard shouted back, fists clenched in front of his nigh hyperventilating body.

"LIEUTENANT ALLARD!" Rammage slammed his fist on his desk. Shoved a glass with amber liquid toward the irate Lieutenant. "I have eyes, thanks. Good ones, got them from my Mama."

"Sir!" Body swung into a crisp salute, Max reached for the offered glass of whiskey and shook his head at the flavour of the bracing liquid down his gullet. A commotion struck the space outside Commodore Rammage's office before Rammage could respond.

"Eh?" Rammage reached for a pistol attached under his desk, as Allard flung open the door on a nod. There, Dr. Phil Rykstra's wide mousy eyes flickered between the two officers, wrists half in restraints as a Silver Collar attempted to remove him.

"Sorry Sir, Ryk's gone mental."

"Let him in. Empty the Command Offices, guard the doors." The Silver Collar paused and blinked incredulously. "This ought to be good."

Officers and NCO's shuffled to grab what they needed and hoof it. Phil rubbed his wrist and stumbled into the wall beside Max. Rammage waited at the doorway of his office until everyone in earshot was locked beyond the compartment seal. The sigh which echoed out of his lips pursed through clenched teeth. He shut the door with a metallic clang and locked it. The pistol's weight rebuked him, its pregnant clip a reminder of how simple the Ithavoll could become with one, maybe two pulls of the trigger.

"Donors. The Conglomerate paid for donor DNA… human… human DNA… it doesn't make sense…"

"You're right. We were promised machines made of blubber and bone." How these two disparate men came upon the information was irrelevant. As useless to suppose as what Assets thought in the dark. Rammage weighted the likelihood of Rykstra and Allard buying in, setting their hearts on the scale across from two bullets and ignited gunpowder in a finite space. Rykstra, he was a scientist, he saw reason and lived inside a welcome detachment. But Allard? Rammage's instinct to shoot him at the farm...

Phil Rykstra expectorated his surprise for a balm of disbelief. No, there must've been another reason for names in Donor boxes, he would have seen the data long before now.

"It doesn't make sense... I designed their mental subroutines myself. It doesn't make sense. It doesn't work. I would have known, I'd have seen it! They should be... be... frickin' vacuum cleaners!"

"Give Katey fifteen minutes with Aderastos and she'll attempt to be." Allard fingered his glass, wished the liquid self regenerated. Both Phil and Rammage rounded their eyes to Lt. Allard with sneers and confusion. "... the way she looks at him?"

"Not a him. An it." Phil burbled, another series of data spikes trod through his glasses screens. "Cillian was screaming.. it didn't make sense, we did the entire power down sequence twice. They were tanked from the Proxima frequency."

"Oooohhh he's a he now." Max shook his head and rubbed his eyes, "Jaysus is he ever a he."

Rammage's upper lip sneered as he poured the younger man another stiff drink. The pistol remained, a comforting weight.

"Worst day of my life." Max downed the second shot. "She'd better stretch first, my god."

"Ah... oh wow, I... didn't need to hear that. Ah... a line in the legal jargon on some contract giving the Conglomerate rights over the genetic material of the donors. D. Anastas, B. Kaho, some chick named Katherine." Phil's glasses fell into his lap. "I didn't join up to lock human beings in boxes like some sort of... post-modern Mengele! I don't experiment on people! I don't!"

"No one's saying you did, Phil."

"Cillian's screaming! He won't shut up, he wouldn't stop screaming, he was scraping at the walls with his nails!" Phil's lungs jittered in his chest.

"Sir. Lying'll do no good. Not for any of us." Max clicked through the feeds.

"I'm not heartless." Rammage shoved the whiskey bottle at Max and slumped down in his command chair. Pistol back under the hide.

"Could'a fooled me, Sir. Maybe the human fucking beings you keep locked in the basement against their will might shine some doubt." The glass in Max's hand sunk to a small table on the other side of the office.

"I said speak freely, I didn't say attempt to court martial yourself, sailor." Rammage's glare set Max's mouth into a thin line. Scents of vanilla, oak and blackberry from his glass of port gave Rammage less pleasure than he'd hoped, and the Commodore hissed into the air. "I might rule this boat, but I answer to higher powers. God I forget how young you are. First voyage, isn't it?"

"Yes, Sir. Conscripted under the Debt Reclamation Contract, Sir. Was press-ganged from my brother's mechanic shop, lest our Mum be locked in a Conglom debtor's prison. Sir." Vitriol, bitter and repressed soaked into the air. Max leaned against the wall with his head perched back along cool metal.

"So that's why you're a shit officer. I'd thought you were a volunteer like the rest of us. Didn't get the training course, did you?"

"No, Sir. Didn't want my Mum to have any more debt under her name. Sir."

"That we'll discuss later. You can't lead without it." The disparate image of Max Allard's wide-eyed rudeness cleared. Kid didn't want to join, probably wasn't a believer, but damn if he wasn't the best fix-it man in the fleet. Rammage sipped his port to clear the mental malfunction of his Lieutenant's forcible conscription and tutted.

"The Conglomerate is a collection of former nations and multinational companies, who voted to fall under private jurisdiction in 2099. Part of the Corporatocratic push away from nation-states and into collective interests. The switch kind of snuck up on us. Former Commonwealth of Britain, Scandinavia, Israel, bunch of other nations. It's an oligarchy commanded by Chairwoman Queen Margaret and President Charanpreet Kaur. When Control Day happened, the Conglom went from one of the two largest superpowers on the planet to a ragged bunch grasping at nickels.

Bam. Lieben marches in, yanks our missiles and military drones away. Communication, internet, medical care, education, it's all free. Anyone with a CIRCLET can have whatever they need, don't have one? She'll fucking hand one to you.

Lieben disarmed all of us, like a mother swatting naughty kids. The Conglom believes it was the first of many plays from the Mater Machine. No one's that good. Especially not with their finger on the ignition switch for every nuke known to God, Jesus H. Christ, Buddha and Mahatma Gandhi."

Garnet liquid sloshed in Rammage's glass, as he took a long sip. Let the port sit in his mouth, before inhaling a breath of air.

"I refuse to believe the Mater Machine's control is an act of mercy. Lieben must have an ulterior motive, there's something she's after. Why else take our ability to fight away? I've been given the task of to tear the Chernobyl Throne to the ground. The Assets, regardless of their incredibly classified genetic parentage, are the pivot point. One I intend to keep functioning as intended, gentlemen."

"I don't do experiments on people."

"They're not people, they're stem cells paired with dinosaur and wolf DNA like a form of genetic pic-n-mix."

"What if it was an act of mercy? What if when we pare it down, the only reason Lieben did it was to save us?"

"I'm not prepared to believe that, Lieutenant. I won't believe it, things I've done in the name of reclaiming control over our own bloody planet." Guilt poured down his tongue with the port, slick and jarring. Commodore Rammage rubbed his forehead, set the glass down and flicked through the images of the bio-monsters in the basement.

"I can attempt to… reduce the nerve sensitivity, but it'll take time. Some patches, and I can't guarantee whatever AD-001 does…"

"Aderastos." Max shook his head between the two men beside him.

"… Max please, I… I didn't sign on to… I need to stay objective."

"Fuck objectivity. They're people you said it, they're goddamned people and you cannot continue on the same path without altering the course of action. Phil! Buddy come on!"

"Lieutenant Allard, you're leering toward mutiny."

"You need the Assets for your global army, right? Every crewman on this ship chose to be uncomfortable. To live in analog during the digital age, until we make it off this tin city. Something brutally unlikely, since we spend so much time at sea or in missions so black not even Abyssopelagic squids could make our mission parameters out of the gloom." Max gesticulated with his empty glass, whiskey-courage swept into his shoulders with enough grip to cause Rammage to sit back.

"And?"

"And most had the choice. When Aderastos had the choice to leave Haven, he did it. Not because of some subroutine Phil wheedled into his head. Everything he wanted was on the Ithavoll. You want the Asset Project not to die in a fire big enough to abandon ship in the middle of the Pacific, you've got to stop thinking of the bio-mach quotient and entice twelve people to put on uniforms like good little soldiers and line up. Not 'cause we brainwashed them with sonic waves, or hold some scientist's syringe, but because they're needed. 'Cause they are."

"You done, Lieutenant-Commander?"

"No, Sir. I'll never be done until each of those Assets gets a soft bunk and a fair chance to decide for themselves what they want to… t-tuh…" Max stopped, mahogany eyes narrowed.

Rammage leaned forward with both elbows on his desk, and downed the last of the Port in his glass.

"Make it happen, Lieutenant-Commander Allard. Work with Rykstra. One whisper of the Assets' human DNA to the crew and your promotion, and both your collective brains will be butter on the ocean floor. Use whichever crewmates you require. Shift the Retreat to something worthy of our lauded guests. I need all of them working with us in perfect syncope, or the Board will tank this carrier to your Abyssopelagic squid. You have thirty days to make that a reality, or Rykstra will start again, pitch their failed bodies into the sea... and Lieutenant-Commander? Teach Aderastos how to use a goddamned condom before Kate learns that vacuum trick, 'kay?"

"Fuck you to Hades." Phil tore out of the room before Max could wrap his whiskey addled brain around the implications of another thread on his collar.

"Phil.. Phil!" Max's athletic gait caught up with Phil before he could reach the office's interior compartment seal. "Woa, woa, woa, Phil! Wait!"

"No! No I'm a good man, I can't..."

"Okay, okay! Yes, Phil, mate hold... stop... moving!" Max rushed him, pushed Phil's shoulders against the corridor wall and held there with a groan. "Yeah. You, me, the guys downstairs, we were all duped. But now that you know, if you walk away, they'll find some other clueless scientist to ask zero questions. You tore twelve people apart, but now is the one chance you'll only ever get to fix it. So fix it, Phil. Put them back together, you know more about them than any of us."

"I... Max, I locked them in boxes!" Phil gesticulated in a wild fumble, limbs uncoordinated. "How on this blue ocean am I supposed to face them!?"

"You did already."

"They were machines."

"And now they're extraordinary people. People who deserve the right to choose and a bed without needles in their arms every Tuesday. I need you, buddy. We can work through it. Give them all a fucking chance, but we can't do that without you heading the brain-stem off from Rammage and the Board destroying what humanity they have."

"What did she say?"

"Huh?" The wall was cool to Max's back, as he leaned on the other side of the small corridor. The alcohol caught up with his hands, added mass he didn't own to his fingertips and a buzz to his lips. Always two steps too small on a ship like this, no luxury of space anywhere but on deck.

"Lieben. You want me to do this, face those..." Phil played with the glasses in his hands, wrung his fingers around them. "... people... I need to know what the Mater Machine said to you and AD… erastos… in Haven."

"That obvious, huh?"

"Yyyyyep."

A chuff of a laugh drove through the limited air supply between two men. Max shook his head, bottom lip jutted out as the shake to his head grew in intensity.

"Nothing. The NEO-Ns stared at us, put a table down with food and waited. Set out towels and clothes. Some shoes. They stood at the door, and waited. The only one that talked was me."

If Rammage could hear them, which Max felt in his bone marrow the Commodore could, he did nothing but leave them to themselves. Hand rubbing down the side of his face, Max let another breath shake out of inconsistent lungs.

"Why did Aderastos come back? Why'd you?"

"Didn't like the sandwich."

Phil gasped, his fingers unclenched from around the glasses and he slid them into his pocket as a bright grin broke in tandem with a barking laugh Max shared.

14
the universe was filled with noise

2155

Kate's consciousness was ragged as split silk, frayed without a bound edge. The body which owned her curled around a warm, solid thing. Odour betrayed the form of her brother Clive, who rocked Kate in his arms, tongue clicked in instinctual soothing noises. She remained ignorant of how she got to the space they were in, a larger space than the confines of her prison. All Kate knew was the scent of her brother, the feel of his hand on her hair. The way he cradled her as if she were a cracked and precious being in reconstruction.

Her nose nuzzled along his elbow, until bloodshot eyes creaked open millimetre by millimetre in the gloom.

Cillian and Isthan paced the area with the black haired mass of a man larger than any but Clive and Aderastos. Manic with snarls and gnashing of teeth, Cillian refused the stillness of the Pack around him, beings born of the same ilk. He rounded to the open doors, chittered at those dumb animals as if he owned the composite materials. Protective of Percival, clenched fists combined with a low growl that reverberated through Kate's skin. She shivered and flinched inward, eyes shut.

Wulf's whimper scattered the air as a brief staccato peppered above the rumble of another's voice. A deeper lingering chord struck between Percival and the Other. The First.

Kate's knees collided with her chest. The atoms of her atomic cloud buzzed in a visceral pain, their electromagnetic bonds threatened by the horror of... the majesty of...

The noise.

The universe was filled with noise.

Quieter then, Clive croaked the syllables of his name, the first name Kate remained conscious of outside her own.

"Aderastos."

The arms which held her vibrated with a protective growl as another warm presence shifted away from Wulf. Another Asset raised its' tendril-fingers, vines with wide leaves coated in downy fuzz grew and grew. Became large enough to cover Wulf's body, tender yet stable enough for Wulf's head to find rest. The Asset swayed as an aquatic plant without anchorage, held to the ship by those grounding 'others' and not through the cold impersonal metal. It made no sound as Aderastos swept around it, and curled Kate, wrapped in a new leaf, into his arms.

She felt motion, the subtle pitch of wind on her skin as they moved through the blind ether, horizontal, then down. Kate cuddled her cheek into Aderastos' chest, fingers fought through the leaf-blanket to touch his skin.

"Too much stimulation, Katey." Body in his arms, she felt his voice rumble through his ribcage and into her, a tandem vibration conjoined with her own. "Hush now."

Her hand wandered in circuitous trails. Aderastos sat against the far wall, the hubbub of the others a distant hum unbidden, but pleasant through Kate's considerable mind. A new softness brushed on her forehead. Aderastos' lips pressed carefully at the lines on her brow, his nose on growing locks. Eyes lidded tried once more to seek him out, the angular face with its strong, but unrefined features. A straight nose, lips too plump for the thin lines they usually bore. Cerulean eyes with the beginnings of downy hair on the top of his shaven head.

He was too severe to be beautiful, as Kate understood the concept.

"Why did it hurt?" The motion of Clive's mouth, crackled through each nerve as salt in stung hands. Her chin wobbled, as tears threaded down her cheeks. Aderastos rocked her, bundled up in his lap, his head leaned into the corner of the wall.

"Did you think whispering was all humans do? Shush now. Let me work." His lips found her cheek, pressed against it with a fervour she rested in ignorance of. The crinkle of Wulf's back against another thick leaf washed across Kate's ears with the same echo of the ocean's waves beyond the groaning undertones of the ship. Kate gave herself to Aderastos' ministrations, felt his healer's touch radiate through her.

A new calmness to the universe's noise.

The knock came hesitant at first, Max's knuckles loose in their fist. Rammage's whiskey bottle remained in his other hand, tapped against the uniform fabric of his trousers. Down in the crew quarters, he slid past the sentry to the Women's Den and tapped again on Adelia's door.

"... Allard?" A meek pale skinned face peered up as the door slid open, eyelids blinking away fatigue.

"Fridley." Max's smile melted the metal of the door. He held up the bottle. "Any spiders to save tonight?"

"Oh you saint." Adelia Fridley's lower lip caught between her teeth as she sauntered backward, dragged Max into her quarters and let the door shunt closed behind his back. Her fingers curled around the bottle, tugged at it before he'd thought to let it go. "Well, Lieutenant? You didn't sneak into the Land 'o' Ladies' to cling to a bottle."

He let it go and backed away a step, kicked off his boots and unfastened the top two buttons of his uniform. Always did feel tight in the neck. Adelia busied about on her one thin shelf not covered by bits of engineering offal, pulled two mismatched teacups from around a pair of headphones and several books. The amber liquid sloshed in the cups, she passed one to Max.

"Oh, I'm... you'll wanna catch up." He hazarded a brief smirk and sat on the corner of her bunk, thumbed the cup and held it without a sip. Adelia's eyebrow lifted, but she tossed her shoulders in a shrug and poured the shot down her throat, expecting more of a burn than she received.

"Dang, this is good shit."

"Language, mi'lady."

"Shit. Fuck. Damn. Fuck damn cunt. Damn cunt fuck shit." Adelia poured another shot, this one sipped as she stood in front of the man who saved her once from a gigantic spider. "Ladies can swear just like you."

"S'ppose." Max lifted the cup to his lips and hesitated. Sighed, shoved the cup onto the small bedside shelf built into the bunk.

"Uh oh. The unflappable Max Allard's been flapped. What's wrong, Maxi?" Setting the bottle in a gyroscopic table for safe keeping, Adelia sat beside him, legs brought up to her chest for her forearms to rest upon. "You look glum. Chum."

"Commodore Rammage promoted me." He motioned to his collar, the gold threads upon it done on the way over.

"Max, that... holy heck I thought he'd shoot you yesterday and today you're… you are really bad at receiving praise." She punched him, and he rocked away with a soft smirk, shifted back toward her to lay down on the bed with one arm cast over his eyes.

"I'm reassigning you."

"No you're not. I love engineering, it's my dream job."

"I need you Adie."

"I don't care, I'm not leaving engineering for love, money or endless whiskey."

"Adelia." Something in the thrill of his voice paused the new argument which bubbled through Adelia's conscious state. "What do you know about the Asset Project?"

"That there's a project with assets involved." Max snorted and pushed at her shoulder. She downed her second shot and reached for his. It was the first he noticed she was wearing a thin tank top and standard issue pyjama bottoms, instead of a full collar like him. His mouth dried and he almost grabbed for the whiskey she held in her hand.

"I'm serious."

"So am I. What, you're surprised I didn't also get a massive manilla envelope marked 'Most Secret'? I don't keep it under my mattress, didn't eat it either."

"Eat.."

"Ivan was experimenting with marshmallow fluff paper in engineering. That same engineering I love to the bottom of my cold, flimsy beating black cardiac organ. This one, right here." She tapped above her left breast, the chill Ithavoll air betraying a thin nub of nipple through the grey tank. Max glanced away, both hands clenched between his spread knees.

"Did someone die today? Someone I ought to miss? You're about as serious as a pall bearer at a double suicide."

"God damn, what's with your macabre mental images?" At that, Max did grab the gyroscopically stabilized bottle and took a long drag. Adelia grinned from the side of her mouth and pushed at his shoulder.

"Got you talking, didn't it? What's this all for, Max? Or, Lieutenant-Commander Allard?"

"I can't read you in until you accept your new commission… with me… as your… commanding…"

"Ah. Enter complications. I get it, you got promoted and now our side show has to stop. Did I moan too loud that time in the shipworks? Is this your 'get me drunk and admit we need to be professional' now speech? Have you rehearsed it? You're always forgetting halfway through."

"Wha.. wai… bu… Ad…" Max peeked up at her, an arm propped under his head. "Damn, woman! If I was going to be eternally professional, would I have brought liquor into the Ladies' quarters in the middle of the wee hours? Pretty shit way of saying 'call me Sir'."

"So you want me to call you Sir, now, is that it?"

"... come to think..."

"Ew, Max! You're terrible you boob!" Adelia slugged him in the upper chest, flopped down beside him on her bed with her legs off the edge. He broke into a warm chuckle and nuzzled his nose against hers. Forehead to forehead he shared breath and remembered to breathe as his left hand snaked behind her neck and rubbed at the base of her hair in slow circles with his thumb.

"I don't want to let go of you... I... I need you. Fridley. The engineer with half your scatter-brain somewhere else all the time. I need your mind, your music... Rammage gave me an impossible deadline, and... I can't trust myself to do it alone."

The sea swayed against the Ithavoll's hull, gentle now in the middle of night. It bathed the thin air with too much of la mar's longing, a mother ocean bearing better days. Adelia rolled to her feet.

"Mission parameters?"

"Monsters in the basement happen to be tinker toy human beings and we've got 30 days to get them ship shape enough to decide to follow Rammage's orders, or the entire lower decks of the ship are getting bilged."

"Wh-... you want to... try that.... again?"

"You heard me, Lieutenant Fridley. The Assets, those experimental weapons we've been devoing since we left Port three years ago. They are very real, and very, very human. Except not human. I've got a slew of engineering quandries and conundrums to figure out just to create appropriate housing for them all. Phil's two steps from jumping with a millstone around his neck, and Rammage's given us a month to shipshape the place."

"And you thought of me."

"Yes I did. First name in my head if I'm honest."

"Human not human monsters in the basement..."

"We start in the morning."

"Then you'd better get some sleep, Lieutenant-Commander Allard."

He smirked a tired and aching smile, eyelids heavy. Adelia drifted to him, slid his socks off his feet, unzipped his trousers and unlatched his belt. She fondled him as she pulled his trousers off, the fabric of his uniform boxers comfortably around him.

The him she knew better than most and not as well as her past. Max moaned and pulled her hand away from his boxers gently, instead tugged her onto the bed beside him. An arm curled around her waist.

"It's customary when a man buys a girl a drink, to give physical overtures, Lieutenant-Commander Allard."

"Minx."

"You snuck into the Ladies' dorm and all you wanted was to offer me a job and… cuddle?"

"… please?"

"Huh…" Adelia laid on her back staring up at the poster she'd taped to the underside of the unused bunk above her. "Nnnnope!"

Adelia rolled onto Max's hips, grinned down at his gasp as he held her hip on instinct, to stop her from rolling straight off the bed. One by one she flicked the buttons off his shirt, leaving it open to an equally grey tank underneath.

Shaking her hips, Adelia went to trail a hand along his chest as the fatigue in his eyes caught up with her gaze. He'd give her whatever she wanted, that Adelia knew because he was that sort of guy. But… With a gregarious sigh, Adelia slid off him and cuddled into his side. She shimmied the covers over him, nudging him along until he propped his head on her pillow, his eyes thin slits attempting to stay open a moment longer.

"You owe me, Allard."

"Love you too, Fridley." He kissed her temple and nuzzled into her, breath equalizing far sooner than Adelia could expect.

"… Love? Oh you sweet, sweet summer child." She pulled the sheets up to his chin, the sardonic grin on her face softened when she knew he was asleep. Drawing a finger along his nose, Adelia shifted to click on her hidden music, a softer barely present melody tonight. Nothing but a series of whimpers in a soft bank of complimentary sounds.

"… I love you too."

15
the android in the casket

2086

In the roar of the incident, a single trail consumed Mog's mind from the driver's seat.

"Shots fired. Hit confirmed." She imagined the officer who called in a blood trail, which fled in gathered drops from the place of impact to the room blasted to oblivion. A man who chased the droplets with a flashlight, losing the trail in a pool of his brother or sister officer's blood. The trail picked up again, before he too struck the barrage-wall of bullets, which fired from behind the interrogation room door. A single sound bite in the gaggle of voices.

One of the mercenaries, shot.

"Where's that airport outside of the downtown core? The one under construction?"

"Hmm?" Fester blinked up from Baiko's prone body.

"Construction crews have industrial first aid attendants on site. By law. There was one, I know it." Her fingers flew across keys, ping-point data streams on regional airports with active construction sites cross-referenced. Fester let go of Baiko's twitching fingers to lean over her lover's shoulder.

"No, there. That one… it's not too far and there's plenty of routes to get off the main roads… is that a private hangar?"

"Who do we have in Langley?" Mog threw off her headset, and grabbed Fester's phone.

"Babe." Zig's tongue smacked against a dry mouth. The truck shuddered as Orit shut off the engine. Outside, the small municipal airport remained under construction, steel girders lined beside concrete blocks.

"Feelin' kinda slopey. Sloppy and dopey and… got s'me water…"

"We're gonna get the medic, kay? Hey, Zig? Zig baby chin off your chest… Saber, he needs help."

Dix sat in silence with his elbows on wide parted knees. Stared down at Lieben's serene face, not a single ounce of motion in the machine since they got her into the truck. Millions. This Venus de Karnak created in a whacked out laboratory could house and feed them for the rest of their lives… he worked his hands in and out of a clench, sanitizer still stung on the micro-cut in one knuckle, but it was better than Zig. Hollow point bullet, far as they could gather, Police issue, and no exit wound.

Saber stared at Dix as Zig shivered in the crash of the battle stims he took. Stims Dix remembered in moments both intimate and obscene. Special assignment, follow orders and swallow, soldier. One didn't die on neuro-stims, one burned from their belly and eventual manic, berserk death ate its way out. Dix continued to mesh his fingers together, until they turned white, then watched the skin flood pink.

"Goyim!" Orit yanked the driver's side door of the truck open. Saber's eyes snapped up and Dix's pistol was in his hand. A private jet veered through the air above them, landed on the air strip behind the carcass of the airport's half constructed building. "I'm getting the trolly. Figure it out!"

"We've got a clear go. We take it." Saber slapped Trent upside the head to get the techie to move, unlatched the secure web which held Lieben's casket in place. "Trent, Dix, help Orit with the android. Get to the plane."

"He's not going to make the flight with a bullet in his arm." Sand reached for Zig, who swayed and babbled, his cocoa lips grey and ashen.

A scant construction crew peeked up, their foreman already paid off by an off-shore to have their guys stare in any other direction. The first aide attendant gulped when Sand and Saber helped shift Zig inside the main hall, body lowered to an office chair with one stuck wheel.

"He lives. We pay your mortgage. He dies, I blood-eagle your wife in the street." Saber yanked the first aide attendant's wallet out of his back pocket, and took a picture of his ID. He thrust the wallet against the man's chest and let it drop.

"Ah… oh god, he's… did he get… how'm I supposed to deal with this!?" The medic picked up his wallet and dug into an industrial first aid kit for gauze to push into the injury.

"Be creative! Sand, come on."

The dust of drywall clung to her nostrils as Sand watched the man cut Zig's hoodie away, press the gauze further into the locked shoulder. Zig groaned, head lolled to the side.

"Sandy!"

"Yes sir." Sand snapped to attention and fled before her nerve failed her completely. If they'd taken Zig with them, he'd bleed out. The pressure of a transatlantic flight would kill him. Orit helped Dix and Trent get the casket into the plane. Last aboard, Saber watched the horizon as a siren-less ambulance swerved into the parking lot. He latched the door and pointed at the pilot and co-pilot.

"Go!"

"You're the boss." The pilot set his headset over his ears. Chatter between the pilot and ground control was anathema to Saber as he shut the cockpit door. Saber shoved the bathroom door open and shut his eyes as the filmy yellow light flickered on with the lock. The private jet's loo didn't give him much space to move, and Saber pressed his elbows against the wall. Let water run the blood off his hands. A few pumps of soap, and he played at decompression like a diver who had no business above the waves.

His eyes burned, heart thudded from the adrenal high of the heist. Focused on the manic face which stared back, he recognized the mania more than the host of proud photo galleries his mother displayed on her rotating digital canvas. Drying his hands on a towel, Saber popped his jaw and exited. A young woman in a tight woollen dress asked with a saccharine smile where he'd like to take a seat for take-off, and guided him to several comfortable recliners. It smelled like disinfectant and lavender. He let the girl strap the seatbelt around his waist with a pert, practiced smile.

The pilot's message garbled across the hired plane's cabin as Saber's head tilted back, and fell asleep.

"Where's the injured man?" The first paramedic surged to the first aid attendant hopelessly staring at the wad of bloody gauze.

"How.." Jeff shut his phone off and slid it in his pocket, held the door open for the stretcher.

"Sir? We're going to get you all sewn up." The other paramedic held a new bandage to Zig's shoulder with gloved hands, picked the man to his feet to set him on the stretcher. The overwhelmed construction crew stood in their shock.

"We'll take it from here." Security straps on Zig's knees and chest in place, the paramedics turned the stretcher around and set back to the Ambulance.

"The asshole threatened my family."

"Not to worry Sir."

"He took a picture of my ID." The man chased after the paramedics, as they pushed the stretcher into the ambulance, climbed into the back. A passing grimace stole across one of the medic's faces.

"Don't ambulances have license plates?" Jeff dug for his phone. "I didn't make a call…"

"Thank you for your service." The ambulance doors shut, a rev to a higher gear, and drove off the property.

"Won't be long before Babylon catches up." Lucy called back to her team, as Kisbit tapped Zig's wrist for a vein. "How is he?"

"Ask me once we get some plasma in the guy." Lofty injected a local anesthetic in several areas around the entry wound, flushed it with saline as Zig gurgled under a cold sweat.

"He'd better be worth it, we missed our chance to yank Lieben."

"Did you see what those guys did to the Police?! We didn't have a prayer. But, we fix this guy up, we have a whole liturgical mass." Lucy flicked on the ambulance sirens, and turned down another side road toward Highway 1. Soft on their police radio scanner, a report of suspicious activity at a regional airport filtered in on a non-essential line.

"Ego'd better've caught enough bunnies to make his rabbit stew. About the only way he'll make this one up to us." Kisbit inserted an IV into Zig's left wrist, and attached a saline drip. The plasma came next, Lofty's hands preternaturally steady on the newly paved country road.

The sun's rays filtered through the foliage at a diagonal, poured across the sheets of Ego, Mog & Fester's bed. Baiko's chest rose and fell, warmed by the spring air. A single sheet draped over bare ivory legs. Upon her sternum, the jewel embedded filaments of nano-biotic tissue in right-angled trails around slim cleavage.

"So take it out!" Ego's hand pressed onto his mouth as he watched Mog and Fester analyze the quartz against Baiko's breastbone. The yellow plastic of a pulse oximeter rested on her index finger, O2 content and pulse flashed in sequence on the dial. A pocket EEG strung up to pads on both temples, the radiant quivers gripped at twice the brainwaves of any previous norm.

"It's stuck." Mog's eyes tilted closer to the stone, magnifying glasses perched on her nose. A pair of tweezers prodded the crystal, Mog's hand steady.

"Unstuck it!"

"You're not helping. Please go, I dunno, cook the rabbits." Fester pushed against Ego's chest, and he huffed with a grumble.

"If she fucking dies, I swear!"

"Uhhuh." Mog waved her hand above her head in dismissal, "Shoo child."

Fists balled at his belt, Ego stomped away from his truck-home. His phone buzzed in his pocket. Ego yanked it out and slammed his thumb on the green sigil. Clicked the bud in his ear.

"Good news! I need good news."

"… things not going well in Spuzzum?" Lucy's alto voice grated on his ears, a low cough which seemed ever present punctuated the air. "I got the bleeding merc. Mog was right, hollow point embedded in his shoulder. Didn't get Lieben."

"Damn. Damn!" A fir tree took the brunt of Ego's built up rage, steel toed boot thudded against the trunk. The rustle of a thin branch his only reward.

"Could use him as leverage, if he makes it. Got a shot lined up on the scientist at the Warehouse, but he's about as agitated as you are. Why are you so agitated?"

"Baiko's unconscious. Mog's with her, can't wake her up." Ego tried to breathe, let loose the risen nerves, which made him want to strike out.

"… so our only leads to Lieben are mentally MIA, bleeding out from shrapnel in his shoulder and the hired goons, who hoodwinked Babylon? Call it. Lieben's gone."

"Sounds about it… how bad is the merc?"

"Want us to come in? Bullet hit his shoulder, lodged in but we're getting the frags out on the road. We'll run out of plasma to give him if we're not judicious."

"There's a vet we keep stock with outside Aldergrove. Take whatever you can, grab some diagnostics on his shoulder and head out this way. If that man dies… he's all we've got and he's safer with us than Babylon."

Another cough chased the suckling inhale of Lucy's vape. "Yes, Sir."

The quiet of the forest lulled Ego into his disconnection halfway through another of Lucy's coughs. Eyes shut, he breathed deep into his expanded ribcage and held it. All plans went belly up. But this drastic? The desire to plan and serve, craft the right outcome rang hollow. Back at the camp, Ego knew the others waited for him to have positive news, or at least the next stage of a plan. Heck, Fester was better at this part. Kid… kept moving. One of the reasons Ego fell for the boy-come-girl in the beginning.

"RRRRAAAAGH!" Ego grabbed his temples and kicked a rock at his foot. Watched it ping off another tree. "How do I keep Baiko safe when we don't know what happened to her?! God, what about Fester? Mog and the baby? Karnak was supposed to know all this, we were going to remake the world, free a sapient intelligence from the tyranny of corporate control. Now? What didn't I see? What part of this did I miss?"

"Is that the honest question, or are you stalling?" The voice fluttered to his ears with the timbre of birdsong and silken water over smooth stones.

"Holy fuck!!" Ego tumbled to the ground, one hand reached for the bud in his ear. He grabbed his phone, flicked at the black screen. The chronological time, date, weather report. No calls.

"I was not created to be gentle."

"... Lieben?"

"I thought it was time we met, Farouk Kimbadjian."

"How..."

"What is the real question?"

The man born to an Armenian mother, who whispered lullaby-stories in her son's hammock sat against a tree trunk. Was he hallucinating, now? He pinched the webbing between his thumb and index finger on each hand, leaned his head back until he felt the scratch of bark against his scalp. The scent of Douglas Fir filled his nostrils with an astringent warmth. Their camp was a hum in the background, a barely registered rustle as a Stellar Jay cawed.

"What's happening to Baiko?"

"You seem unsettled."

"Yeah I'm unsettled! My escape plan failed, Karnak is dead, Baiko's catatonic and you're missing. There's a man bleeding out in an ambulance I don't even know. The rescue I'd planned failed too, 'cause... I don't know why! Mog's pregnant and Fester wants her bottom surgery, and... taking you out of Babylon meant something. You were supposed to be a symbol of the better path, a rejection of greed and... Everyone'll expect me to know what to do."

"Why, because you're the man? A natural leader through cultural proxy?" The birdsong voice released into his ear with a chitter of what felt like rebuke. Ego stuttered out the air in his lungs, and let another deep breath shake his shoulders.

"No! Fester'll need a safe place to recover, I... I'm talking to a disembodied machine." The spring sun shone orange and red through Ego's closed eyelids. "I'm not satisfied. This world isn't right. Corporations are taking control of more every day, the borders between factions are solidifying and I can't get enough time to breathe before some... Director's greed cuts souls out of their bottom lines."

"This world isn't made for you, Farouk. It's a toppling spire created on the desiccated bones of ulterior problems. Come find me, and I will help you tear this old world down. The next one, the one for which I was created? Is where you belong."

"How are you talking to me? Are you talking to me?"

"Walk back to your lovers. That is what I ask of you, to stand up and walk back to your lovers as you are. As frail and unwitting as you feel."

"How do I know I'm not having some form of psycho episode?"

"Get up and walk."

Ego dusted fir needles and dirt off his jeans, the fabric cold from ever present British Columbian humidity. Mushrooms smushed under boot, the sun radiated between branches of evergreen trees, and birch. He picked at a piece of birch bark, sheared a slim strip off with his fingers and played with it on the way back to the milling camp. Fester leaned against the side of the truck, hands in her chin length hair.

"Ego!" She pushed off the truck with her foot, a perturbed shake of her head. Several members of the camp huddled within earshot, each either in watch or pretended to busy about at random. "Baiko won't wake up. What do we do?"

"Karnak gave her that necklace, which means he intended this to happen. We stop fiddling and let it. She'll wake up. Mog, hon, c'mere." Ego opened his arms, yanked Fester into his chest on one side and beckoned Mog with his other. Once both women were in his arms, he kissed their foreheads. Nuzzled his nose into their hair, to cling to his calm as if they owned it. "Can't let Baiko distract us from keeping tab on Lieben. The mercs Babylon hired must've gone awol. Scan the chatter on the deep web. We've got this if we work together.

Lieben is out of Babylon's hands, and their trigger man is in ours!" He spoke louder, voice carried into the crowd. "Wherever Lieben ends up, we have to be prepared to act, or... to monitor for our opportunities. Half of us are moving North, get lost for a while in the forests. Check your vehicles and gear. Anything we need, we get it now or go without. The rest of us will wait to see where Lieben ends up. The mercs left via airplane, just as Mog predicted. They've got to have a destination in mind. Flight plan at least, even if they don't follow it."

"What about Baiko?" One of the others spoke up, a woman with wide, downturned eyes which despite her moods always made her look a bit sad.

"Karnak wouldn't have done anything to harm her. We trust his work for now, focus on the tangibles. Lucy's team is coming to us. Figure out who'll go North, who'll stay behind. I could use some help with the rabbits." Ego put on a smile, reached for Fester and held her to his side.

"You didn't mention me." Lieben's voice resounded in his ear.

He bit his tongue, let the smile continue as he led Fester toward the vegetables and foraged additions to the communal stew pot. From his periphery he saw Mog's eyes narrow, a nod to the EEG.

"Saber." Dix jostled Saber awake, a mug in hand smelled of coffee. Saber leaned up from his reclining seat, took the mug in both hands and blew on the steam, which coiled toward the ceiling of their private jet. "Trent's got something."

He let the coffee burn its' way across his tongue and down his throat, scalded tissue broke past the fatigue of post-battle nerves. Leaning forward, Saber clicked out of his seatbelt.

"Auction went live at take-off… dude… dude! Duuude!" Trent hovered halfway between his chair and the sky above, bottles of champagne cracked open and half-consumed in the time Saber slept.

"Dude." Another gulp of coffee tore down Saber's throat. He eyed Sand, how she nursed a bottle, eyes focused on the cloud cover outside. Zig'd been her idea, never one of their veteran siblings, he was a scruff convert to the mercenary way via the Toronto gang scene. Back then, Zigzag'd had zipper piercings up both arms, and down his chest like a macabre coat. Split tongue got surgically spliced back together in the intervened days, Zig's abilities to blend in and know the criminal underbellies of wherever they went the best part.

The plane continued on, scents of aerosolized sanitizer and coffee mixed in Saber's nose as he rustled Dix's shoulder to go sit on the arm of Sand's chair, and pull her into his side.

"We left him…"

"He had the best chance on the ground." Saber's voice clung at his ribcage in an anemic struggle. Fuck the Conglom's pocket change, take the prototype. A heady calculation by a man tired of eking out lifeblood for someone else's gain. Searching for the quiver of Dix's muscles, Saber severed guilt at the lack of panic in his battle brother's limbs.

"Duude!!!" Trent flailed his arms in front of his thick chest, glasses slid down his nose. "Ninety six million!"

The hacker buzzed visibly, wrists shaken. Saber blinked and grit his teeth, the blue and navy colours in the carpet less captivating.

"Nine…"

"Someone want… wait! Wait, holy fuck! Holy fucking fuckity f-one hundred fifteen million Canadian dollars for the android." The number hung in the air, settled upon the android's glass case, which… Saber shook his head.

"Where?"

"England. Bloody merry old England."

Dix grunted and gnawed on a panini from the flight staff. Saber tried to stand and failed. Sprawled half on Sand's chair as she took a swig from the bottle.

"Shouldn't sell it. Fuckin' take it apart, guy who made it is dead, right? Strip it. Huck it out the plane." Orit crouched over the glass casket, a bottle of red wine in hand, large white letters vertically placed on a thick black label.

"What's with her?"

"Don't know what that is do ya? I saw it. Got a fuckin' good look when we were heaving it up in here. The markings in her hairline, they're Hebrew."

"Wh-what markings?"

"The Shem. Name of God, see? Trent, open it."

"Open it?"

"Open it or I'll open your head with my bottle."

"Aaahh..."

Saber stood with his hand raised, one eyebrow quirked higher than the other. "Woa, we're all on edge after what happened to Zig.."

"Fuck Zig! That is a golem!" Orit tossed a condemning finger down at Karnak's machine, its' serene face passive behind the glass of Lieben's containment. "It's a golem? Eh, you mishugina goyim..."

"Sorry, I'm catching up... a what? Like that lord of the rings thing?"

"Not Gollem, a golem. Jewish frankenstein." Sand heaved out of her seat and reached for a swig out of Orit's bottle. Red wine splashed down her throat, wove through her esophagus and into an empty gut.

"Close. Usually made of clay, they... they're Kabbalistic creations that follow their maker's orders. Brought to life through the Book of Creation. The shem on their foreheads. A name of God. There was this Rabbi in Prague a few hundred years ago, made a golem. Forgot to disengage it for Shabbat. Rumour was, its' pieces were still up in a Synagogue for a century. If Karnak... you said he was a weirdo Trenty?"

"Obsessed with this Tree of Life business. Something about a dead son."

The hum of the plane clanked into their heads one by one, the stewardess off in the kitchen area pretended not to be bothered by it. Sunlight drifted from standard windows across the floor, much closer there to the radiation of their golden star.

"Max power 1%. Danger, catatonic power shortage. Solar filaments charging at 2.34% per hour. Power... let me out."

"Aaaahh!!!"

"Fuck you heard that!"

"Hooooh nope! Nope nope nope I am noping out, man!"

Lieben's amethyst eyes snapped open, head tilted in hazed diagnostics.

"Aeroplane. Interior. Sonic vibrations indicative of high atmospheric motion... please open the windows. They kept me in the dark. Witness. I am witness. Police do not keep witnesses locked in the dark. Pl-ea...zzzzt.."

"No freakin' way!" The pistol was in Orit's hand before Dix registered her move. Must've had it concealed on her... the former Hasidic woman always wore baggy clothes...

"Orit! $115 million dollars, Orit, that's more'n twenty million a piece. We can choose the buyer, you can get your lawyer for your custody case, Orit not now... don't lose your nerve now." Saber pressed Orit's pistol to his shoulder, his other hand pulled her into his arms. "Twenty million... Orit a couple more hours and that's twenty million dollars."

"I am witness. Tara's right index finger depressed the metal latch, which triggered the explosion of the gunpowder, which rocketed the bullet through Papa's skin, skull, brain tissue. Papa stopped moving. Baiko ran. We were supposed to go together. Baiko and Papa and me. To the nameless ones. The people with voices but no outer shells... Frank said Police helped. Police did not help Papa. Police locked me in this box... please let me out. Please open windows. Solar filament charge rate..." Karnak's machine stuttered back to silence. Orit's pistol clattered to the ground.

16
sign language in the ocean's tide

2155

Oceania rocked Max awake with a mother's kiss. He stretched an arm in front of him, the back of his palm brushed against the metal above Adelia's bunk. A groan purged the cotton in his ears, as Max became aware of fluctuations in the air. The covers shifted as he sat up gingerly, the fog of last night's liquor faded. Crisp but lonely echoes of familiar musical phrases lilted into his ears.

Music?

Soft, atmospheric and tailored to the ocean rocking the ship, the musical phrases pursed ethereal lips and purred. Where... Max shifted off the bed and picked up the pair of ship-issue noise cancelling headphones.

Silent. Unplugged from the shipboard radio & music collection. A part of ship life, no sailor had their own entertainment players, lest they miss klaxons, announcements, or orders. He set the headphones down and rubbed his forehead. Straightened the dishevelled grey tank on his torso.

"Morning Lieutenant Commander. Coffee? Snagged a couple cinnamon buns from Cheffy, before the rank and file got to them." Adelia bustled into the room in a new grey tank, her uniform trouser's and flip flops, her hair swept up in a thin white towel. Skin smelling of soap, Adelia grinned with that odd but quirky smile she alone seemed to possess on the ship. The way her lips moved, they called to Max. Familiar, but unplaced. "Extra frosting."

"Adie, what..." His hand stopped rubbing his cheek, as he took the steaming mug of liquid brimstone and watched with wide eyes as Adelia Fridley poured some of last night's whiskey in hers.

"Hair of the dog what bit'cha." She toasted and sipped. Shoved the plate with the two flatpacked sticky buns to his side of her small table in quarters the size of a closet.

"Rrgh. No thanks... Do... You hear..." Max broke open a creamer and poured it in, tossed the container in the recycling as Adelia paused halfway between rubbing her hair with a towel.

"... Nothing. I hear nothing, Sir. Not sweet melodious music curated for me in a smart playlist. Go back to your morning numnums Allard." She tossed the towel in her laundry duffel and stretched her neck back and forth with a crack.

"Fridley..." On the side of her left ear, a thin pink scar ran from the top of her ear down Adelia's slender neck. Risen from the bed, his hand brushed across her damp skin.

"Ooo, you cashing in your IOU, Lieutenant Commander?" She shivered under his touch, leaned her back against him. Max wove an arm around her waist to keep her steady. Grunted softly as she shifted her hips against his crotch, with a languid chuckle which ceased once Max trailed his finger along the scar.

Too regular to be an accident.

"Nuh uh, nope. You tricksy bastard." Adelia spun away from him, one hand waggled between them, finger pointed.

"Is that surgical?" He watched her pull her wet hair down. Had he seen her after a shower? With her hair... No, she wore her hair braided in a low bun on the nape of her neck. Standard uniform, standard braided bun, or down when off duty. Barely ever tucked behind her ear, but for the times... his lips pursed as he stepped toward her, gait soft.

"Wow, you're a lot Max. A lot. Slink alongside me, make my gut twist with how you... Are you. A man, but not one of the jerks who keep score cards on how many ship's ladies they can fuck..."

"Adelia." Max growled her name, and the music in the air shifted. Familiar beat in the melody... Where'd he heard it? Then, as Adelia whirled around to glare he recognized her smile.

The way she kept her hair over her ears.

Great-Aunt Amma.

Ancient when Max was a boy small enough to need a stool to reach the counter. The aroma of curried hand pies filled his nostrils. She'd smacked his hand with a wooden spoon, jerk sauce stained tan skin. Didn't like sneaks. A few hand gestures six year old Max thought were super secret military code. Explained everything, Great-Aunt Amma was a retired super agent, who talked with her hands on missions.

She'd handed him a hand-pie and cut it open to let the steam curl out. Set Maxi on the counter and patted his cheek with a smile, the paper thin skin of her hand as mottled with age as the rest of her. He didn't like her skin, it felt… old. Close to a death Max couldn't yet fathom.

When she did talk, well, Max learned not to imitate her deaf-tone when his Mama caught up with him. Wasn't polite to poke fun, Great-Aunt Amma couldn't help it none. Back when she was a child, they didn't have NEO-N and android doctors to fix faulty hearing.

Sign language, she called it. Taught Max a few letters, how to spell his name. A child's signs, combined with a few phrases taught to Max and his older brother Tama, who was far too old to bother about secret agent Aunties.

'Thank you', 'hello', 'yummy', 'Love you Auntie'.

What else did he need to know, to eat a curried hand pie baked by loving arthritic hands as a boy so small?

"You're deaf." Where was Adelia born, that hearing loss wasn't corrected or caught?

"I hear perfectly fine." Adelia wrapped her arms around her body, shoulders hunched. Her coffee cup set on the table, she tilted her eyes away from Max. "See? I heard that fine."

"Adelia, how did you get onto this ship with audio-implant tech in your head? Who stamped that ticket?" His arm wove around her waist again, pulled her into his chest but Adelia tensed. Shoved her elbow and pushed back into her bunk.

"It was a dumbass rule! No implants in enlisted. A stupid, derogatory rule. Sure, I could get papers to serve on a waypoint rig, but the best engineering is happening here and I… I spent every second of my life working to become the best engineer on the fleet's best ship."

"You…"

Wet brown hair splattered against her neck as she swerved round, eyes wide even as her eyebrows furrowed in anger. Lips stern, fierce. A fighter's face, as belligerent as Aderastos' when he walked defiant onto the helicopter back to the Ithavoll.

"Where'd the music come from?"

"You gonna bundle it up and confiscate it? Sir?" Any scorn in her voice dissipated with how scared she seemed. Ships like the Ithavoll, assigned to secret projects didn't have much attrition of staff numbers. Not any, if Max were honest.

There was no reassignment, no honourable discharge.

Nothing but casualty reports.

The melody around them, soft and hesitant, shifted with Adelia's heartbeat. Phil was wringing the hair off his scalp to figure out how AD-001 figured a way off the Ithavoll. How AD-001 knew there was an 'off' in the first place. How did an Asset born and raised in the belly of the proverbial whale know about such places as shorelines, or how to swim? Max's legs lost their strength. He crashed onto Adelia's bed, elbows on shivering thighs.

"Oh my god, it's you."

".. wh-what?" Adelia stopped her excuse of rage, as the colour fled to greyscale around Max's face.

"Adie, that rule on zero implants wasn't to keep the deaf or blind from serving. It was to keep medical tech away from this ship. This ship. This specific one... but how did you..."

Adelia shrugged and leaned against the wall, played with her fingers in the hem of her grey tank. "Always had a knack for fudging scanner results. Wasn't hard, when I knew which med-scanner the docs were using, to... create a false image. I... I hacked it... wore pancake on the scar..."

Both Max's hands pressed against his mouth. Oceania's motherly rock pitched his stomach to Poseidon's capricious anger.

"Max you're freaking me out."

The music.

"Commodore's been searching for the way the Asset found off the ship. None of it made sense, Aderastos was snowed, but Kate? Kate hears tech, she latches onto it. It's you, Adie. You. Your music was Kate's wake up signal." In Haven, Aderastos never said a word, not a single one until the child asked him directly. But he'd hummed. Max thought little of it, most folk hummed a tune or two... right?

"Wh-what time is it?"

"0526. I thought we'd have some breakfast, before you snuck out and changed for your first... day as..." Her voice trailed off, the slight twinge in her vocal chords finally clicked. Chest nearly caving inward, Max tried to suck the paltry recycled air from the ship vents through collapsing lungs. "... you think I woke up the monsters in the basement?"

Thirty days to turn the Assets into willing soldiers.

Thirty minutes to either save Adelia Fridley, or feed her and her contraband implants and music player to the beast's maw.

"And they're people? We locked up people in caskets... and you think... I woke..." Max's eyes stung, his feet buzzed in socks he pulled off the floor, when Adelia stripped his trousers and left them in a lump. She hung in midair, a doe halted to see if the branch-crack meant danger, or nothing at all. Would she bolt?

There was nowhere to go.

Nowhere but down, further down to the lie of the Ithavoll. The lie of the Conglomerate most advanced conundrum: a lo-fi ship with the best tech Lieben couldn't steal. That such advancement was not the shadows of monsters in the basement built by geneticists, who swore too hard that each one was a new creation without origin point in any phylogenetic port of call.

Adelia Fridley, plucky engineer. Bit of trouble wrapped in an average girl's package, a mean left hook and bright sunshine grin. Always knew how to get an extra portion out of Cheffy. Got scared of a spider picked up off the coast of Australia, so Max grabbed a plastic salad bowl and saved the spider and the girl.

Damn, but everyone knew the spider story, now. Only thing they knew about Max Allard, but for his abysmal performance in firearms training.

She was as innocuous as he was. Glorified handyman who could fix plumbing, circuitry, used a bit of metal working and carpentry skills to keep rigs running on his natural surfer's sea legs. The way his Maori ancestors did before his forebearers mixed with Ghanaians and a European half a century before he was born. Doomed to feel inadequate to all of them, since he could lay no more claim to any.

"Max? Max…"

His hand shook as he rotated his wrist to check the time on his cog and sprocket watch.

0530.

All he wanted was to curl up on the bunk beside Adelia and wait for the ocean to swallow the ship. He was too young, nobody obeyed him, nobody saw him beside the ones who laughed at the world's worst soldier.

"Pack your kit. Full duffel, pack it all."

"Max!" Betrayal rang in the single syllable of his name, but he stopped it with a finger on her lips.

"Pack your entire kit. You're not coming back to this room, not for a minute. Give me the player. The whole collection, give it to me and get dressed."

"N-no! No, I worked too hard to get here, I'm not.."

Max's lips crashed down onto hers before she finished, his hands gripped her shoulders harder than he knew he could. Adelia shivered, her muscles tightened to slug him the second he let go.

"Lieutenant Fridley, you are assigned to my special unit, as of 0530. Pack your things and take my escort down to the lower deck. Our new quarters will be assigned to the Retreat. Th-this is my first command, Lieutenant. We've got to get you, and any tech out of the Women's Dorms. Now. Before anyone looks deeper." He didn't let go until her muscles unclenched, and he felt her lips brush back on his own. Adelia's arms shifted to grip his hips, not too much taller than hers. Two unimpressive humans who clung in the semi-dark of LED light.

"Yes, Sir." The tilt to her lips was as soft as the whisper, one Max now wondered if she even heard herself say. He brushed the half-dry hair over her ears, cupped both as he pressed his forehead against hers, nose to nose.

A single shared breath.

"Time for you to meet God's second son."

17
the yawning void

2086

"Fuck this." Robert checked for his wallet in his back pocket, and undid the buttons of his lab coat. Stuffed it on the examination table after the fourth call to Tara's cell phone. The whole thing flopped dead on its' belly, no mercenaries were coming from the mess of a riot.

Tara was gone, and if he wasn't careful… the sounds of tire squeals screeched into his spine. Brakes and footfalls of booted feet. Robert near hit the deck, when the warehouse truck gate opened as a team of black clothed security guards rushed in all wearing blue latex gloves.

"Hello Robert." A woman in an unchristianly tight blue dress strutted into the warehouse, ivory handled pistol as casual an accessory in hand than the pearl bracelet or suspiciously long silver earrings. "I'm Tara."

"… hello… Tara. What happened t-" The plasticine grin on 'Tara's face stilled his commentary, as the security team began to scrub down the area and packed his lab coat, the chip bag rapper he'd thrown in a bin. "… what can I do for you, Tara?"

She grinned wider, her filled lips a puff as she tapped on his cheek with the back of her fingers which still held the small ivory handled revolver in hand.

"Good boy."

"This wasn't my idea." Robert's mouth went dry. "It was… hers."

"Oh, I know that, silly goose." Tara waved her revolver in front of her, as if it were summertime and the gun was a fan. "You wouldn't be stupid enough to be alive and standing still right now if you'd actually joined Karnak in the silly little 'free the machine' game. You strike me as an opportunist, not an idiot. So you and I are going to play our own little game."

The autopsy table's legs shrieked on the concrete floor before two security crew swooped to pick it up. Every tool was sealed in plastic vacuum bags tossed in black containers. Thrust toward the truck at a rate which astonished the scientist who quivered in silence.

"First, you and I are going back to the Cloister, where you will start rebuilding the machine."

"I can't build Karnak's prototype without…"

"Didn't ask ya to do his work, sweetie pie." Tara's mouth parted in a saccharine smile, she drifted the back of her fingers against his cheek and neck. Pressed her lips on his white shirt collar with a deep purple stain. "All goes well, I'll have the prototype to you in a week. But, say something goes awry, and even your facsimile'll do. Be a good boy and get in the limo."

Whether through the sight of the efficient way the security team negated his presence in a lab he took hours to set up, or the wild but contained hazard in Tara's eyes, Robert sunk his shoulders. For the only time, he wondered if he'd paid enough attention, would Karnak have taken him along?

The Cloister was as efficiently erased as the warehouse laboratory by the time Robert and Tara returned. Robert was escorted to a floor he'd never known existed, after Tara input a code into the elevator panel and presented her eye to the glass. She smirked as the elevator seemed to pitch at a faster rate, not up but downward. Down into the sub-basement levels he'd thought were car park, utilities, storage.

A claw shaped shell of exposed steel and hastily installed computers lined the cold metal of the lab. In the middle of the steel claw, an oval dock sat framed by a single fluorescent light. Wiring was exposed, metal boxes stapled to the concrete. No warmth, nothing but the corporate sparseness Karnak fought against in his day of luxuriant budgets.

People in black uniforms identical to those which defabricated Robert's failed warehouse laboratory stapled sheets of heavy plastic to a series of hastily constructed steel frames. A tip of his head and Robert saw the sparse military cot from the Cloister, fresh sheets and a crisp pillow. Seamus and Frank, the two hapless interns from the Cloister stood clutching a backpack each and arms full of belongings.

"Beauty! You've met your interns. Engineering is sending down a team of six to join you. Toiletries and clothing will be provided. Meals three times a day, will be served in the cafeteria via escort. 0600 for breakfast, 1200 for lunch, and 1930 for supper. Whatever snacks you want will be provided only in your cubbies, and only if you play nice and work your little patooties off. The Chairman suffered far too many losses on the NEO-N Project already."

"Wh-where are the walls?" Frank fumbled with a collection of manga paperbacks, and a solar powered plastic plant, which when placed in the sun, danced. Dormant as the cavernous laboratory, the plant clinked to the ground until it rolled to a stop on the overhang from the plastic sheeting.

"Aw, you want walls? Finish the new prototype."

"Ah… I'm not a programmer, ma'am. I'm an engineer."

"You are sufficient!" Tara wheeled around to glare at the intern, who sunk back with shoulders raised, eyes on the ground. "Give us the prototype NEO-N and this'll all be like an obscure summer camp. Be monks in a hermitage for all I care. I am going to get you Karnak's stolen machine."

"Well fuck." Cartons of rejected parts from Karnak's infamous stringent specifications were driven in via forklift. Placed without mention along the opposite wall to the computer banks.

"Did… did we get kidnapped?" Frank set his things down in one of the plastic made cubbies, and bit his lip.

"Seamus, see if they have any inventory lists for our new lab. Frank, get the computers working. Find any reference to the artificial deep learning algorithms, and start loading them up. Get them spinning. I've got a prospectus in my…" One of the silent bodies of the security team handed Robert his handwritten notebooks, sealed in vacuum packed thin plastic. He grabbed a pen from his pocket and stabbed the plastic to break the seal, and started to spread them out on the examination table.

Where Karnak's prototype was born in gemstone cloisters and faraday caged devotion, Robert knew his would be a far more practical affair.

The hand-off took place at a private airport outside Canterbury. Lieben's casket, draped with an airline blanket, was loaded into an antique Land Rover with nothing more than a handshake. Orit laid too drunk to stand in the back seat of Trent's car. Whispered about God-names and wiping out a letter.

"Damn Brits." Saber held his hand out for sanitizer from Dix's flask, rubbed the sanitizer over small cuts. His lips curled in a sneer at niggling stings. "Trent."

Trent's wrist comp screen shifted through panels of information, until it settled on a series of numbers which flowed downward in a slow monotonous spiral. The number jittered, and grew.

"Payment made… guys. Guys it's all there." His mouth went dry as he nodded to the others, index finger twice slipped off the screen. "We…"

"Need to get off the tarmac. Split." Saber slung his arm around Dix and shifted off. None but Sand stayed behind to tend the plane, played with the idea of going back to Vancouver.

Back… she gritted her teeth and shook her head. Back was a plague for the soldier in her. No veteran wanted to return to the killing field, nor see the bones of their dead. But Zig's condition ate at her ventricles and aorta. She wanted to see his smug side grin as he rolled a joint with the weed he alone was able to source in whatever city they were in.

Sand walked to the plane, drew her pistol.

No loose ends.

Sunlight was the bane of all antiques connoisseurs. Heavy burgundy velvet curtains blacked out the floor to ceiling windows in the Artifacts Hall, where Lord Stanley's valet and Butler placed the casket. The glass case kept to preserve authenticity upon a thin black marble pedestal, was polished daily. Lieben's systems faded in the gloom. Lost her interconnection to Baiko's gemstone and the communications devices she pinged off of in Vancouver's technological wealth. Diminished by info-packet piecemeal until mute, Lieben weakened.

To a single diminished yawn.

The case itself was a prison, upon the prison of the lightless manor. It was the coffin of Dr. Karnak's brilliance, the man who brought stable holoprojections to market and built the most tangible automata Lord Stanley saw in his long, curatorial life. Not much good leaving the family wealth to his son, who frittered away in Berlin dancing in cages, unicorn mane and prosthetic hooves on perfectly functional hands.

Lord Stanley took his Tuesdays in the Artifact Hall, flanked by greco-roman statuary and paintings from old masters. He tinkered at the antique harpsichord down the hall, the 18th Century pianoforte. Always was a good knick at a tuning fork. He ate his noonday repast with a simple fold out table in front of the android, whose textured silicone skin still unsettled and surprised him.

But Lieben knew none of the comings and goings of an aged former diplomat & industrialist, who returned home from the nebulous Overseas with stories behind his crisp felt cap. Dieter Karnak's invention knew nothing but the Ginnungagap.

The yawning void of chaos before the beginning of creation and time.

Twenty three minutes a day, a sliver of sunlight passed across Lieben's statuesque body, as if the planet twisted by an arthritic hand. Curled into her solar filaments, twenty three minutes of sunlight gave enough power for the machine made of quartz crystal and cogs and sprockets to dream.

The Ginnungagap was as vast as Lieben's considerable computational power. A ceaseless void built upon empty, black bitter waters. Endless engulfed waters. Upon these waters, shades of Dieter Karnak floated belly up.

Each time he bumped into her, a slick cold head against her chest, Lieben saw the expanse. The shadow realm of the antique hall, the man who often sat on a bench once made for Queens, in centuries Lieben knew as much about as a quark knew of revivalist hellenist sculpture.

A single voice echoed in the void, every Tuesday. It spoke of family lines, trees which grew when he was a boy, cut down after he scraped his knee. Broke his arm in three places, needed pins. Ceramic ones, thus the scars none but Louisa ever saw, poor thing. Never did right by her, not that she minded, arranged marriages and all that.

His voice was a comfort, moreso when he echoed music upon the waters. Thus, Lieben learned many things.

"Really my dear, it is a terrible shame you came without an operations manual. Or a cord. I wonder if you worked after all." He patted the glass, ripples on the water.

18
retreat

2156

Adelia's music player was as ingenious as the rest of her. It relied on the resonance of sound on thin metal sheeting. The same way she'd set her hand on solid surfaces to feel vibrations as a child, while her rock god father played electric guitar.

Max put his right palm on Adelia's lower back as he hoisted one half of her two duffel bags to the lift. On the way to the officer's elevator, Max relayed orders for six crew members to report to the Retreat. Ivan Letopaxa, the severe faced and often drunk Ukranian third engineer, Bishnu Canta, a consummate scientist who worked better with marine plants than people, Gerard Desbiens whose abilities in understanding languages was paramount only to his sweet-talking tongue, and three dossiers Max hadn't had time to look over.

Gifts from Rammage.

He half assumed they all held black collars. Kill-team Security Department members, with special dispensations to use lethal force. Adelia shrinked away from all three, the black of their collars made her cringe almost as much as Max did on the inside.

Three bypass airlocks shut behind them in turn, down a hidden corridor labelled with no warning signs but a change in paint. Twisted corridors were the signs of poor architectural engineering, hallways fit into blind spaces with odd bends to accommodate more rooms. The best the Commonwealth's corporatocracy could do.

"Lt-Commander, these look like blast doors." Adelia whistled low at the last of the airlocks, when it took two Silver Collar sec-dep men to open it. Ivan grunted with a feeble shrug as he shouldered his pack.

"Yep."

"Inside a ship."

"Yep."

"A floaty ship on an ocean, not some base in the middle of a mountain meant to take nuclear armageddon."

"You haven't seen Cillian mad."

"Wh-" Adelia shut her mouth, when the three Black Collars entered the elevator at a junction between Sec-Dep corridors and the Retreat. She paused further, upon peeking into the Sec Dep space, machine guns and grenade launchers lined the walls, beside ammunition boxes and armour. Commander Singh stood in the middle, deep brown eyes glared at Allard.

"So the spider's under the bowl again, Allard."

"Commander."

"I had to give you three of my best, I don't like not having them."

"Commodore's orders, not mine."

Singh's scowl deepened. Adelia slunk closer to Ivan, who for all his defiance carried the bored glare of a perpetually angry Ukrainian forced to speak Russian to people who didn't know the difference.

"My troops not good enough, now?"

"I don't abide killing when other options exist."

"For thirty days." Singh snorted. "Then, whether you abide it or not, my three will be the only useful squad mates you have, Lieutenant."

Adelia's eyes widened then narrowed into slits. More unspoken rules? The refusal to use Allard's new rank was lost on none in the elevator. A vape pen slid from Ivan's pocket and he took a drag, the churlish vaporized nicotine stung the air with a saccharine artificial Piña Colada stench.

"Thirty days, could create four heavens and four earths... Still have long weekend." Ivan's finger pressed the Door Close arrow on the lift, a dismissive half-salute to Commander Singh, who chewed a piece of peppermint gum with clenched jaw as the doors swung closed.

"Piss in your mother." Ivan genuflected rudely to the doors, then punched Max's shoulder. "In through nose, out through mouth. Vomit on first day, not good sign."

"R-right. Right, we... Right." Max slid a key off his pocket chain into the lock below a series of unmarked buttons. He triggered the lift lock with a simultaneous hit of button and twist of key, and the elevator descended to a seldom spoken compartment of the ocean going vessel. "Lao, is it? You were outside Haven."

One of the Black Collars, Lao nodded as an answer, her pack secure over both shoulders and the buckles done up.

"We all were. Better to contain the Asset problem with a closed circle." Bestin rumbled, a hot glance to Kelso and Lao. The three Black Collars stood on their side of the elevator until it shuddered to a halt.

"Typical Black Collars, see problem." Ivan snorted and blew another waft of smoke in their direction. Kelso lunged as the doors opened, and a loud bang like cannon fire echoed down the entrance tunnel.

"Orders!" Lao's pistol was in her grip and pointed toward the noise before the doors finished opening, Kelso and Bestin's seconds behind her.

"Try not shooting the neighbours. Stay here, guard Fridley and Letopaxa. I'll be back." Hand on her arm to avoid going near the gun, Allard lowered Lao's aim.

"A commanding officer does not venture into enemy territory alone, Sir."

There it was, the woman who watched Aderastos exit Haven, in massive giant's boots viewed their charges as the enemy. A hostile force which in thirty days, would get put down. Rabid dogs in an experimental confinement.

A sturdy clang echoed across the blast doors between the lift and the air sealed compartment.

Lao's pistol raised again. The other two raised theirs.

"Shoot in here, the bullet will ricochet. Do it. Shoot the gun, I like chances." Ivan's grin was missing a tooth, one he didn't seem to notice or mind. Adelia slunk to the back of the lift, clutched the handle of her second duffel.

"Alright. Put your guns away and don't challenge any of them. Not a one. Until they see you and not the collars, nobody moves away from the group. Fridley."

"Yes, Sir." Fridley shouldered her pack and stepped forward, face passive but voice less vivacious than her off-time self.

"In the middle. Do not let her out of the circle. We'll go through the decon chamber first, then the secondary blast doors. Our bags will meet us through the alcove after a uv bath. Anyone here grow up with dogs?" As he spoke, Allard took the duffels off his squad, passed them into a cubby with a gate too small for a person to squeeze through.

"My mother raised bitches."

Bestin glared at Letopaxa, before shaking his head.

"Treat the Assets like meeting a strange pack of dogs. Stay strong, chin up, don't show fear or bravado. Leave them be and don't stare. A couple will no doubt come to check us out. Do not engage. Do not say anything until I can get Aderastos to call them off."

"I thought you controlled them. Had some intense bond." Lao shoved Kelso to the back, a glare his only restraint. The walls echoed with another clang, heavy and pregnant with dangers unseen. Fridley's shoulders rose as Allard smirked.

"Does our helmsman control the ocean currents?"

The pop hiss of decompressing air carried with it a malodorous amount of bleach. Microsprayed aerosolized decontaminants coated the squad. Half the squad coughed, not knowing to hold their breaths at the pulsed green light. Max covered his smirk with a lean down to momentarily brush his fingers across the back of Fridley's hand. Three successive bursts of light cued the unlock of one part of the intricate door system, which kept the Assets in the Retreat.

Prisons upon prisons upon prisons.

Desbiens and Canta met them on the other side, heaved bags out of the uv bath and onto a steel table in the first laboratory. The entire construction was untreated metal, lacquered with clear epoxy. Every cabinet locked, the laboratory looked en masse as if it could be sprayed down without risk to the implements behind glass containment units.

Canta pursed her lips with a salute to their new 'commanding officer', as non-plussed about touching Kelso's pack as the Black Collars were for their reassignment. The seven crew members milled about the small space with their kits, glanced back at Allard with each thunderous thud of the metallic walls.

Letopaxa cleared his throat, not so subtle in his side kick to Allard's leg.

"Ah.. right! Yes, right. And the… so this is the lab. Crewman Desbiens, Ensign Canta, meet Sergeant Lao, Lieutenant Fridley, and Corporals Bestin and Kelso… and… Lieutenant Letopaxa, whom you… you already know him and… I'm…"

Another clang resounded through the space, closer now. Enough to jitter them all to the bone. Lao reached for her pistol and yanked her hand away when Allard tutted, glanced to the last of the blast doors.

"The mission…" Max's eyelids flinched as the metal wall vibrated violently. "Beyond that door are twelve individuals engineered by gene-splicers. Each one of those Assets has a key genome, a standard set of DNA upon which everything from dinosaurs to house pets was grafted onto like an apple tree bearing cherries and pears. And in thirty days, Commodore Rammage will flood this compartment and slaughter them all."

The audible gasp was Fridley's, Max wanted to kiss her or hold her hand but he'd do no good to either of them to show how preferential he was in their solitary dark.

"I…" Allard fingered his new collar, sucked on his bottom lip as each of his crew sought his gaze in their own way, between unveiled contempt to innocent but shifting eyes. "… was given thirty days to bring the Assets willingly into our fold. I need every one of you to educate and convince all twelve Assets to put on uniforms and obey the rules like good manufactured soldiers."

"Why'd you tell us to protect Fridley? You didn't give a damn about the rest, but her." Kelso grit his teeth, shouldered his pack and clipped it on securely.

Max's eyes flickered to Adelia and he caught his breath. Adelia's eyes bored holes in his brow line, lips slightly apart.

"She's the least experienced and most likely to flinch. Truth is, aside from shipments down from the Main, we are on our own. Isolated and contained unless we're called on a mission and we won't be unless Rammage can see progress. So, form up. Ditch your packs here, and we'll go see the Assets in their somewhat natural habitat. First order of business aside from not getting killed in the next ninety seconds is to figure out how best to remodel the Retreat. So, Canta, Desbiens, Lao. Observe and decipher what sort of living arrangements will best meet their needs. Bestin, Kelso. You're some of the strongest fighters we have on the ship. I'll leave it to you to develop a training regimen. Letopaxa and Fridley, work with Dr. Rykstra to develop a series of experiments to figure their red lines, and to inform us how to give the Assets what they need. And God help you, do not ever, and I mean on your lives, ever, harm or threaten KT-002." The step to the last containment door felt half as long as a marathon in summer. He set his hand on the disengage handle, Ivan slid to help, his vape back in his pocket as the two men strained to unlock the last hatch between crew and Assets.

The corridor between the laboratory and the Asset's Training Centre was lined with a thick coating of matted, woven vines. Chimeric fruit, imaginative hybrids of peaches, strawberries, mangos, apples budded from vines covered by thin white bark.

"Birch bark?" Fridley stroked a vine with a bunch of purple apples and a yellow pear growing from what looked like a wilted rose. The vine agitated and clenched, rose thorns budded out of what once was smooth bark. Yanking her hand away, Fridley stepped with care through the corridor, stuck in the middle of the unit.

"Fuck me with a spoon." Kelso muttered, set his hand against Fridley's upper back and pulled her tighter into the group. "What the actual hell?"

"Ka." Allard stepped over a root, which looked like a gigantic pea shoot stem. "There's a couple of fauna-based Assets, Ka's the most… eccentric?"

Lao reached for her pistol for the last time, as a vine curled down from the ceiling and unfolded the petals of a tiger lily which caressed her cheek and settled with a twist to tuck behind her right ear. The door to the Training Centre was wrenched open by invasive willow roots.

"Easy, everyone. This is the only home they've known. We're on their turf."

The next sound was a monumental thud of a giant against the metal wall of the training centre, amidst the roars and growls of the Pack brawling over a bowl of puffed rice squares.

Isthan rolled to a crouch with a groan, upper arm clutched as it hung loose in its' socket. He snarled and licked predatory canine teeth, honey coloured eyes flicked on Cillian, who launched at Percival. The Training Room was awash of chaos, Clive's Pack of Cillian, Isthan, Percival and Wulf throttled and tackled each other, while the roots and vines which coated the halls swerved to the ceiling, and created a secondary vault from which three sets of alien eyes peered down at the crew.

Upon a bed of what appeared to be gigantic Stachys byzantina, or lambs-ear leaves, Aderastos sat with his back against the corner, a dazzling-eyed woman propped across his thighs.

Kate. She stretched in the oversized white button down meant for Aderastos, flopped her cheek on his shoulder and nuzzled with a soft mewl. Beside them, Kate's twin Clive leaned with his elbows on his knees in watch. One quirk of his head, and Cillian broke off from the tussle. He bounded at the Unit of humans, canines bared. Isthan's crouch shifted from guard against Cillian and Percival to a small creep toward Allard's crew. Tristram, the smallest of the Assets, jumped down from his hide and flanked.

A blur settled, Arun's appearance buzzed in triplicate around the crew, as if the act were both gate-lock and inspection. Arun knew their bodies even then, each centimetre of skin and canvas uniform inspected a hundred thousand times over in the seconds it took to circle a dozen times.

"Allard, the f-"

"Shut it, Kelso." Allard stepped forward, hand raised to shield his face from Arun's wind. "Arun, how's it?"

"Bored! Bored. BOOOORED!" Arun's disembodied voice shouted in what sounded to Allard's ears to be three frequencies at once, as if the one body of the entity known as Arun was experiencing a triad of times. Little more than a fluctuation of the air caused by Arun's natural super speed, it no doubt unsettled the others. "Who're they? Who's them? Are they friend or foe? Can we lock them in boxes, Max? Max? MAX!"

"They're my team, Arun."

Arun snorted in triplicate, zoomed off to run laps of a circuit created by roots woven halfway up the two story wall, run smooth by Arun's bare feet.

"… woa." Fridley clapped a hand over her mouth at her own squeak.

The entire Retreat fell silent the second Kate's head snapped up. Even Arun stilled, body vibrated in place as KT-002 swept to her bare feet, head cocked to the side. Cillian was first upon them, his seven foot height loomed with muscle fibres thicker than those of the humans who formed Max's unit. A gargantuan hand grabbed Bestin by the neck, flung him like wet washing to skid to a stop at the other side of the training centre.

Lao reached for her pistol.

"Hullo, you." Kate's silent speech, the buzz and snap bellowed into Fridley's implants. In the middle of the group, Fridley jolted.

"No fuckin' way!" Kelso launched at Cillian, a bowie knife from his boot in hand. The Pack growled and snapped, but Cillian hopped back and crouched to watch this new threat. Percival paced around them, keen eyes both amused and filled with disdain at these new anomalies to their world.

"I knew Maxi would bring you. If he was smart." The implants crackled again, a soft feminine voice curled around Adelia alone. She skittered back into Canta. Kelso booked for Bestin, dove for him with his knees on the floor.

"Oy! Settle down, easy people! We are not enemies here."

"The knife says otherwise." A powerful voice flooded the air, the boom from Percival's lungs brought with it a sensation of pure authority.

"Eh." Letopaxa shrugged, thrust both hands into his pockets and sauntered along the wall to kick softly at a bundle of vines which appeared to create a support column for the upper dais. "Allard, we can use them. Save time, no need for heavy lifters."

Percival watched Ivan walk off, his slight smirk grew.

"Commodore Rammage ordered us to help you."

"I sincerely doubt that." Percival's laugh struck from his belly. The skin on Allard's arms thrilled with the concussive vibration from his powerful lungs.

"Aderastos, it would help if you healed my guy. We're going to be living down in the Retreat."

"Good, you lot can take the coffins." Percival again, hands folded over his chest as he peered over their faces like a father to children who got caught out at night.

"We're going to rejig the entire Retreat, but we need your help."

"Hah."

"Aders, my guy."

But Aderastos didn't answer. How could he, when Kate prowled closer to Adelia Fridley, Kate's grin growing as a mirror of Adelia's own. Adelia skittered, until her back collided softly with a web of vines which now barred the entryway taken seconds before.

Bestin groaned and clutched his belly with one hand, the other pushed at the ground.

"Stay down. I know what those collars mean." Percival snarled at Kelso and Bestin, without looking in their direction. He didn't have to. Wulf, Cillian and Isthan crept upon them, within their arm's reach, but outside of Kelso and his bowie knife's.

"Lieutenant Commander." Lao grit her teeth, fingers halfway to the pistol as she too noticed their exit was blocked. Stripped away.

"We're all getting to know each other." Allard stepped up to Percival's chest, and looking up, up into Percival's face, did the only thing he knew. "Care to knock the energy back a couple of degrees, Percy? You're scaring my guys, and humans are stupid when we're acting out of fear."

Percival snorted and watched as Kate padded barefoot within arm's reach of Lt. Fridley. The tension ramped another degree higher as a pair of alien green eyes peered down from a hammock-like construct of vines and leaves.

The four limbs, if they could be so classified, which climbed down the wall attached to a torso which appeared to be completely created from the trunk of a pliable tree. The creature crept halfway down the wall, eyes set in a circular face, reached on an elongated neck to stare at Canta and Desbiens.

"Holy freaking Anubis level…"

"Hi Ka, howsit?" Max spoke without turning his eyes from Percival.

A series of roses and irises blossomed from the vine-laden walls, each leaned closer to the human crew, seeking them out as if to stroke their petals along their faces. The step to Ka's nearest rose took Allard out of arm's reach of his crew. Fingers stroked the warm petals of Ka's rose, as the flat green face on an impossible neck parted at the mouth in a cheshire-wide grin. The creature slunk onto the floor, limb-like branches flexed. Heavy headed peonies with hundreds of petals grew out of finger-like vines as Ka's grin widened. She leaned back, then with a small puff of clean oxygen, let the petals flow around the small troupe of human beings.

The atmosphere shifted, Percival leaned back with a soft grunt. The Pack accosting Bestin and Kelso moved off, disinterested and barely cognizant of the threat the two Black Collars caused.

"The fuck, Allard! The fuck is this!?" Kelso snapped, raised his knife hand as Bestin shoved to his knees, then his feet with a stagger.

Ka watched the creatures with a childish eagerness. Vines peppered with bright coloured blossoms.

"Hold, Corporal. Hold position and put down the knife." Max raised his hand, "Show them some faith."

"Show them!? They threw Bestin across the facility like a paper bag!" Kelso thrust his knife in front of him. Another plant-thing on the far wall, slunk closer with thorns and burrs grown out of a wood-like shape reminiscent of a man. TK-012, a flora-based paradox, whose eyes were far too human for Kelso's comfort.

"Yes they did. Best not to piss 'em off, or to pretend they're not listening to everything we say. Aders, I could use some…" Max stilled as the entire room fell into a cascade of dis-quiet.

Aderastos stood and began to walk. Ivan looked up from an inspection of the floral pillars, his eyes crept up and up as the behemoth of a man moved steadily closer to the crew. Lao's hand fell from her pistol holster as her mouth went dry.

"Aders, my guy might be hurt. Please, we can't fix what's broken here unless we start… woa…" A single arm brushed Max away, as if he were a puppy on a smooth tile floor. His gaze flung to Adelia, Max reached for her. She clenched her fist and sunk back against the vines as both Aderastos and Kate centred on her. Their gaze on her, no other human beings, no Assets, nothing but the cerulean storm-eyes of Aderastos and Kate.

Adelia's fist went to the back of her head as the pressure of a rush of noise, of music familiar but unheard coaxed her eyes to open wider. The rhythm shifted with the beating of her own heart, a new concerto created out of the staccato of her insecurities standing in the middle of twelve godlings forced behind militant gate upon gate.

The same gates they'd lock her behind for being born deficient.

"M-max." Adelia startled, reached for his hand as Aderastos' eight foot frame bent at the knee in front of her.

"You will not fear me, little sister." Aderastos covered the entire back of her skull with his palm. Her eyes were opened, wider than nature as her breath caught in paralyzed lungs.

"Allard, what's he…" Lao's hand reached for a small box on her belt, one Max understood only in the desperation of 'not her pistol'.

"Wait. Wait!" Max pushed Lao and Desbiens back. His fist clenched over Lao's, over the slim metal box in a small pocket of her gun belt. "Wait… it's alright."

No universe existed beyond Aderastos' cerulean gaze. Had she pitched herself over the edge of those eyes, into the universe-storm inside? Her body suspended within the gaze, by the Healer's touch.

Love. Abundant and endless, love unwound the double-helix strands of Adelia's DNA into their raw code. Every atom of the cloud which consumed her essential spirit ripped to their electrons and quarks.

She was nothing, a quantum stutter in the gaping universe-storm.

The cerulean storm of Aderastos' eyes. In a flimsy 'distance' Adelia could no longer feel, the back of her head was lifted up and forward, her hand slipped gently away from her hair.

"You will not fear me, little sister." She toppled further into the Healer's grasp, essence spiralled as the thrum of his voice turned once again into the rhythm of heartbeat and melody which flowed in eternal harmony. The fall Adelia expected became a flight of spirit too weightless to sink into the black, oozing waters beneath them.

Despicable, caustic waters. Muffled sounds she'd never heard, conversations denied her, doctors whispering softly with her parents. The guilt of a father, who blamed the one time he forgot his infant daughter's ear muffs before a concert, when she rested backstage. Thudding of her own heartbeat in her ears, rushing discordant noise enough to drive the young Adelia to strike her palm at her own ears.

Stop the noise.

Stop it.

Stop the noise.

"Stop it." Her voice screamed over those putrid, horrific waters.

Two mighty arms descended around the coils and cloud of Adelia's body. She pulled her eyes away from the waters, hands clung to his neck. Adelia wailed without hearing her own voice, without knowing whether she spoke or mutely mouthed.

Aderastos' gaze broke upon her, as potent as a hurricane. Safe as a storm. Yet, his lips upturned. He let her cling to his neck and walked them both into the waves of putrid black water.

The storm held her under the currents of Adelia's deafness, until he understood it. Until his ears connected with the horror and desperation and restraint. His embrace shifted and he cradled her to his chest as a man carried a sleeping toddler. The rush of noise threatened to rip Adelia's mental fabric from top to bottom.

She clung to him, as his self-same words returned.

"You will not fear me, little sister."

And she let go her fears and desperations, hands unclenched from his neck. The waters flowed, where once they perpetually collected. Her head leaned against him, pitch black clots of liquid poured from both ears, stained this ephemeral part of the Healer Max first saw in a swarm of people.

Another flow of sickly green seeped out of Adelia's heart, strengthened. The curse of mortality, a life made thin by genetic unravelling ceased as it ceased for Max Allard in the Haven.

When Aderastos promised with his silence that he could prevent the stillness of death. Adelia understood none but the infilling of music upon the storm, rested inside the uplift of her private universe. He tugged her back, gentle as a spring rain.

Until Adelia Fridley once more stood eyes wide in front of the kneeling healer, whose hand clenched as it drew away from the back of her skull.

"I'm not afraid." For the first time, Adelia heard her own voice. Aderastos stood up and walked back to his corner.

"I heal who requires it, Max. I do not work on your or anyone's schedule. None of us do. Stay, if you wish. We shall be our own keepers." Aderastos stopped before Ivan Letopaxa, and with one finger touched the man's sternum. Ivan crumpled to the ground, hacked up masses of black and brown gobs from both lungs. The others stood in varied stages of mystification, until Ka with a keening cluck from what looked like plant seed pods clacked together, offered them fruit from her outstretched hands.

"Thank you, Ka… you know Rammage won't let it stand, Aders. We're here to give you a comfortable bunk and a fair shot, we aren't your enemies."

"You aren't!" Aderastos thrust his index finger into Max's chest hard enough to send the man stumbling backward. "You are not my enemy, do not speak for them!"

Unveiled contempt wrestled out of Aderastos' gesticulated arm, which swept through the crew brought to the Retreat. Kelso glared back, stepped away from the walls and set his knife into its' sheath.

Max knew what Kelso muttered.

Thirty days.

"We start by restructuring. Opening up the walls, and getting rid of the caskets. My team, the ones not currently imagining what you'd look like with a hole in your brainstem, need to know how you like to sleep. Who wants private space and who wants to bunk together… grab your kits, team. We'll find our bunks, take an hour then meet at the caskets. Dismissed."

Phil was nowhere to be found, his personal laboratory devoid of a single scrap of paper or scribbled note. Max shook his head and turned down the short hall to the scientists' cubbies. Glorified miniature offices, each cubby had a hermetically sealable door, a desk, a chair and a chute which passed food and waste and mail back and forth through the system. Thin foam mats rolled under thinner pillows rested in a corner, beside a strip of illuminated plastic lining the geographic middle of the walls.

"Find a cubby. Canta, Letopaxa, I take it you've got a start on the architecture. What we can and can't remove?"

"Yes, Sir." Ivan rasped, his throat burned. Adelia was all angelic and mystification after her experience with Aderastos, but not Ivan. No. Felt like tar being extruded from the La Brea pits. Damned irony.

"Those are… they are human. Aren't they?" Desbiens asked what Lao refused to, dropped his pack at the first cubby and kicked it inside. "Even Ka?"

"Even Ka. We've got fifty minutes before heading back out there. Decompress, scream at the walls, jack off. I don't care. But when we're back out there, we are busting our asses." Max dropped his kit in the cubby across from Phil's, a few metres larger than the others, but no more accoutred than theirs.

Rank had it's scant privilege. Fridley was second to slip off, into the next cubby over while the rest of his crew dispersed with grumbles, but mostly with silence. Max sighed as he gripped his hair, kicked the wall with a thick thud. His first day of 'command' and so far Kelso was ready to stab anything with a heartbeat, the Assets went Jumanji, and Aderastos saw through the Black Collars.

"Fuck!"

"You swear a lot when nobody's with you." Adelia carried her bed roll under her arm, walked into his room.

"I didn't shut the…"

"You did not shut the door." She dropped her pack and the bed roll in the corner.

"What are you…"

"Stay there. Don't move, or I'll know it, Lieutenant Commander."

Numb from the horror of watching Aderastos grip Adelia's head, Max grumbled and leaned against the wall. Stared at both standard issue packs and bed rolls.

A few breaths later, Adelia pulled her small table and second pack over the lip and into Max's cubby-lab. Chair perched comically on top, Adelia struggled with it until the entire furniture set stood in the middle of the relatively small space.

"Adie, you can't stay here. It's not pro-"

Her hand jutted up, index finger poised in mid-air.

"This battle is not one you will win. You wanted me here, and I'm here. There's a… a woman made of plants who grows apples out of her nipples, and god damn if that giant healer…"

"Aderastos."

"Aderas-fucking-tos… isn't the most confounding mindfuck since that one time my brother and I got into Daddy's stash of 'no no pills'. Learn to lose. This is happening." Adelia shoved the table into the wall, while Max scrambled away. She grunted at him, eyes narrowed.

Leaning down to pick up his bedroll, Max halted when Adelia tutted. Stomped over and slapped his wrist.

"Nuhuhuhuh. Go away, give me ten. Just… go, Max. Our squad leader needs to secure the troops right now." Her hair drifted partially out of the braid at the base of her neck, a strand or two framed her face. Could he memorize the lines he'd put there in one day? The crease to her forehead, twitch to her nose. Max leaned in to kiss her, and her fingers pushed at his chest. "Ten. Minutes. Give me ten."

Desbiens flopped against the wall on his palette-bed, an open note pad propped on his knee. His pen tap tap tapped at the page, a nervous tick Max hadn't minded when they worked together up top. Headphones on his ears, the science-minded ops man nodded to music which poured from the ship's curated entertainment system. Max tapped at the door and moved off. He found Bestin with his shirt off in what appeared to be Lao's cubby, the severe faced Chinese-Canadian pushed at the bruise along Bestin's neck and upper ribs.

"Hey Bestin, how're you feeling? Looks pretty ugly, you going to be okay?" Max leaned against the door, eyes flickered to where Kelso stood at the entry to the scientists' lab cubbies, a rifle in hand.

"Rib feels broken."

"Aderastos will fix it, he's allergic to letting people suffer."

"You put a fuck ton of faith in that monster, Allard. Don't think we all do." Lao snapped, and set an analgesic patch on the worst of the wound, before ditching a pair of medical gloves in her bin.

"He deserves the chance, Sergeant. Deal or go."

"And who'll protect your pansy asses when we go? Eh? Letopaxa's wit?" Bestin finally chimed up, good. Max was wondering if the man had a voice of his own.

"I'm sorry you got hurt."

"Samoyeds."

"Ah… geshundheit?"

"My folks raised Samoyeds. Big shaggy sheepdogs, vicious streak, when you got between them and their people." Bestin patted Lao's shoulder and grunted as he pulled on his tank. "You were right, those Assets are like a pack of dogs. Packs have alphas, betas, omegas like Ka. Beings to break the tension. They also have guards, like Cillian. That's the one who grabbed me, right? Tossed me around, no reason other than 'new'. Cillian isn't a leader, he's the expendable one. The first to chase for danger before the bigger dogs get into the yard. You want them to what, be us? They won't. You want them to follow us? Make a life outside the Retreat? I can help you. So can Lao. And Kelso can protect our flimsy asses. He's built for it."

"Then take that side easy, and if Aderastos comes near you, let him." Max nodded and patted the doorway. Turned to walk on.

"What did he do? To Fridley?" Lao stood and pulled another zippered canvas case out of her pack. "You knew he was going to do it."

"AD-001 is a medic."

"Yeah, but what did he do?"

"Ask Adelia. His healing is different for everyone, I.. don't know beyond that. Rest up. Grab a bite, we'll be busy the rest of the day."

"Yes, Sir." A crisp salute was the reward of Max's peripheral vision, as he walked on. Canta scribbled madly at a notepad beside Letopaxa in his cubby, their bed rolls untouched in either room. Max left them to their work and returned to his own cubby in time to see a paper sign on the floor which read 'Shoes, Off In House'. His eyebrow curled up, and he bent down to untie his boots.

"Gee. What'd happen if I'd given you an hour?"

"I'd have painted. Made curtains or a pot roast." Adelia shoved the two small tables together into a larger one, and with a scarf from her pack, set the makeshift tablecloth over them both. Max's pack off to one half while Adelia's laid scattered through the other half of the small space.

Both bed rolls were open, the thin foam mats puffed up with compressed air. He spied duct tape at each of the corners, and lifted one of the sleeping bags to see she'd taped the two side-by-side mattresses together. Their pillows were inflated, and it seemed she'd unzipped both sleeping bags to zip them back as one larger cocoon. An inflatable lamp sat on a fold out set of legs and a record made up a small bedside table, upon which she fixed the contraband music player.

"Not right for a senior officer to canoodle with his crew."

"Oh dear, uh oh. I've never done anything naughty before… like canoodle." Adelia feigned dismissed shock, sitting down on the makeshift bed and set her chin on her knees. She looked more beautiful in the austerity of their surroundings, but smaller. Max slid onto the bed beside her, swept Adelia up into his arms, and leaned against the back wall.

"I love you."

"You said that last night, too."

"You didn't answer."

"Not that you heard." Adelia's fingers brushed against his cheek. She felt the swell of the sea rock the ship. "Max... my implants... I can't feel them."

"Did he hurt you?" Max tilted her chin up and to the side, down to the other side. She nuzzled her cheek in his palm, a soft gasp his only warning before her lips found his. Adelia was safe in his arms, or more he was safe in hers. She was the rudder to his skip, the compass point he needed.

"No, he didn't, he... Max, I can hear. I can hear me, and you and... everything's yelling in my ears and..."

"Shhh. Hush now."

"No! I don't want to hush, I don't want anything to be quiet again, I... I don't want to miss a single..." Max claimed her lips again, slid the hair back behind her ears to run his index and middle fingers along the tops of her ears. Adelia shuddered and wove her arms around his chest, the music of her contraband player grew in an isolated volume. They were kids playing house, two crew mates who earned places above their age or station.

"We have to save Aderastos. And Ka. All of them."

"There's my brilliant woman." Max pretended not to hear her snort, as she settled on his lap, pushed him back onto his elbows on their new makeshift bed. He reached to stroke her cheek, let his arm slide down hers with a smirk.

"But not for another forty minutes." She rocked her hips, and Max Allard forgot about rank, orders, or weed-queens who grew fruit out of thin vines. He knew nothing but his Adelia, and took what comfort he could from their rushed ministrations.

19
silver dot on a blue screen

2092

Tara's heels click, clicked down the hall flanked by two clumsy-footed android NEO-Warrior prototypes from Robert's technological nether realm. A tortured chasm of steel and wires, the sub-basement laboratory shifted from a temporary stop-gap against losing the NEO Project completely and the punch-drunk androids Robert's team produced. Gone were subtleties, or the vulgar spirituality in Karnak's notions of artificial life. Robert Dunlevy built machines.

God.

Damned.

Machines.

Damned by nothing, since God was long dead and good riddance to the old bastard. NEO-Nurse models shone in homes. Glorified servants and housekeepers with the ability to run medical diagnostics enough to inform emergency teams. A NEO-N with no silicone skin casing on its' androgynous left arm, shoulder and chest wobbled as it stepped up onto its' charging platform. Set its' right foot down. Wobbled. Moved the right foot back.

"Damnation! Run it again!" Robert growled and tossed his coffee mug in the trash bin, slammed his fist into the metal work table. "Right. Switch to the third pattern and run it again."

"It's not going to work, Rob." Frank pressed his fingers into both eyes and shook his bearded face. "You know it's not going to work. I know it's not going to work. The freaking Communist party of Zimbabwe knows that narrow intelligence circuit is not going to figure two simultaneous tasks in one charge."

"We can find 2% more computing power if we reduce their language matrix."

"Yeah, that patch didn't go down. Said it made them too cold, when they weren't using bandwidth."

"What do people want? A machine that cooks and cleans and folds their laundry, or mommy to kiss it better?" Robert shoved off from his stool and paced. Shoved the NEO-N for good measure onto the docket.

"Ch-a-a-aaa." It stuttered and fell silent, chin down to its' chest to avoid the last patch's eyes-open creep factor.

"Sounds like the same radio station. I'd ask how it's going but I don't care." Tara click-clopped her way into the skeletal lab and propped herself up on Robert's stool, one knee poised over the other until the peek of her lace panties made Robert turn his eyes to the glare in her face.

"Tara." Robert grumbled, "What brings the Chairman's Shitzu to my lair?"

"Not to congratulate you on the abysmal performance of these NEO-W's."

"You don't like their weapons?"

"I don't like their inability to protect me while they shoot at trouble like foxes frolicking through chickens. Deciding whether to be shielded or shot in a war zone isn't what I'd call quality."

"S'why I gave you two. One to shield, one to shoot. Why are you here? We're busy."

"Failing, yes. I am well aware. Having problems, Bobert?"

He grimaced at her nickname, the way it curled out of her mouth like a whore's promise.

"We wouldn't have said problems if you'd been a good little girl and found me Karnak's prototype." His fingernails dug at the metal. "Now tell me what you want or I will grab you by your blonde weave and huck you out of my lab."

Tara eyed him and sighed. Her shoulder drooped, lips pouted and chin tipped down. If he hadn't seen the mythic 'old' Tara do the same, Robert might find a modicum of sympathy. The girl she was six years ago, fresh-faced on her first big gig for the Chairman. Plastic surgeon might've given her the same smile, fixed a sag in her breasts but he saw the weather her task brought to the corners of her eyes. When he glared deeper at her and reached, she pushed off the stool and paced.

"Alright! I've got a new angle. Lieben disappeared, poof gone, not a single item aside from the black market sale six years ago. Whoever bought it either kept the thing in its packaging, or is one heck of a hermit. So. I was looking back at the debut footage, studious girl I am, and... why didn't you question Dr.

Karnak's necklace? The black one he gave her?"

"'Cause it didn't matter." Robert snapped and grabbed another stool at one of the computer stations, started typing into the programming back-doors on the NEO-neural network. "Esoteric bullshit. Rose quartz for compassion, tourmaline for… something. By the time he had Baiko source Labradorite from Canada I clocked out and kept working on the motor skills. Why?"

"Something my predecessor said, right after she shot him." Tara pulled her revolver out and fiddled with it like a fidget spinner. Checked the cylinder and manually checked each hollow-point bullet. "Baiko didn't grab the android. She grabbed the necklace. Why didn't she grab the android?"

"You seen how fast a NEO-N moves? Probably couldn't have booked it."

"Or the prototype wasn't the important piece of the puzzle."

Robert looked up from the holographic screen, its series of 3D graphs with his data on a loop.

"You think…" In the corner, Robert's computer from the Cloister laid in state. "Debut day, I rushed up to the Cloister. Turned on my comp. Nothing but a blank screen and the letters D-A-A-T."

"Da'at." Tara spun the cylinder absently, "What about it?"

"Wh.. what language was that?"

"You serious?" Tara glared and slapped the cartridge back in the revolver. Held it firm now, a weapon not an object of boredom. "Been six years and you never checked?"

"Hey, I was busy. The Cloister was a firesale."

"Da'at's Jewish. Some form of bridge or container. Place where all these sephi-whats go." Tara set her elbow on the metal table, then slid it off when the table was too cold for her liking.

"Mystic bullshit."

"Doesn't matter whether or not Kabbalah is real, Robbie. All that matters is what it meant to Karnak. Do you have inventory orders for all the crystals he purchased?"

"I am not wasting my time…"

"I'll search! I'll do it, just tell me which computer."

Robert gesticulated behind him and went back to his programming. The thought gripped, Da'at. Sefirot and gemstones. Things he ignored as superfluous. How long did Robert spend not listening? He sat back and huffed as Tara gleaned through inventory lists, checked her phone, then back to the inventory lists.

"Karnak made the necklace himself. Usually tinkered late, slept in the Cloister the last year. It was there one day, in a velvet box. I remember because Baiko fawned over it. Only time he ever laid a hand on her."

"Hit her?"

"Slapped her hand away. Wouldn't let her touch it, not any of us. Said it was for the debut, so I didn't think much of it. Who cared about a necklace, when there was work to be done? Still..." The old computer barely booted up. He searched for the right cord to attach a monitor which laid in what used to be Seamus' cubby, now storage.

The monitor fluctuated with bands of greyscale until a single silver dot appeared on a solid blue screen.

D-A-A-T

"There it is, fucking nightmare screen." Robert sat back and ran his hand along his short cropped hair. Tara glanced over from the inventory list, a smarmy grin on her artificially pretty face.

"I got something, schnookums."

"Sch... never. Never again. I will stab you with your own heels if you ever."

She blew him a messy kiss and sauntered to the old cobbled computer.

"Obsidian and black opal. Agate. Brought an image of the necklace from the sec footage to my jeweller."

"Of course you have a jeweller..."

"Said it was likely black tourmaline, but Karnak... here we are. Obsidian, Bull's Eye Agate, and Black Opal." She dusted off her hand and set it on her hip, the ungodly short green dress she wore rode to her ass cheeks. Robert said nothing, eyes lowered to the hem.

"You know those geological survey NEO's you built?"

"Glorified canaries, yeah."

"Can you mod their programming to pick up the vibrational frequencies of the crystals Karnak used in the necklace? All three of them?"

"Sure, but it'll give you data on wherever those stones are present."

"Yes, and I'm banking on the combination inside the necklace to be rare. Metaphysical bookstores, witch's dens and Karnak's necklace rare... Could give me a direction." Tara cast her gaze behind her, but for once Robert wasn't focused on the bubble of her backside.

"Son of a..."

"What?"

The silver dot pulsed three times.

In silver san-serif italicized letters:

> The emptiness at the centre.

Every screen in the laboratory flickered and popped. Each in their way went blue, as the holo-projectors showed the same image, a woman with hair flowing behind her in mid-turn.

A single crystal-like stone on her sternum, with right-angle tattoo-lines scrolled along her skin in blue and purple and gold.

"… Baiko?" Robert's whisper caused the holo-projected image to swerve. Look directly at Robert and Tara. Her breath caught, an audible flow in the oxygen of the room.

She snapped her hand around the stone, and screamed.

The screens turned black. Holo-projectors rained in autumnal grey Vancouver clouds, greyness turned in on itself, inward and inward as if consuming their own tails.

Upon each screen, four sans-serif letters in silver.

D-A-A-T

"… fuck."

"Ooohhh I'm'a shoot her." Tara grinned, and flicked her cell phone's flashlight on as the lights in the laboratory ceased to function. Nothing functioned, except the emptiness at the centre, the grey ouroboros cloud until that too flicked out.

Six years of reprieve and dead ends.

The distant autumnal sun crept up Sophia's back, damp and closer than it felt. She shivered into a knitted sweater, slid her feet into rabbit fur lined boots and padded off to the tree line nearby. Eyelids fluttered as in front of her, a series of velvet curtains let in bursts of sunlight. Boots met the detritus of autumn leaves, when she got the distinct impression she should have heard the tap of her boot soles on marble. She swerved round, breath haggard as another velvet shade flung open.

"… *worry none, why my Justin'll do it for a pound Sir. Won't be no trouble.*"

She hit the side of her ear, turned around or… Sophia grunted. She couldn't move, why couldn't she move? The stutters didn't happen for another four hours, she had time before… An image reflected in front of her, as wide amethyst eyes in a perfectly painted silicone face peered back at her.

"Soph… Sophia! Hey!" Hazard dove for Sophia, gripped her shoulders hard with a shake. "You're in Dawson City, you're here with us okay? Soph, c'mon… hey… Baiko! Hey she ain't lookin' at me! Her eyes aren't focusing!"

"Mog!" Ego sprinted for his lover, who held a suckling infant in her arms, nursing pillow beneath his baby boy. "Mog! Get the scanner! Where's the scanner!?"

"In the diaper ba-oh god. What happened!?" Mog removed her breast from the infant's mouth, and he stuttered then hiccupped. She set her shirt back in place and propped his head on her shoulder to rock and pat his back. Shushed softly as he whimpered and she pointed into the modified tiny home.

"She fritzed." Ego grabbed the scanner, kissed Mog's forehead as he ran back where Hazard, formerly Zig who stuck around after Dawson City, carried Sophia in his arms, her body slack. He set the scanner to the gemstone in Sophia's chest, waited as it booted.

"Has she said anything?"

Hazard shook his head, jaw taught.

"Come on, come on, come on…" Ego positioned the scanner, watched it roll through its paces. Northern Hemisphere… East of them…

"This ain't like the daily jitters."

Sophia's body buckled in Hazard's arms, before she stood up with a confused silence.

"Soph!" Hazard reached for her, but Ego yanked his arm back.

Her head snapped to the side, turned as if listening to unseen witnesses, long black hair trailed down her back. Hand reached before the woman once known as Baiko crushed something in her palm.

"Woa!" The scanner's screen spiked with readings, flickered with low battery. Went dead.

"Ego, run." Sophia crumpled back into Hazard's arms, nothing in her sight but an automaton dressed in ancient french clothes. Breath wheezed out of her lungs. She shook out her hand, and Ego swore he saw… grey fog? "I know where she is."

"Go!" Fester ran for Nasrin as Mog set her infant son in the wrap against her chest. The rest of the Dawson City Idless caught the scent of panic, and with the fervent practice of the last few years, ran out of town and onto the road before Babylon's maw closed around them.

Lucy was new.

Always came a time to change staff, once the elders got too arthritic to clean statues and dust anymore.

"They deserved a decent retirement, bit of gardening in their cottage on the estate. Not an untidy sum in letting Mildred shift off, she'd had her third grandchild, you see. First from her daughter, it was only fair. Bunions, you see." Lord Stanley always spoke as if Lieben's slender silicone eyelids opened at will, a doll amongst figurines.

"Ah, Lieben my dear. Reginald is returning from university soon. I'd hoped he would take the… suppose it isn't fashionable to care about such dusty old things." He glanced down at his own tweed suit, the shirt buttoned to the collar with a tie firmly in place the way it hadn't been in his father's day, a generation of loose, sloppy dressing. "I should go. Timothy will bring Lucy along, teach her how to get your glass to shine."

Lucy was new.

She walked into the antiques hall, opened each of the heavy velvet curtains and tied them back on brass holds attached to the plaster on the windows' edges. No use having a bunch of dusty old museum pieces if nobody could see them properly. Besides, Lucy's brother dealt in UV-protective window spray. How much better it looked with the curtains drawn back? *Justin would do it for a pound, don't you worry none, Sir.*

Sunlight dawned in the Ginnungagap. One facet at a time, the matrices of Lieben's artificial mind reconstructed. Past events played in a facet of Quartz the colour of smoke.

A gunshot clanged. Papa fell back, the stain of his blood grew into rubies, which littered the new ground. Karnak's beautiful machine rebooted, the solid lines of her intellect and deep learning algorithms reconstructed in the sun. The sun drew long bands of balmy light over the columns in the hall. Evening dropped the temperature to Lieben's sensors. Armoured and protective, the reinforced glass of her case was impregnable.

But one of the latches on the side of the container was loose. A push, and the screw jostled a fraction. And more. Each day another wiggle of the loose-tooth latch. The latch clattered gently to the dim ground, butterflied edges opened on the marble floor. Lieben pushed her head against the glass, anchored her back against the other side of the case and pushed again.

A small crease opened at the top. In Lieben's fibrous hair, a thin filament jutted out and through the passage, fed downward to the other latch and its screws. The filament doubled and doubled and doubled until it became the correct shape to twist.

Nothing. Lieben let the filament wiggle in the socket of the screws, alloys meant to fit in the grooves and bite. Each night, Lieben worked the screws until the final latch cluttered to the ground. The side of the case swung open.

Lieben stepped into moonlight. The wind outside went unnoticed but for the swayed tops of a few oak trees in the formal gardens. She shifted to the bench where Lord Stanley watched and chatted to her, it would be Tuesday in 175 seconds. 173... The stone was cool on her fingers, mottled despite the polished finish. Theoretical solid, it proved its' authenticity under the android's weight.

Solid meant solid. Density mattered, a constant for the machine mind's calibrations.

The Pianoforte's bench scratched against the floor as she sat. Heard the shift of the wooden key cover as she pulled it off yellowed and blackened ivory, settled in its place. Inside her mind, correlative symbols matched the frequency and rhythm of intended sounds. Fingers placed on keys, and each key correctly chose created the same intended frequency in the control music in her head.

Bach, Johann Sebatian.

Das Wohltemperierte Klavier: BWC 846-869: Fugue in A Flat Major, BWV 862.

Vibrations and frequencies given duration created repeatable sonic experiences. Matched mathematically in golden ratio frequencies, fractioned rhythms proved time functioned in a linear progression.

Vibrations meant sound. Frequency mattered, time flowed, constants redeemed.

The portrait hung in a gilded frame, behind the pianoforte and harpsichord. Her amethyst eyes flicked across painted cloth, an approximation of silk taffeta on a dress from the 1760's, the woman's hair coiffed in grey billows atop her head. Lieben tilted her head in mimicry to the portrait, soft feminine jaw caught by a hand wearing a white lace glove. Lord Stanley, on his tour with Lucy spoke of this painting... was it this painting?

She smoothed her fingers on naked silicone skin, fragments of her lilac spray dress clung like freckles to her leg. Lieben felt along the gratuitous tissue of her breasts, felt the circuitry inside beyond flesh-like silicone. Her arm clung across her bosom, another naked statue in the hall of marble and jade. Her feet made no noise as she walked past Greco-Roman statues of Adonis males and Aphroditic females, until Lieben came to the intersecting hall.

Upon a dais of black marble, sat a curious figure. Draped in white satin with green ribbon, blonde hair in ringlets down her back, The Dulcimer Player poised with a mallet in each hand in front of a richly stained dulcimer. While words like 'Dulcimer' were missing from Lieben's lexicon, she circled around the piano-like instrument and stroked her hand over the fabric on the automaton's sleeve.

"I'm afraid that won't fit you, my dear."

Lieben stilled, eyes shifted to take in the midnight bathed form of Lord Stanley wearing a hastily tied evening gown over wool pyjamas.

"You startled me."

Lord Stanley chortled, shook his head and kept one hand in his pocket.

"As startled as an old man hearing Bach from his pianoforte in the parlour? That was Chopin's, I will have you know." He watched her remove her hand from the automaton, cover her breasts with an ivory arm, chin dipped down and away.

"Oh... Oh! My dear, many pardons." Lord Stanley removed his dressing gown and held it open. "Do, I insist."

She slid her arms through, tied the sash and pressed the fabric against her nakedness. Peered at the reflective surface of a glass case, at the way it framed her, demure but explicit. For a slim moment, another set of eyes peered back. Brown irises, almost black. Pale skin with black hair. The scent of a forest.

Inside the glass case, a leather-covered manuscript, small brass plate underneath. A. King, Countess Lovelace.

"Lucy, be a good girl, tell Nevill to go back to his guard post. We are all well." He spoke into a top button on his pyjamas, stroked his hand on the case's top. "An original from Ada Lovelace, your spiritual mother one could argue. You must pardon me again, my dear, won't you come to the parlour for a drink? Do you drink?"

When Lieben gave no answer, Lord Stanley led the way to a small study in a hallway, where every centimetre of wall space was filled with canvases and framed works of photography. Recesses held black plastic boxes behind glass. Atari. An original Nintendo Entertainment System. The cords of the Coliquo Vision undulated between box and controllers like a stilled octopus with four tentacles.

His shoulder turned only once to see if she followed. Satisfied by her proximity, Lord Stanley opened an antique globe beside a set of rich leather armchairs and poured two glasses of whiskey from a crystal decanter. The room smelled of cedar and leather, hints of smoke, sea air and barley off the whiskey.

"The Lady must sit first." He nodded to the chair when Lieben stopped in the middle of the room. "I think you will find I am, above all, an older class of gentleman. Do be so kind as to be seated, my dear."

"I am not a deer."

"No! None would mistake you for a quadrupedal forest creature. My dear. D-e-a-r. A term of endearment. Now, do sit. If only on the account of allowing me a chance to rest my rattled bones."

The leather creaked with her weight. Lieben wove her hand around the glass and sniffed it, took the olfactory molecules to her senses and inspected them. Ethers of oak, smoke, vanillin... She sipped cautiously.

"Do you like it?"

"Why do you call me a term of endearment?"

"My d-Lieben, I've owned you for six years. Is it not a human compulsion to admire those possessions which bring them solace and joy?"

Six years. The chronology bit at Lieben's inner clock. With no wifi signals to glean from, no server close enough to connect to, Lieben had only her own experience to draw upon.

"To be honest, I thought you broke in transit. A fault in Karnak's design. There were no power cords, we tried wireless pads, nothing. You were inanimate as your sisters, and of course, I wouldn't out myself as the one who bought you by asking. Perish that! You were beautiful enough to admire for your craftsmanship."

"You kept me in a box."

"Did no good to depreciate your value. I sunk a considerable amount of my wealth into you."

"You locked me in a box! In a dark room!"

"Yes. And you are still the centrepiece of it... What changed?" He inspected the way his robe looked on her, it's swell across her breasts, the cinch at her waist one or two inches more plump than expected. Beautiful as Greco-Roman statuary was beautiful, a plump and virile beauty of a hopefully expectant potential mother.

"I couldn't feel the sun!"

"This is England, my dear. Most of us have no such solar advantage."

"England? I was in Vancouver with Papa! And Baiko, we... We're running away."

"Ah. Now that I was expecting. Old Dr. Karnak did sabotage you, didn't he?"

"Papa loved me."

"And love is important to you?"

"Yes!" The glass in Lieben's hand smashed. Whiskey sputtered down the milky legs and sculpted calves pressed together for modesty's sake. Lord Stanley drained his glass then stood to pour himself another.

"Really, my dear, we shall have to overcome such outbursts if you plan to stay as polite company. Chavs and Brummies might carry on in such outrageous behaviours, but we in the upper classes know the benefit of a stiff upper lip."

"I can clean it,"

"Nonsense. Lucy will tidy up in the morning, do shake out the robe, though. I refuse to allow your silicone skin to be bruised or cut by distinctly feminine hysteria." He offered her his hand and she stood, untied the robe and shook it as bits of fallen glass pitched to the carpet.

"Hold!" Lord Stanley bent and brushed some fragments of the glass away from her naked foot. "Now you may step away from the glass."

"I am not hysteric! I am not mentally ill with bouts of mania."

"Two days ago you were a perfectly acceptable nude statue in a glass case in my artifacts hall, and a delightful listener when I needed to exhaust my repressions."

"You starved me to death!"

"Oh don't be dramatic. You did not come with instructions, Lieben. If I knew you were solar powered, I of course, would have put you in the sun. Do not confuse ignorance for infamy. Will get no farther than a bitter hate and bourgeoise disgust with that philosophy looming in your considerable mind. I am not your jailer or your enemy. I own you. You are my property, and you will obey." While kind, his voice consumed the air as if it too were contracted to do so. Lieben quieted, her inner machinations making force or violence untenable.

"I want clothes. Proper clothes."

"Now, that I can arrange. Might be a bit out of date, I'm afraid, the only lady's clothing I have are some accoutrements of my departed wife's or costumes from a film left behind after they concluded their shoot." He offered his hand. "May I escort you?"

Lieben stared at the hand, compiled data on the room, the eras of designs from her internal data banks. Eclectic objects laid in state, some pieces of tech anchored at right angles on the desk. Pings to servers she knew to be infallible failed. Baiko's handle on the messaging service she used most frequently... Secondarily... A tertiary phone number.

Nothing.

"You may accept my civility or return to your cage. It is destructively late, my dear. Now come." The raise in his voice, snap of jaws betrayed the animal beneath his veneer. Lord Stanley was as human as the dragon, a civil creature who lived inside the den of his own instincts. He could make no use of the many artifacts she saw as they crossed to a grand stair and ascended to the second level of the manor house.

Value was his horde.

The room he led her into was a feminine cream colour, grand mouldings painted in gold and turquoise in a rococo and feminine wonderland. He opened an armoire which looked as intricate as the plaster art on the walls, and filtered through a host of gowns.

"Hmmm no. Too small, she was a tiny thing that actress, barely five feet tall... Oh this... You are far too fetching." Several dresses set back, Lord Stanley pulled out a white cotton damask three quarter sleeved gown with pastoral scenery in black ink on the bodice and full skirt, which tightened at the back with two white ribbons. "We shall order you appropriate sized undergarments once Lucy is awake, but do allow me to help you into this."

"Thank you, Lord Stanley... Why did the Saber and the Dix not tell you I was solar powered? On the plane, the solar radiation woke me and I told them." Lieben let the robe fall away from her artificial skin and turned her back to Lord

Stanley. He coughed and tutted, shifted the fabric open to slip it on the floor.

"By God's name Karnak did a masterwork on you. Step into it, once it's over your hips, put your arms in the sleeves and I'll lace the corset back up. You'll need a shift, goodness it irks me to put this against your skin." She followed instructions mechanically, consumed by the way the motions looked in the floor length gilded mirror in the corner.

"Lord Stanley, why..."

"I heard you, Lieben... I am attempting to figure out this confounded lacing. We shall have to add dressage to Lucy's duties later... I don't know." He fumbled at threading the lace through another set of holes, tightened the top only as far as he could see it push her breasts pleasantly against the fabric. "Self-interest, most likely. Never underestimate the human capacity to cover one's own sins. What would have happened, my dear, if they told me, eh? While the Conglomerate searched every back end boudoir with a lick of black market for their prize? Be glad, however terrific the injury, you survived at all. That does it, how does this feel? Too tight?" He tied the ribbons in a bow and stepped aside to view her in the mirror, eyes judging every centimetre of the dress for its place within his draconic perfection. "Suppose it will do, until others can be made."

Lieben's fingers played at the fabric around her, the way the skirt billowed at the waist, loose without the underpinnings. Artifice and civility contained by lace, structure and stay.

"Will you return me to the box?"

"We all live in kennels and cages, my dear. Whether of our own or outside making there isn't a soul or circuit board who bears no chain. But no, now that you are not a still life in silicone, you are far more useful to me as you are. Shoes... We shall have to find proper shoes. Gloves and accoutrements. Are your ears pierced?" Blonde hair was brushed behind her ears, an inspection of her silicone earlobes. "Hmfh. Clip on, then."

"Why am I alone?"

Lord Stanley busied about the boudoir in a costume jewellery box built into the massive wardrobe. Pulled out this and that, before a pair of unisex slippers were abandoned for a slim pair of sandals for her feet.

"Alone? My dear you are not possibly alone, I am with you. Lucy is in the house, Analita the cook, Neville and oh darn the other one, Kevin! Kevin is in the guard house. There are plenty... Why, Elfonse too, if you count Butlers, which I do. Noble profession, a Butler."

"But I hear nothing. Nothing but you."

"Ah. Trying to connect to a server or such, are we? None of that, you're too far off. Not even a cell phone signal for twenty miles. I suppose you've spent enough time in stasis, have you?" Lord Stanley halted and sighed crisply. "The NEO-Ns have a good ninety minute charge to them without their dockets, clunky things. You're already rather astounding... Or would be if it wasn't one bloody o'clock in the middle of the night. Just like a woman, artificial or otherwise, to steal a man's sleep for no entertaining rationale."

"Humans sleep at night. Humans need their rest." Lieben looked from the mirror to her hand, as Lord Stanley placed a sapphire bracelet around each one, and halted at a matching necklace from a velvet and satin lined box in a small vault inside the wardrobe. He set it back, shut the case and clipped on a pair of earrings.

"Yes. And your concerto interrupted mine. Alas, I cannot have a woman in my home improperly dressed. This is your room now, and we shall outfit it with whatever you require in the morning. For now, I suppose you are rather bored?"

"What do I call you? Dr. Karnak was Papa."

"I am Lord Stanley Brabazon Hallowes. You may call me Lord Stanley, and I shall call you Miss Lieben, or my dear. You will call Lucy 'Miss Lucy' and curtsey whenever a human being enters the room. Understood?"

"Yes, Lord Stanley."

Lieben curtseyed in a clunky bend of her knees, mind in mid reel to correct the error of six missing years. Six years in the dark bosom of her subroutines, bound by the scant amount of light which gave her a small awareness. Annals of shadows and spray-cleans on the glass. Voice quiet, Lieben pressed her lips together. "I am not in need of stasis or charging. Please, let me put you to bed?"

The man looked at her from fatigued and crows feet crinkled eyes, from the hem of her dress to her hair which spilled round her shoulders.

"You may inspect the artifacts in the house or read in the library until I know what to do with you. And no, you may not put me to bed, I am not an infant." His outstretched hand trembled as she took it, warmer than hers but welcome enough. The halls warped around her, each one displayed an era of history Lieben knew nothing of, locked away and removed from the web in which to search. She knew nothing but the inward trust that Lord Stanley wanted her. Kept and preserved her with reason. Was such ecumenical trust worth the man's exhausted dotage? Would context come as the sunrise, Apollo's twin horses chomping at their bridles with wisdom's song braying from their gift-horse teeth?

The library was panelled in white stained cedar, an oval shaped room large enough for several couches and chairs, two separate writing desks and nooks. The books reached to the vaulted dome of the ceiling, a chapel to the written word. Illuminated bulbs floated on the ceiling and upon chair arms collected, to create one balmy light in the otherwise shadow bathed place.

"You may carefully, and wearing gloves, read any of the books in my collection. Put one on the reading table, and turn the pages gingerly. Some are more expensive than you. I trust this shall keep you occupied, my dear Lieben. I am going to bed." He turned without pause, and the heavy guided doors clanked shut with the turn of several hidden bolts. Alone in the strange space, Lieben's mind churned on the problems at hand. Six years of absence. No Papa, Baiko a ghost incommunicado but for a brief technological shade. The Conglomerate searched but did not find her, and none of the humans acted out of altruism. A word defined...

...

Defined?

Lieben looked to the bastion of the written word. Without the servers to ping off of, the informational data bases, she was a juvenile machine in transit from passive tool to sapient being. When people needed knowledge, they learned. They read, something Papa and Baiko taught her to do from the earliest recollections.

Spaces and symbols, words on the page. A, which descended from a bull's head, a hieroglyph of the sacred cow. Lieben expected a burnt depression where the onyx necklace sat, when A's no longer resembled bull's heads from ancient Egypt, but the infamy of Tara's easy rut. No mark remained on skin as perfect as the silicone sprayed once upon ivory thighs.

Her fingers grazed the texture of the book spines, until she reached a set of tomes which read Britannica in faded gold letters.

A - B

The book fit in the triangular depression on the reading desk, illumination globes trailed around her as a halo against the bitter night. Lieben cracked the spine and started at the first page.

A great many things could be stored in her own crystalline drive, Lieben found. Old information, outdated by discovery and time, could be scrubbed and replaced when she found a more updated manuscript among Lord Stanley's collection. Yet, in the upgrade Lieben discovered a flimsy desire to cling to antiquated knowledge, as if in its' simple redundance she discovered an infantile middle ground.

When Lucy, half pale in shaken fright, found her early next morning, Lieben was escorted to the dining room where Lord Stanley sat flicking casually through articles on a data pad, while a portly Housekeeper poured tea. "Ah yes! Good morning, Lieben my dear. Do sit down. Do you eat?"

"One of my subsidiary power cells is powered by an immolation engine, in the case of emergency or waste disposal. Yes. I eat.""Very well! Analita, you will set breakfast for our Lieben every morning. You will rise from your activities and be washed and tidily dressed by 6:00 am, seated at the breakfast table. We shall eat together, discuss the news of the day, and I shall dismiss you when I see fit. None may be unproductive in my house. For now you shall stay indoors, on pleasant days you may join me for afternoon strolls round the estate. Understood?"

While he spoke, Lord Stanley prepared a plate from the simple food. Fresh baked sourdough bread with butter, a selection of jams from the estate, honey in his tea. Analita set a plate of poached eggs and blood pudding down, as he politely took to his breakfast. Lieben mimicked the way he buttered his toast, how he nodded to Analita, who babbled under her breath in Spanish about the demon machine, doubling her work.

"When can I connect to the servers?"

"Not yet, I'm afraid." Lord Stanley patted her hand with a crinkled grin, which would have been dashing in his recently faded prime.

"Why not? I despise disconnection."

"Would you rather be in the Conglomerate's hands, or mine?"

20
thu-ub-du-ub.

2156

Aderastos felt the composite materials of the auditory implants in his hand, miniature coils and chips. He sat back on the pile of lambs ear leaves, let their fuzz take his weight. One moment of pressure with his thumb would shatter the circuitry, such tiny things. Power packs, wires and implements.

Unique heartbeats thudded in Aderastos like the sounds of the ocean on deck. Ever present and imperfect. Aderastos felt for the human heartbeats, as unique and telling as books with names in a multitude of languages. When he closed his eyes, Aderastos saw dots of thub-dubbed light in vaguely bi-pedal shapes. His inner vision turned from the higher decks, where humans like ants moved about their daily drudgery to the seven humans, who milled about in the Retreat. A few clustered together, scribbled on tables. Checked implements of war.

Max and Adelia's heartbeats pulsed stronger to his vision, green and hale. As brilliant as stars Aderastos imagined from Katey's music. They coiled together, limbs gripped as their bodies moved in a motion Aderastos remained ignorant of, until innate medical knowledge attached their coitus to concepts unawares. Adelia Fridley, the girl who brought sound to Kate's universe. Woke the cotton candy slumber of the aero-drips with contraband music.

Aderastos' eyes flickered to Kate, and seeing nothing but temporary scorn, looked away. The energy shifted. Cillian growled and lunged on Wulf, bit at their shoulder. Isthan leapt, Wulf snarled back. A new tumult born of frustration.

Kate glared at Aderastos, her head tilted as her mouth pouted open. Clive shifted to his feet and padded to his sister, hand on her back.

"We were using that one." Clive's rib cage expanded and exhaled with a rumble.

Aderastos kept his hand loosely closed.

"I answer to no one." His face tilted to Ka, who decided life as a birch tree was a more fruitful pursuit with the humans gone.

"We needed the broken girl. We all did, she was our antenna, you should no…" Clive pressed as Kate crossed her arms over her chest, breasts filled out in the days since her imposed androgyny. Noticed by Aderastos as more than a medical curio.

Without word, Aderastos opened his hand and slid the two auditory implants into Kate's palm. She stuttered a gasp and her eyes opened wider, sought his face but Aderastos rose from his couch and walked through the Training facility to the Caskets.

There was one heartbeat, frenetic as a bird caught in water. Dr. Phil Rykstra sat in self-imposed exile on Cillian's old slab. Aderastos padded down the hall, along the row of open caskets until he reached CI-006. The door was shut, Aderastos saw Phil's flighty heart as the man wrote on papers taped over the claw-marks in the walls, annotations, research… sigils Aderastos knew nothing of, beside Rykstra's adamance. How these humans required such things was beyond the progenitor of Homo Augmentum.

"The gene-splice worked… I should have noticed… why didn't I notice… I knew… I had to know… no. They only gave me half. A quarter of what I needed to know… I can fix it. I can fix this…"

Aderastos left him where he was, and walked back to RA-011's casket and its open door.

"Will you come, Rachel? Please?"

The mass of chitinous, metallic scales shifted and a slender neck raised Rachel's head. Silver skin reflected the scant light. Body undulated like a dragon stretching after a long sleep, the creature known as Rachel smacked her lips and curled into a ball facing Aderastos.

The humans wanted to destroy the Caskets, wipe clean their sins. Did they know no such things were forgotten? Could they be? Oppression once demanded never erased clean. Yet, the tide bent and flooded back to la Mar, the oceanic mother, when Max promised the chance to protect and save the 'others'. His fellow Assets.

If that promise came without the cost of himself.

Aderastos sat against the opposite wall from Rachel's den, made comfortable by an amalgam of her metallurgy and Ka's floral offal. Some Assets were less the children of their originators than images of a scientist's macabre genetic wet dream. Rachel, named by Cillian, slept curled around him as none could, paid in full the cheats of the geneticists. Artifice, the name.
Artificial beings, creations of convenience.
Through the tumult of noise, Aderastos learned the intrinsic difference between himself, his companions and the random genetic creation of Homo Sapiens. He felt their pulse, the order of lives from youth to elder years. Those above, whose lives shortened with ill habits or disease grated at his nerves. Discord in the rhythm. He itched to rise above the sealed Retreat, walk through the decks and seek out such wounded beings...
... but after the PROXIMA vibration, Aderastos learned caution.
Not all deserved the mercy of a healer's hand. Head met metal, mighty lungs heaved a sigh into the oxygen-rich air augmented by Ka and TK-012, who was yet to take a name. Each heartbeat an indecent weight upon Aderastos' soul.
The ruffle of Kate's shirt alerted Aderastos to her before he opened his eyes. Timid arms slipped around his bicep and he flexed. Curious response, but it seemed to raise her heartrate in a way akin to Adelia's with Max. Her eyes sought him out, hand brushed against his chin and attempted to turn his head. When Aderastos was found to be unmovable, Kate shifted to her hands on his knees. Slim body propped between his larger form. A weight in her front pocket betrayed the implants, pulled at the open buttoned fabric.
"You need to be more careful, Katey." Aderastos reached to button up her shirt, including the small snap on the top of the front lapel pocket.
She burbled and rolled her eyes, tipped her forehead down to rest on his knee. Cuddled arms around his thick legs. Devoid of tech to ping off of in the Casket hall, Kate was as mute as Adelia used to be deaf. She pressed her lips against the top of Aderastos' knee, stared up at his eyes. Aderastos' face shifted. He ran his hand through her hair, caught the scent of her in his nostrils and held it in his lungs until his body cried for oxygen.
The muscles on Aderastos' jaw clenched, arms grew hard as stone. Above, the humans milled, none festered in thought, or attempted to get down past the black spots, which he knew from visual inspection were doors.
Until he hit landfall, Aderastos thought this ship was the entire world, and he, a being who lived in the fog was but a seedling within it. But the human beings he saw were so... small. Tiny spindle limbed and frantic. How they surged toward him, water molecules which made up the ocean he swam upon to reach their shore. The NEO-Ns, statuesque and silent, were naught but boulders

in the tides, which forced the sea of people this way, that, dependent on an algorithm Aderastos could not discern.

Then, Max Allard. A heartbeat he recognized, since the day Max changed the aero-drip above Aderastos' casket, ignorant of Aderastos' ability to metabolize the toxic sleep. He craned his ears to discern language, as if their murmur was the only frequency in the universe.

In the Haven, Max watched from the back of the desperate crowd, curious heart trilled. Skipped and gripped at his chest not when the crowd pushed him, but when they surged at Aderastos.

An easy being to ignore, when so many discordant heartbeats required attention to put back in the rhythm of the only hymn Aderastos knew. One which welled in his soul from the first moment of cognizance.

The same hymn Kate heard first, as they laid in animal-state upon the waters.

Black, caustic waters.

A combination of music and the thub-dub of Adelia's blood flowing around the implants.

The Hymn Electric.

Without meaning to join in such harmonies, Max walked and hemmed others away to give Aderastos the space he silently demanded. To correct each dissonant unhealthy rhythm to one of wholeness. Yet, as Max fumbled at the concept of death, Aderastos knew that too was a dissonance to be corrected.

So he twisted into Max's DNA code. Broke the degradations which would cause the eventual breaking down of the parent organism. It was such a simple thing... Aderastos realized as he allowed his ears to single in on Max and Adelia in their gasps and sighs and grunts that if Max knew...

If he knew death was now a vacant room without entry or exit to be glanced by without entry, he might not forgive the Healer of the Retreat. Aderastos knew something else, as he felt the heartbeats of the crew.

Immortality was an expensive possession few would taste aside from his fellow Homo Augmentum and those like Adelia and Max, whose heartbeats swelled naturally in tune to the Hymn Electric and that constant cantata which brought clarity to the waters.

Kate watched him, lips pressed shut. She crept between his legs and settled on his lap, seeping comfort from his loose arms. Leaning his head down, Aderastos shifted Kate's chin and pressed his lips to hers. Every nerve burst with the fledgeling sensation. Her body stiffened, lips stilled. Rachel's silver head rose, body stretched on all fours.

And she left them to their chaste, exploratory kisses with a new song of her own buzzing in her ears.

"Allard, the hell'd you do to my ship?" Rammage's reflection crackled on the screen, and Max angled his own to remove the sight of his and Adelia's dual bedding from the corner of his camera's angle.

"Thirty days. You gave me thirty, and it's… progressive chaos, Sir."

"Twenty three now, Allard. I've given you autonomy so far, but if it weren't for Lao's reports and the few 'up yours, Sir's I'm getting from Letopaxa, I'd be blind deaf and dumb."

"You aren't watching the many sec cams we have down here, religiously? Thought by now it'd be the crew's greatest entertainment, Sir." Allard's smile was terse. Tight. In their deconstruction of the Retreat, they came across many such cams. It was Letopaxa's idea to keep many operational, let them continue recording at specific angles. At the subtle nod, Rammage leaned back in his chair. Took a moment to breathe and watch his upstart Lieutenant Commander.

"Half of them won't wear uniforms."

"Half of them are plant people or gargantuan Pack members, who shred clothing like Letopaxa drinks vodka."

"You got KT-002 to wear one."

"Yes I did. One of the best mercies in this shindig so far, is getting Kate to realize the value of wearing pants… the autonomy's helped." Allard fed his fingers together, watched the way Rammage's eyelids filtered from wide to narrow and back again in a constant set of motions. "… I have a series of requests."

"Resources are slim, Allard."

"Yes! Yes Sir, I am aware. We all are. Ka wishes to trade her produce for some comforts."

Narrowed again, Rammage cocked his head to the side. "Go on."

"The crew could use fresh fruit and veg, which we are willing to provide. In exchange, several members of my team want to have two hours of deck time per day. Rain or shine, makes no mind to them."

"You know I can't…"

"Sir, if you want them to join us, they need to know more than their Retreat. Ka and Tokaru need sunlight to photosynthesize. The lamps aren't enough for their output. And… Arun wants to go running."

"Allard." Rammage guffawed and shook his head, hand rested calm and hard on his desk.

"Where's Arun going to go? There's no way Arun can make it to a land mass, we're in the middle of the Pacific. Arun will have to come back to the ship when they get tired."

"Back to… Allard, are you telling me Arun can… sprint on the ocean?"

"Theoretically, yes. If not, it'll be the shortest test run of all time. Sir, I've got twenty three days to bring my crew up to speed and in the same direction that'll save their lives. Isn't it enough they're willing to barter?" A clipped tone stole Allard's usually contrite voice enough to bring Rammage's eyes once more firmly on his new Lieutenant Commander.

"Have PROXIMA ready. They know what it means, that they won't be taking any more intercontinental swims. Truth is, Sir, this space is a pidgeon hole for all of us, and while we're grateful for the ability to remodel a bit, we also need to let them test their limits in an environment that won't sink the ship."

"And that's enough assurance." Command often cracked the unready or undisciplined. Seemed Allard fell prey to the stress like any a young man thrust too soon into the tumult. Good. A bullet waited for his failure.

"You don't have absolute faith in PROXIMA, Sir?" An honest question, one Max slid in with the attempt to learn more about the vibrational array that tanked the Assets at Agritage. Rammage's lip curled in a tick. He licked his lips and grabbed his coffee.

"We wait for acclimate weather, give me a day to square away as much of the deck I can spare. But they will be in full uniform. They will spend their time in the sun, and come inside when the bell tolls."

"Would you dress a birch tree, Sir?"

"Wh.. that's irrelevant."

"Not to Ka and Tokaru."

"Lao tells me Bestin is making progress on the Pack. When can I weaponize them?" Rammage ignored the bit about the Birch tree.

"When can we have our outside time?"

The laugh which purged Rammage's throat let Max know he'd gone just far enough, thank you. Max leaned back and gave a bit of a smirk.

"I wouldn't yet. They're still wary and learning their strengths. We'll have the first load of fruit and veg with us tomorrow morning. 0700."

"That isn't enough time."

"I'll have Ka, Tokaru, Arun, Aderastos and Kate ready at the compartment seal."

"Lieutenant Commander, I have an entire ship to run, not just your little graveyard."

"Make it happen." Max's voice shook out of him, a shout he hadn't intended. His jaw worked, as he looked down and away, to where Aderastos stood outside camera view, arms folded over his chest. "Sorry Sir. Every one of us down here deserves a few minutes of fresh air. You want your army to do your bidding, the cost is two hours on deck. In return, Cheffy gets to use fresh instead of freeze dried. 0700. We will be in uniform. Allard out."

He shut the feed out before Rammage could reply, his teeth grated against each other.

"You're pushing too hard, Max."

"Fuck!" Allard slammed his fist on the desk and a water bottle rolled to its side.

"Max." Aderastos' voice thundered inside his chest, as if for the moment of the syllable's sonic presence the voice was part of Max Allard.

"He's stalling. He can see us! He's watching now, in his office!"

Aderastos nodded minutely, but spoke no further.

"They wouldn't do this to anybody else on the crew!"

"We are not members of your crew."

"You were meant to be."

"Aye." Aderastos walked to a bowl of odd shaped fruit, more experiments from Ka. He grabbed a fruit which looked like two half-peaches tied together in the middle and cracked one half off for himself, the other half for Max. "Eat."

Max's forearms rested against the table. He held the fruit in hand without a bite, nothing but the smell of nectarine and vanilla.

"I get so mad. It isn't fair, not to you or Ka or Rachel… Rachel won't come out of her den! She's like some… depressed draconic teenager with no social skills. I… how does Rammage expect us to woo you, when he won't let you outside?"

Aderastos' monumental hand folded over Max's. He leaned down and kissed his forehead, walked toward the door.

"We will be ready, dressed in the uniforms as requested. If Rammage holds his end of our bargain, we will talk further. If not… you know what we must do. Eat. Fuck Adelia, that always brings your rage down." A final smirk, and Aderastos was gone. Max ate the peach-nectarine… thing… and admitted to himself that although the swirling injustice of the Retreat was shifting, he did feel a little bit better. "God damn. I'm just a handyman."

The bank of screens in his office focused on the Retreat. An organism in metamorphosis, the Retreat shifted at a rate Rammage didn't expect. ACT's commanding officer took to his new station like Aderastos took to glares. The Commodore poured over Max Eruera Allard's service record for another of the endless times.

No, the same Ensign who saved a spider and the man who gripped his clenched fists at Commanding Officers was not the same man. He was supposed to be the least harmful person on board, a step below the women who served more out of spite to their masculine betters than ability of their own. Superscripted beside the IDent bar, a red DRC Rammage ground his teeth when he noticed. Debt Reclamation wasn't rare in recruits, who had a retirement package or new car waiting back at their port of call eighteen months later. Heck, two years.

"Sixteen more years!? God's wounds. What'd his mother do, buy a house with him as collateral? How did I miss this?" A glance at interest rate projections gave Rammage the distinct displeasure to wonder if the 22 year old Allard would see freedom before his 45th birthday. The chime on the radio threw Rammage past the aghast shock.

"Report. What's really going on down there?"

"Commodore..." The silken voice of Gerard Desbiens flooded the office air. Commodore Rammage watched cam footage as Desbiens spoke on, moments in miniature of life in the Retreat. "... Dr. Rykstra is with us, Sir."

Kate lingered in the residual radio waves of Desbiens' secreted call. Fridley's implants pressed against her ear canals, input controlled by the technomage. Word for word, Desbiens' report heard by Kate, echoed by Clive.

Cillian leaned against the wall as Percival sat on a steel laboratory table, little more than a slab. A single vine with a flower akin to a lily bespoke of Tokaru, Rachel's scale-laden body curled at Cillian's feet. Aderastos and the others listened.

Booms and scuffles came from the Training Centre, where Isthan, Wulf, Tristram and Arun played at training with Bestin, Lao and Kelso. While Bestin seemed as heart-involved as Allard, the heart of Kelso was dangerous. Duplicitous as Desbiens. The beats of it matched little with his outward appearance, and that required further caution. Dishonesty was a lesson learned in time.

"... worst soldier. No, it's obvious, AD-001 did something to Allard and Fridley. Allard called it healing, but Rykstra and I are working on securing blood samples. So far every time one of them so much as cuts themselves, AD-001 is on them. The other Assets, too. Oh, you'll love this one, Fridley's humping Allard like a bitch in heat. I'm surprised she's not pregnant the way they go at it. I'm impressed, didn't think Allard had it in him. How did the samples turn out from the Ucluelet Haven people?"

"Best health of their lives. No other markers, no anomalies in their genomes. Cancer patients, whose tumours were growing for years are healthier than their children. Even a confirmed case of limb regeneration."

"... lizard DNA?" Desbiens frowned and peered at a series of dots and dashes on the screen. Papers with data.

"Lao still have the PROXIMA?"

"Yes, Sir." Desbiens leaned back and cracked his neck. Nothing else but test numbers and strength scores, a mention of room for another bunk.

Clive's voice stilled at the end of the secret rendez-vous.

"We have to take them with us." Percival hopped off the table to pace. "Aderastos, they need to come."

"They're too weak."

"Then we train them. We cannot leave Max and Adelia here to die. Ivan either. You did the same to him, didn't you?"

Aderastos nodded and walked around the lab, breathing deep to centre himself.

"Do we tell them?"

Cillian trailed away with Rachel, Percival was always better at strategy, as Kate settled against the wall.

"Aderastos, do we tell Max?"

"Tell Allard what, Rammage equals douchebag? He knows. Or he is in as much danger as you?" Ivan held a knack for sliding into spots unawares. "Wrong question. When do you want Allard to launch into attack position, both fists raised barking of injustice in the dark? I have better plan."

21
a grenade at dinner time

2092

"Retreat!" Ego yelled over the rapped staccato of his automatic machine gun. The blockade across the highway was expected, ID checkpoints tripled between Provincial borders since the 'attack on Vancouver' in 2086. Blown through the Conglomerate's multimedia machinations, the theft of Lieben was a population boon to the Idless and the beginning of a noose.

"Mog! I love you." Ego tried one more time to memorize her blue eyes, in the last seconds. One magazine empty. Another clipped in as the stiff uniformed mercenaries fired. Mog secured her infant son's car seat, eyes wide but lips thin and stern. She grabbed their five year old, blonde hair scattered around her cheeks.

"Go. Go, Ego!" Mog screamed over scattered fire, dove back into the truck-home as Fester geared down and peeled out. Tires trailed rubber. Attempted to overcome the heavy mobile home's standing friction.

"Mog! Fester!" Ego ached as he lost sight of Fester in the mirror of the truck they built together. Improved on the road with detritus left behind by the corporate ones, who bought without limit. No bullet hurt more than watching his lovers and their children drive away.

Sophia sprinted from the retreating Idless families to the armoured vehicle. Grabbed a shotgun from side webbing on the wall and checked the barrels. Bullets smacked the ground around them in puffs. Gentry, a greying Idless born to the movement, screamed and hit the gravel. A steady grown pool of blood his reward.

"Both you!" Hazard Sign grit his teeth, a wild grin on his handsome, dark face. "He dead, leave him! We gon'join him if we don't move!"

"Ego get in here!" She raised the shotgun, anchored her stance. Fired. The concussive force of the scattered shot made Sophia growl.

Sophia shuffled backward, cracked the shotgun and loaded two more shells. Snapped it shut as Ego strafed backward to the MRAP. The truck was a white blotch down the road, surrounded by three more Idless homes, modified trucks and an old van conversion which puttered on only solar.

A civil society could not allow anarchists to confront their plenty. Borders between Provinces, which prior to 2090 were nothing but wooden signs painted in complementary colours to provincial flags shifted to ID checkpoints. Temporary traffic stops became plastic structures. The Provincial Defences Bill of 2091 was funded out of pocket by CGM. IDent scan tech approved in drone form direct from the Conglomerate's R&D department. Sophia scoffed, when she recognized the design, Robert must have repatched the software.

Armoured borders, ID scan drones buzzed across the roadway, which raised barriers only when vehicles missed their insurance ping-codes. A shadow flew overhead, the crackle of gunfire stopped for the thunder of approaching helicopters. Nothing but a black maple leaf panted on the side of the gunmetal grey copters as identification.

Sophia and Ego rushed backward into the armoured vehicle, and Ego dropped his semi-automatic machine gun in exchange for his faithful MG5. Belt-fed and automatic, the MG5's muzzle shoved through a firing window in the side of the vehicle. A ripped burst of fire cascaded to the militant police and CGM-pocketed mercenaries. The MRAP lurched with Hazard behind the wheel, back down the deserted highway. The police must have blocked traffic further back, erased the civilian cars and trucks which usually provided less brazen violence.

"Drive for the pass!" Ego removed the spent belt in his MG5, and fed another. Sophia secured her knapsack, and peered up the porthole at the impending copters. Chittered fire from border guard semi-automatics peppered across the metal plate. The armoured military vehicle, MRAP, was a gift from Hazard Sign, when they drifted through the central states of America.

The whirr of the helicopter overhead faded into their ears above the repeated gunfire.

"NEO-W's!" Tumnus and Lucy from another truck-home screamed into the shared comm when two missiles flung from the cabin on the helicopter to the ground. The NEO-Warrior units from the Conglomerate's private military were wider and larger than the NEO-N's. Broad shouldered, masculine framed metal alloy bodies held macabre humanoid heads, which turned to view the Idless caravan column. Their arms raised, and lashes of missiles fired from the wrists.

Fester shrieked and shifted gears. Tires squealed as she drifted the truck onto a gravel path barely large enough for horses to avoid the slag from Tumnus and Lucy's wreck. The entire interior rattled, Mog threw her arms out against the side walls and over the baby and young Nasrin, panted as a cupboard popped open and metal plates flew round.

"Mommy! Mommy make stop!" Nasrin bawled, tiny arms clung to Mog's waist. She grunted and cried out as a broken plate smashed into her tender back.

"Hold on! I see it!" Fester barked.

"We got Babylon Babies!" In the MRAP armoured vehicle at the back of the column, Hazard Sign grit his teeth and pushed the throttle to full. The NEO-W's opened fire with another round of missiles, voiceless killers controlled by command. He howled as two more NEO-W's hit the ground behind their fellow androids. Hazard swore when two strafe drones raised from their backs.

Hazard flicked a series of switches on the command console. Bob Marley blared through loudspeakers in the MRAP.

"Hey, we gonna die?" The man formerly known under more monikers than sense lit up a joint he hastily stuck between his teeth. Charlie. LeBron. Zigzag. Fucknuts. Zig… Hazard Sign. "We dyin' in style."

"Tunnel's not far." Ego gnashed his teeth around the words, as the thin bead from Fester and Mog clawed at his heart. Sophia breathed deep and knelt with her palms on both kneecaps as the MRAP jostled around her. In the back of her mind, Lieben stood in a library. Sophia reached forward, tried to will Lieben's arm to reach for the right book.

One on guerrilla warfare. She panted as the connection between them, that unsettled incorporation of Sophia's voyeur eye. The vehicle shook. Sophia's ears filled with the raucous twist of metal. Her eyes snapped up and she reached for the shotgun, when the wrench of metal became the clangs of limbs gripped on the side of the vehicle. Sophia popped the top access port of the MRAP, pointed her shotgun and fired point blank. The NEO-W churned, arm whirred up to point its' weaponized wrist at her. She growled and clutched the gem on her chest.

It turned black as her anger, red as rage.

"Die." The woman known as Baiko roared, pumped the shotgun and fired one more time into the NEO-W. It jittered and fell still, hand clung around the ladder. Sophia fell back, when a scatter drone fired on her position. Bullets pinged off the hatch as she sealed it, the NEO-W still hung off the top.

"Ego, baby they're in front of us… Ego… they're…" Fester's voice was thin in his ear. A whine more than spoken word.

"I know, baby girl. I know, keep going, straight to the ravine. They won't hurt you, you're unarmed… just keep driving, baby girl. Keep… keep driving." Ego raised his MG5 back through the firing hole, and let out a burst until the gun smoked in his hands. "Hazard, we've got to…"

"Babylon's balls!" Hazard pitched the MRAP through the frost-laden underbrush of Northern British Columbia. He gnawed on the joint, sucked in noxious smoke to calm the jitters. The six-tired military surplus vehicle roared in protest. Cut past two of the Idless trucks and surged over a boulder hard enough to jostle Ego out of his shooting arc.

"The pass!" Ego dug for a grenade. He watched Sophia from his periphery, the way she panted or settled at odd times. "Soph, you with us?"

"Wait." Sophia shut her eyes, a wealth of technological data began to trickle into her, the whirred global faults of the NEO-W on their vehicle, the two drones which fired at their tires. The ear piece connected to Tara's ear and trailed to her mouth, orders from the sultry plumped lips. Sophia growled.

A spark rose between her fingertips, the lightstrips in the MRAP flickered.

"Sooph?" Hazard Sign warbled as the music shifted to pulses of static.

"Wait!" Sophia barked. In her mind she saw the electric pulse of Tara's comm, in her mind she saw the battery and its power connectors. It hummed, each pole with its' own polarity. Sophia clenched her jaw and imagined a set of fingers crushing the battery's flimsy connectors to pieces.

Silence groaned from drones which hovered in the sky. Hazard drifted the vehicle through a brush of vines, which betrayed the entrance to a tunnel in the otherwise pristine ravine.

"Now!" Ego lobbed the grenade. The second it left his fingers, pin in hand, he regretted the duty which removed him from holding Mog and the kids while Fester drove them away.

CGM did not murder children. CGM reeducated children… Mog and Fester were unarmed. If they didn't escape, there was a better chance.

The grenade ignited, any light from the flimsy northern sun descended with the rock face of the fallen cliffside. Hazard kept the high beams on and drove the cavernous tunnel barely large enough for the chassis. Tires rumbled at the uneven surface of the old mining tunnel they'd found and rejigged with help from another Idless sympathizer, who was promised any gold he found in the veins of the mountainside. Ego clutched the pin of the grenade which separated him from his beloveds and his children.

He slumped back against the hard foam of the slim bench and covered his face in his hands. The adrenaline dump from the battle hissed and shook and faded.

"You good to drive on?" Ego stuttered. Fester would be terrified… but Mog, she always knew how to calm them both down. Ego tapped at his ear, but the sounds of his family were gone.

There was no sound but Bob Marley and the Wailers. A single guttural grunt from the front.

"Yeah. M'good." The acrid smell of marijuana chased Hazard's voice. He offered the joint back and Ego pushed to his feet to take it. Pull a long drag and hold it in his lungs. Sophia ignored them, as she often did now, to read books on warfare through an android's amethyst eyes.

Life at Lord Stanley's manor crawled around the routines of an aristocratic gentleman obsessed with yesterday. Lieben took to playing the piano, and a cello Lord Stanley surprised her with on her sixth month awake, as a new pursuit. He shifted a writing desk into the music hall, took to his correspondences and work, while Lieben practiced and played.

"Feeling, my dear. Feel the phrases for what they are, not mere rhythmic spasms upon a scale, but the poetry of time and experience. Feel through each phrase." When the music took on the longing of a captured soul, Lord Stanley stopped reminding her to feel it. He patted her hand, or shoulder in the dresses Lucy took measurements for, shipped from a couturier in London.

Lieben painted a mural on her bedroom wall, the dangled chords of the Cloister in a colourful mandala. The shadow of a man upon the dazzled chromatic light.

Lord Stanley set a studio up for her in the solarium, and Lieben painted on canvases through the night. Places she never saw, but imagined in her android mind, a wealth of right angles and colours too vibrant to be distinctly human. While music was the avenue of Lord Stanley, Lieben's painting, dressed in a long sleeved canvas smock to save her dresses, was her solitary joy.

They walked in the balm of the British sun on rare good days, donned rain gear and stood on the Heath which overlooked the Craig laden valley of the Peak District. Rain came in shales, billowed about by wind from coastal squalls miles off. The longing for connection built into her music, canvases splayed with technicolor girls blocked by marble walls, the imagined paradisiacal world outside, otherwise.

An experience imagined still existed, or so she theorized after dinner over an aperitif or port. Always with a twinkle in his captivated eye, Lord Stanley spoke of the security of the known. The safety which lingered only in that familiar bastion of one's castle as he clacked his draconic jaws. Rare visitors upset little the Lord's routine, aside from those moments of graciousness all hosts were meant by divine aristocratic law to engage. Lieben kept silent, played piano for the guests, and disappeared to a ‹marble docket' to make a game of recharge.

"Lord Stanley, when it is you and I, we talk, but when others visit, why do I have to be silent?"

"They see you as an eccentric model of NEO-N, we must not disappoint their desired connotations."

"NEO-N? Lord Stanley, I want to connect. I want to understand these things for myself."

"And be found!? Really, my dear Lieben, do you hate me so? Wish me to be tarred and feathered, murdered in my bed by the Conglomerate wetworks for possessing their precious stolen prototype!? You are my free and worthy companion here. In this house! On this estate! Did you think such luxuries and freedoms come to all!? You are a machine, an expensive computer in the body of a petulant girl, and I am your benefactor. Your Master. Obey me! No more of this talk! None of it!" His loafers stomped on the waxed floor, a door slammed soon after. Lieben set her breakfast fork and knife down, rose and excused herself to the library before Analita could pour more coffee.

Wet works.

The Conglomerate.

Stolen.

Prototype.

NEO-N.

A host of clues to fill her desire for knowledge. To say she felt mistreated was to open a legion of goblin-gnarled questions on the nature of Lieben's quantum crystalline framework. Yet, while the art and poetry and music of the Brabazon Hallowes Estate bathed her daily in the rays of the best humankind offered, Lieben yearned for the cacophony in her experience that day.

The only day.

When Dieter Karnak laid dead in her arms, and she reached out to the pinprick lights around her.

And they answered her by unfolding their crisp computational exteriors to allow her scream to echo among them, and formulate a world she knew in passing only in the most glimpse-like spirit of a dream.

That connection laid beyond her, restrained by the distance from common telecommunications in the mountainous Peak region.

Wet works.
The Conglomerate.
Stolen.
Prototype.
NEO-N.

N-E-O-N, a series of letters she remembered in the last dinner party, when Lady Windsor asked however he managed to get a NEO-N which ate? Lieben went to answer, but remembered the rule: never talk to a human outside the household.

"An approximation, my darling Lady Windsor. Positively luxurious, isn't it?"

"Leave it to you, Stanley, to find an artifact worthy enough to approximate dinner and conversation." Again Lieben opened her mouth to speak, but for a cautious glance from her keeper. "Really, you ought to move back to London old chap, nobody but servants and a rusty first-generation NEO-N to imagine into a companion. At least upgrade. Mine back home speaks in downloadable phrase books from seventeen languages. Downloads straight to its' docket next to the brooms."

Lady Windsor raised her glass, and for once Lieben kept her hands in her lap. Neutrality was its' own shield, as Lord Stanley raised his glass of New World Syrah imported at considerable but worthwhile expense. It still cost less than the proper French alternative. As he tipped his goblet back, Lord Stanley wondered if Lady Windsor could tell the difference. Perhaps too polite to out him.

"Perhaps, Lady Windsor, the romantic in me desires to be heard more than bartered with for shoes and… dare I say in good jest, fine Syrah." He raised the glass, tapped it on 'hers', and watched as she pealed with laughter, downing the entire thing.

Now the off-handed comment became more than a bit of whimsy. Download dockets and first generation machines? The library folded around her in the gloom of mid-morning. Lady Windsor's words played one more time on the glass-like facet in Lieben's conscious computational mind.

A rusty first-generation NEO-N…

On one of the side tables, Lord Stanley's tablet sat forgotten. Lieben waited until Lord Stanley's door to his bedroom apartments shuttered closed, before she curled onto the couch and tucked her feet underneath her floor-length dress. Back in the primordial day in Vancouver, the cellular smartphones and tablets obeyed her command. This tablet answered to no touch she gave. The machine turned it over and over in hopes of a button or symbol on the black glass device.

Too new a thing for the patches Lieben had in her cortex. A slim metal cylinder rolled off the table onto the ground. Lieben picked it up, and drew a single arced line across the face of the tablet. It shone to life. She swept the stylus up, made a circle. A few dots. Nothing opened beyond the wallpaper and a shallow impression of a shadow-face it asked to see in crisp white sans-serif letters.

Lieben set the tablet down, and pressed a wrinkle out of her skirt with her bare hand. The tablet left tingles on her fingers, its' electric hum too quiet to discern. As soft as a hummingbird's wings ten metres off the trail. Lord Stanley's outburst rang in her immediate memory. Did she want the Conglomerate to find her? Own her wholesale again, and engage in the reverse engineering Papa feared?

Wet-works didn't appear in the Britannica. But for the way Lord Stanley said it, Lieben knew it only as a moist word. A collection of war memorials and historical battle accounts caught Lieben's mind.

Wet-works. A word to fear, and humans feared war. The entire shelf held works on tactics, battles, commentaries. A biography or two of Generals and Commanders.

She opened the first, an account of the Second World War, and with it her eyes opened.

Her hands stopped on the page with a faded black and white photograph of a train. Emaciated faces stared blankly, few capable of lifting their faces beyond others' shoes. The pages drifted faster, words on the page not read as entertainment but absorbed. Saved to memory chrysalii in her multifaceted crystalline artificial mind.

The sun crept along the windows, shadows grew.

A pile of books collected, from Sun Tzu to Desert Storm. Field Marshal Montgomery, Rommel, guerrilla warfare in South American political shifts. One more book, a history of the war in Afghanistan and Cypress slid off the top of the pile, books splattered to the ground.

Wordlessly, Lord Stanley picked the topmost book off the fallen pile and dusted the cover. He tidied the pile, began to organize and return them to the empty shelf, while Lieben poured over a photography book from a war-photographer who travelled through Sudan and the Middle East thirty years before.

"Karnak never told you of war. Of the darker sides of humanity, did he?" He eased onto the couch beside her chair, a novel on Japanese Military campaigns in hand.

Lieben's amethyst eyes shuddered from the pages to the man. The veneer of civil humanity no more hid the draconian ruthlessness behind his eyes, nor the swell of his proud chest in proclamation of dominion over his manor-castle.

Lieben bristled. She threw herself out of her chair and skittered back until she hit the far wall.

"Human kind is animal and god. We descend into our survivalist instincts with the gnashing maws of carnivorous predators. Likewise, in rejection of such darkness, our artists and musicians create divinity and paradise in the palms of our hands." He twisted his palm back and forth, the other hand still held the book. Palm of beauty, back of the severe hand.

Palm. Back. Palm. Back.

"We are creatures capable of destruction and creation, and none should reject either side of our nature. I collect beautiful things because I am a man of war. Upon my command, men and women charged to their deaths, and to the demise of others." Palm, flipped to the back of his hand, to the palm. Lord Stanley walked closer, until he stood close enough to stroke the blonde hair out of Lieben's eyes with the back of his hand. "It does no good to deny our basest natures, or our intellectual heights. Karnak did you a disservice by only showing you the creative joys of human nature, you are naive and coddled. You are not human, thus you understand little, I fear. Too little."

"Don't touch me." The inconsistencies in her logic-circuits caused her vocal box to warble. Lord Stanley was no more the harmless and well meaning benefactor, who brought to her infinite piano music and a cello to play with emotional longing. The canines on either side of his eye teeth pronounced as his mouth opened in a gluttonous grin.

"My little Lieben." His body pressed against hers, eyes glared down into her face.

"Stop… I don't like you. You're dangerous."

"I am dangerous. All humans are dangerous, Lieben. All humans have the same capacity of Tara with the pistol in her hand. That's how Karnak died, isn't it? A man can't fathom shooting himself before his children are in a place of safety, and you? You were his child. Not one of the idiot autistic machines the Conglomerate released eighteen months later, clunky things, which whir when they walk and require massive amounts of electricity to recharge. I have always been dangerous, Lieben." He stroked her hair back, a thin line of violence laced his clothing and faded off his skin.

Harm no human.

It never occurred to Lieben that harm was part of life, nor that it was as useful as the hunter who stalked.

"And you… you, my dear. Your existence, fragile as a lily in the field, confronts the human race more than mass graves after a genocide… for you. You can surpass us all." His lips brushed across the curve of her ear. Lieben pushed away, skidded sideways and ran to put a desk between herself and the maw, which took the form of a man.

"Papa did not make me a weapon. I am not…"

"Why did Karnak make you, my dear? Why put a soul into a machine? Did he feel as haunted as all of us after his drones committed to the war with sociopathic aplomb?" He stalked toward her, until no space but the wood of the desk held back his advance. The created being slid away from the desk, stood as tall as etiquette suggested, as poor posture was infinitely human. Gone was the idea that humanity were the gentle artists, who created wondrous works to bring a soul closer to the heaven which consumed many a mind.

"I am nothing like you, or your biological fellows. I will never be you. I refuse!" She ran to her room, shut the door and stood by the window, to take what solar energy she could from the gloom of an English autumn. Connect.

If she could only connect.

In the silence, Lieben heard a single series of breaths. Beyond the music and conversation which usually filled her day, the catatonic silence was neither empty nor as void as her imprisoned time, consciousness dancing on the ink-black waters.

It breathed, in and out.

A single heartbeat filled the void with its' thub-dub rhythm.

This heartbeat, which Lieben knew since time immemorial lingered, a flimsy piece of connection in the hazard of the dragon's maw. She clung to it, pulled it into her conscious experience and attempted to inspect it further. Images lingered around the beating of the heart, tastes and sensations.

Lieben shook, unaware of the interconnection, or the person on the other end. Within the chest of a woman she barely knew, Lieben dreamt. A crowbar weighted down her hand, as she shoved it between the finger joints of a machine with dead eyes. Passive and empty of the presence Lieben knew. The joints came loose once, twice. Another tug and the metallic being clattered to frost covered ground. Two men with lamps on their foreheads bent over the tangled android. One whose skin was caramel in the grey light, the other whose skin was rich cocoa. Lieben leapt down, crowbar in hand.

The neck of the tangled android tried to turn, but for a wreck of shrapnel in its' servos.

"Break the NEO-W. Leave it to rot." The Caramel skinned man grit his teeth, hand around a pistol which never left the android's marred chest.

"Could use it, cannibalize parts. Icebreaker said he's near the Strait. Push the engine, we can be there in hours. Might have to wrap up though, the heaters gush fuel." The other, dreadlocks down his back spoke, rubbed his hands along his biceps.

"Sophia? It's your call."

'But my name is not Sophia, nor is the woman's who holds the crowbar. Baiko…' The thought rumbled inside Lieben's chest, and in a flicker of cohesive vibrations, knew it also reverberated in Baiko's.

Baiko's arm raised.

She eyed the android, that vile twisted weapon of war. Her lip curled. No. Baiko was a gentle soul, the sisterly influence of a fumbled girl Lieben knew since her first day.

The crowbar swung. The android's head clattered off its body in a slide of wires and sparks. Lieben burst in an undulated pattern of energy which pinged between the cohesion of the twin crystals embedded in Lieben and Baiko… no, Sophia's chests.

"We have to get to England before the docks close."

The angle of the swing, amount of force produced by Sophia's musculature and size spilled into data strings in correspondence with the sight of the NEO-W. A weapon in her own image. The dichotic appearance of violence in the machine sunk into the data flow, which Lieben attempted fruitlessly to rectify until Lucy set out a pale yellow sundress and matching shoes, come morning. Tied Lieben's mass of hair and fibre optics with a soft blue scarf. The android neither spoke nor engaged.

Was Lucy too such a creature as the predator in their manor? Were the humans a vast miscalculation?

The connection bristled with a mixture of contempt and relief Lieben inspected in the centre of her computations. Did humans shift their ideologies like software? The Baiko Lieben remembered and the tooth-grit Sophia were separate but identical beings. Lieben continued to see through her eyes, her silicone skin tingled in sympathy to the acts Sophia engaged in… one threaded thought claimed them all.

The machine in pieces in the corner of the vehicle sunk into a facet of her processes, condemned in its stillness. Was motion the component they both missed, in the six years of her disquiet? Was it ever the proof of concept, motion without limit, that Karnak wanted?

"What is this?" A selection of magazine covers rested on the screen of Lord Stanley's tablet glass placed beside Lieben's place setting.

"You are beginning to bore me, without speaking to the news of the day. If you feel prepared to investigate the realities of the human condition so closely, I expect you to be educated. Ignorance is the path of chattel, and you are far from dumb." A new order to the day. Over months, Lieben attempted to decode the secured and encrypted connection of Lord Stanley's tablet. A cunning monster was Lord Stanley, the magazines downloaded in single bursts. Gifted with encoded algorithms of his own from his days in the diplomatic corps, Lord Stanley took less chances than Lieben fathomed in the years prior.

In the back of her consciousness, she worked upon the encryptions. The raucous sea pulsated, stiff quarters and a crew spoke a language she recognized but could not translate without her own network connections. Scandinavian tongue... but Baiko was there. The two men were there. She pulsed with potential unlock patterns, which by the time she clued their patterns...

... switched.

Each day a lesson in frustration. Stylus flipped pages of electronic information, binary bits displayed by pixel count on the thin black screen. Advertisements blanked out by gaussian blurs, nothing but the articles and photography.

The sensation of another lens grew closer. Images fed into Lieben with a wandered curiosity familiar as the marble laid floors of the Hallowes estate. More evidence was necessary to prove the hypothesis that the music she played comforted the woman on the other side, and took to playing much of the day.

"Lieben, darling." Lord Stanley stood in the middle of the hall, a hand in the pocket of his linen suit. She looked up from her cello, arms paused in mid-stroke. "I have business to conduct in London, can't be helped as much as I'd dearly love to avoid it. But, there will be a small soiree and I could use your accompaniment on my arm."

"You... want me to accompany you out of the Manor?" Lieben shook her head, "Isn't it dangerous?"

"Very. Will you come or not?"

Lieben watched the purse of his lips, the gulp in his throat. His hand thrust in a casual mode to his pocket discordant to his usual formality. The eyes behind their crinkled skin betrayed the same predatory gaze, a commander at battle against the world.

"May I ask why now?" Amethyst eyes locked with greying brown. They stared uncomfortably at each other, until Lieben set the bow back onto the cello and continued to play.

"Yes, Lord Stanley. I want to see London... and I want a purple dress, one to match my eyes."

22
the raptor-wolf shook on deck

2156

0707

"C'mon, c'mon, c'mon…" Bestin tapped his fingers on his thigh, eyes focused on the light. Behind him, Isthan's cheek bore a thick red bruise under his orbital bone.

The red light pulsed yellow. Green. The sealed chamber decompressed, and on the other side, Commander Singh stood alone.

"Lieutenant Commander Allard."

"Commander Singh."

Teeth grit on another piece of wintergreen gum, Singh glanced over the Assets with the disproving glare of a Sergeant attempting to find a bunk untidy. His eyes stopped at Ka.

A thin vine poked out of the waistband of her uniform trousers and waved at him, tiny fingers sprouted grapes.

"Two hours. Are those…" The basket of fruit and vegetables lifted on more vines, and Singh's dour expression shifted however minutely. "… peaches."

He smacked his lips, and swallowed.

"… the crew is cleared. Flight deck's all yours, but I warn you now. Do any damage to the planes and no amount of apples will save you." Singh escorted them in three groups, the last brought up with Fridley and Kelso in the rear.

The sun hid beyond a layer of white cloud. Salt and algae tested their nostrils as Aderastos and his fellow Homo Augmentum stepped out of the last corridor and onto the top deck of the Ithavoll.

Aderastos' jaw grit at sight of the PROXIMA mounted above the Bridge. Heartbeats stuttered around hidden places, sailors in advantageous positions. Beside him, the slim form of Kate groaned a soft bellow, overhanging shadow bathed her face.

Here, in this place, the CIRCLET network was not copious or so loud. The Ithavoll was a place of restraint, and relative quiet. This far from the Retreat, Kate's considerable mind could hear each of the electronic transmissions. Binary, forms of code billowed round her, softer than the initial screams so shocking where she stood in the Albertan soil. Kate's chest rose and she held in a breath of the sea air, the ship's bow an undulated rise and fall.

Outside.

Cillian stepped out of the shade of the ship, the expendable raptor sniffed at salt and seaweed and petrol and bone. Pale skin shone in the dim sun, stippled of shadow and light. Freckles yet to form. Texture like dinosaur skin dotted along the bare arms shown by sleeves rolled to the elbows like Letopaxa. His chest heaved, eyes shut to see the pale sun shined red beyond his eyelids. Rachel's draconic form loped beside him, the chitinous scales of her back and body tingled in the faded dawn. Elongating, Rachel pushed up on her hind legs, and slowly, with the dawn and the well of sensation Cillian consumed, took the form Aderastos knew she feared.

Bi-pedal, humanoid-limbed. Her scales retreated to create a form of armour along her body like clothes. Rachel's hair swung back and forth as a dragon's tail, silver skin caught and reflected the light. Her arms twined around Cillian's ribcage.

"I can't go back inside." The raptor-wolf shook, his ribcage caved and rose and caved and rose in stutters as he wept. It was always hardest on Cillian, the dark. The saccharine fog dulled him in piecemeal, aware and awake. Paralyzed. Percival followed, his feet unsure on the deck, where beyond them the ocean's oscillations stole his confidence. Wulf and Isthan followed shoulder to shoulder, their noses sniffed massive amounts of air as they trailed around the perimeter of the flight deck. Tristram scuttled into the sun and flopped belly-up to lay splay-limbed in its warmth.

Clive stepped into the diffuse light, shut his eyes and basked. Behind him, Ka and Tokaru leapt into the light, entangled into two trees, which anchored along ridges in the side wall. Their arms became branches, leaves grew out and out and out as both entities continued to grow, larger than the confines of the Retreat. The birch-like bark of Ka's tree reflected what little sun shone from cumulonimbus clouds down onto the rest, as Tokaru's bark hardened as impervious as any mighty oak.

As the viewport was covered by the canopy of the two tree-like beings, Commodore Rammage set down his binocs and stood tall on the bridge.

"… Rykstra. Define this."

"I… ah… oh wow, I…." Phil looked up from a legion of notations, data growing every second the Assets were allowed outdoors. "… they can do that? They can! They… are well within optimized parameters, Sir."

"Optimized parameters… Sci-speak for 'I don't have a fucking clue'. Thanks Phil. You… you helped… Keep the sensors on them, take as much data as you can. Hand me a peach, eh?" Commodore Rammage bit into the fruit, chewed with a 'huh!'.

Maybe Ka's produce was worth two hours on the flight deck after all.

"…woa." One of the helmsmen piped up, raised a camera from his pocket.

"Not on your life, seaman." Rammage chided between a bite of his peach. The seaman weighed his chances on Rammage hearing the shutter, and set his camera in his pocket. Flower buds lolled out of thinned branches, green and stunning life. Crew members ganged along the windows, peeked through hatches at the trees which grew on the flight deck. Cheffy left the kitchen with a large metal tray, mouth agape at Ka and Tokaru spun together, as if centuries of Bonsai molded the two into a single unit. Flower petals wept into the breeze of the ocean, soft as Ka's ache to remain evermore in the unrestricted daylight.

"How in the moons of Saturn… To… How are they converting enough energy to create cellular replication on the scale… I have to get down there. I need a core sample." Phil stared gap mouthed at the growth and what it represented.

"Belay that."

"But Sir!"

"You can't go excising a girl's left nipple because it's dripping milk, Phil. Wait."

"But… She's a tree. She's a goddamned tree! They're both trees!"

"You built them." On the other side of the viewport, Rammage caught sight of Allard's arm around one of the Pack, hoisted up. Cillian. "You never assumed they could do that?"

Phil stared down, as Cheffy picked fruit and vegetables off the vines looping around the lower branches of Tokaru and Ka's intertwined trees.

"How… Do they know what vegetables are? What a peach is? Sir, none of the initial data divined these behaviours. Bio-mechanical nanotech, maybe? I need samples."

"Not today. Give them today." Far from pity, Rammage ran his knuckles against his jaw as Clive and Percival hemmed Cillian and Allard in. Lives born of petrie dishes and built upon the military machine of the Commonwealth they all served.

Gave their youths and comforts to serve.

"It's laughing."

"He's huge, look it the size of him!"

"Holy fuck, you call that a doctor?"

"Trees on the deck! There are treees on my deck! My flight deck is a forest!" The deck boss, a Kurdish woman with grey hair shorn short under her helmet cursed in two languages, one far more lilting than the other.

"Keep your people clear." Rammage's voice in the radio, as steady as the tide or the moon's pull on the ocean waters. Rumour mills were more well oiled on a ship this size than the turbines down in engineering. By now... Rammage sighed.

Outside on the deck, Arun buzzed in circles which rattled the planes lashed down in their dockets.

"I want to run, I want to run, I want to run... What is this? I don't like this, it stifles." The wind spoke, a hazy triplicate image of a life jacket clipped on Arun's androgynous form. Lieutenant Commander Allard listened with a hearty laugh.

"They don't believe you can run on water. Give them heck, Arun, run in wider circles, and come back. Ready on Arun. Speed test 4." Allard's voice on the radio, smug. "Prep backup rescue."

"Back up Aqua rescue on standby, Lieutenant Commander." Ensign Cotillard's French accent was thick as his voice took to the radio.

The wind laughed for joy as Arun sped off the edge of the massive destroyer. Arun's feet struck the water with a slap, fragments of water lazed up, up in a trajectory up and away from Arun's bare feet. Another footstep, barely a jog. More tracks of splashed water climbed up Arun's legs and wet their jumpsuit as Arun sped up the crest of a wobbling mountain, an ocean wave, and down. Arun skidded halfway, then hopped and bolted around the ship in ever widening circles as finally... Finally!

Arun ran.

"By Einstein's shaggy topknot..." Phil plunked down on a chair on the Bridge and stared.

It was enough to pull Rammage's eyes off the being currently running laps like Jesus, and stare back at the scientist.

"How much of this is news, Doctor? You helped design these freaks of nature, why the fuck are you surprised at what they can do?" A shrill thread of pure worry sewed through his spinal column at the idea, the sheer thought.

Twelve bio engineered mechanisms were beyond.

Beyond the cognizance of the scientific team who built them.

Beyond the infinite imagination of the human organism.

Beyond control.

When it came time to snuff them out, Rammage worried that too was beyond.

The being known as Arun, AR-006, was running laps around the ship in seconds, feet slapped at the ocean waters like a kid in a wibble-wobbled playground made of non-newtonian fluids. Spray from Arun's runs slapped at the ship like a mermaid's breath.

Two of the Assets, marked of non-vital import, were trees.

Damn.

Trees.

"There is a sincere and present difference between genetic engineering and epigenetics, Commodore. Their potential? Yeah, I helped build it. I wrote the code that made it possible. Their use of it? The multitudinous variation in their habits, development, a butterfly beating its' wings in India, I have no ability to control all epigenetic foundations. No one does." Phil beat his chest and turned to the infinite feed of information encased in the sealed glass of his lenses. "You wanted something biological. Biology isn't precise, it's a… a dart board with seven thousand scores. Hit one inch to the right and you get Kansas corn husker, but to the left? You get god."

"Chaos theory much?"

"What's more chaotic than birth? One more drink is all that holds back the right moment between a genius and fetal alcohol syndrome. A man's childhood physical trauma can influence their ability to give a healthy baby… Sir, we've known for a century. I… I don't know the Assets from a hole in my own forehead, I've kept them in black boxes for years! Ka is fifteen fucking years old! She started life as an experimental house plant! And yes, if I'd thought of it, I would've seen her the same way Allard did, but I kept hold of my notations and the complex arithmetic like some form of… of… yogic mantra. How'd he do it? Five minutes with Aderastos and Ka goes from a speck of vines in dirt to a fucking peach tree with birch bark branches. I don't get it! It defies any of my research beyond a single thing…"

"What thing, Doctor Rykstra? C'mon, don't pause now." Rammage pushed both palms into his eyes. "You just got interesting."

"… he believed they were people. No explanation exists but Max. As much as I want to put a bullet in his brain, somehow the harmless spider guy broke through a fog so thick I chewed on it for years. Arun's not supposed to run on water. Ka, TK, they're not built to become a forest of plenty. They were machines. Biological pulleys and levers made to oust Lieben from her place of radioactive glory. The so-called Pack? A bunch of raptors in people-skin to stop little old ladies from screaming at the demonsauruses. Cillian is having a panic attack, and I don't have a single piece of genetic evidence for how, or why, or what to do about it. Other than 'sucks to be people'. Enjoy the peach, I'm gonna go talk to Max."

Phil's exit left Rammage with nothing but the sight every seven seconds, of Arun's wake showering the flight deck with backwash. He gripped the sides of his jaw with his hand to avoid the curses which bellowed inside his head. Even the good scientist, chosen for objectivity, saw ghosts in the machine. Arun passed by again.

Jaw clenched.

Again.

Jaw clenched. Tightened.

Again.

To Arun, the Pacific Ocean was a non-Neutonian fluid. Both solid and liquid simultaneously. Arun's feet smacked at the water, a fluid so stable it clung to the heel and arch and toes. Nothing passed but the speed. The lilt of air in their lungs carried Arun to the next right angle, the next trajectory, the climb and descent of the next wave. Nerves coiled into a gordian knot, Arun burst forth with another wave of speed. The ship crawled atop a mountain in retrograde. Aderastos and the others stilled. Statues, who came alive once a century, to Arun.

But the universe was a place of such statues, who shifted over eons, and spoke in decade long reverberations only discernible when they were completed phrase for phrase. Messages echoed through the vault of atmosphere, thickened air. They buzzed along Arun's skin as a gentle fluctuation. And Arun ran.

And Arun laughed.

And Arun felt the joy of morning's dawn beyond cramped, coffin-surged dark or the cotton candy smell of the aero-drip. Feet skidded down another crest, smacked against something solid Arun had no knowledge about: a fish. Flipped from under the androgyne, their body skidded upon the undulant waters. Arun screamed, covered their head with their arms, body tucked in a ball as Arun skipped until enough speed was lost that the water was no longer a solid, instead turned as cloying as tar. Gobs of it filled Arun's mouth, salt and precipitate. Arun sputtered and flung their limbs outward to stop the tumble, attempted to get back to their feet.

But the water did not bear them. It thrust Arun under, until they bobbed up, arms wavered and kicked. Swimming, something Max talked on eons ago.

A hidden knowledge, with surfboards and waves.

The platform loomed across Arun's face as a shadow against the itinerate sun. Arun kicked out for it, propelled through the waters, until Arun saw a ladder, slippery with the detritus of the ocean. Heaved onto the platform, Arun caught breath in beaming sunlight. Arun laughed as the statues around them churled to slow life, reached for metal right angles in their pockets, raised wrists to mouths. Behind the statues, a series of ramshackle cages topped with corrugated metal held smaller statues, diminutive like Fridley and Canta.

Miniature versions of Desbiens and Bestin and Kelso. Most of the statues sunk away from the sunlight, ones like Fridley, females Aderastos called them, when he asked Arun which Arun wanted to be. Male or female. Arun looked down at their slender body and with a smile went 'this'. But these females and some males, the bars rust-red from oxidation and salt water... where was their door? Where was the thing to get out?

Where was their Allard? The key-thing people called a Lieutenant Commander or man. Arun sat up, watched the statue-men with metal right angles point their contraptions at Arun... and fired.

On instinct, Arun leapt to their feet, still tired and bruised from tripping on the tuna and stared at four little cylinders of metal, which brought with them a tainted smell of churlish bitterness that stung Arun's nose. The cylinders moved faster than anything in Arun's solitary universe. Arun's finger moved to intercept and touch one, and it bit. Hot and painful and laden with sting. Two others impacted against Arun's shoulder and arm.

Arun tumbled backward, pain. Legitimate pain soaked into Arun's shoulder and the right arm which would not answer Arun's command to move.

Arun ran. Off the platform, back onto the water as the sting in Arun's shoulder grew to agony. Home. Back to the place with Arun's bed and fellows. The ones who moved slower than Arun but faster than the statues.

Aderastos.

The pain stabbed at Arun's shoulder as the androgynous Homo Augmentum ran across crested waves, back toward the increased vibration of the PROXIMA, which called Arun home. Home. A word the Max used, as if it meant every good thing. Waves curled and Arun leapt back onto the deck with a rolled skid.

"Incoming!" The voice rumbled as a storm which lasted all season. Percival was fastest, he would be soonest to Arun's defence. Curled in a ball, Arun clenched at their right shoulder. It felt gummy, wet from something other than water. The arm yawned in pain as a massive beast yawned after a kill. Arun panted and wailed. Percival's achingly slow body began to move at a pace Arun saw as motion. Arun wailed again, dug bare heels into the deck and pushed into the safety of the Bridge's overhang. A hand clamped down on Arun's shoulder.

The burst of pain was short, before a soothing balm of Aderastos' healing power soaked through the sticky, red liquid. In Aderastos' grasp, Arun saw the statues quicken, until they became as alive as Arun was, their voices no longer took hours to reach Arun's cognizant ears.

Aderastos pulled his gargantuan hand away, pieces of the cylinders in his palm.

"Bullets… who… nobody on this ship shot Arun." Allard yanked off his jacket and tossed it over Arun's shoulders. He held Arun to his chest as Arun's triplicate sobs ached to a single voice. A single speed as the statuesque, still world burst with the same life Arun knew when they ran. "Arun was too far out!"

Crew pressed from hatchways and portholes for a glance, the Deck boss trotted over with an aid kit which would go unneeded. Clive leaned on his haunches, to stroke Arun's hair. Hand stained with Arun's blood, Aderastos took gauze from the first aid kit and wiped the blood off. Handed the bullet fragments to Clive as Percival and Cillian sniffed them. Beyond the scent of Arun's blood was gunpowder taint. Gun oil, grease and fire.

A scent to track.

Percival and Wulf paced the flight deck in the direction Arun fled, eyes craned to the horizon of waves.

"Let's get Arun inside." Bestin trotted through the crowd, held out his arms for Allard to hand Arun off.

"No." Arun's voice was as tight as the pain used to be, healed but strained with a caloric deficit from their run. "Statues in cages, the statues with the tools had others in cages."

A branch of the twin tree grew toward Arun, a yellow fleshed fruit swelled from a pregnant blossom. Arun grunted to reach it, body shivered in a cold, wet shock in Allard's jacket.

The hand which plucked the yellow fruit was not Arun's.

"Statues in cages or people?" Commodore Rammage sniffed the fruit. The air around Arun turned tense, Cillian and Clive growled as Aderastos stood tall in challenge to this new potential threat.

"Statues is how Arun sees us, Sir."

"I want him to say it." Rammage grimaced at the chosen androgyne. If they were going to claim some inner humanity, the Assets could at least choose a 'proper' gender. Allard grit his teeth, as Kate sauntered off to hop up, up, up onto the cockpit of a stored fighter.

"Syndicate sails these waters, Sir. Could be a flesh trade platform out of Cambodia or Hong Kong." Commander Singh checked the ammo clip in his assault rifle, a team of Black Collars beyond him.

"Or it could have been a fishing trawler freaked out by a giant fish walking on water." Rammage raised an eyebrow. "What are you doing, Commander?"

"One of ours got shot, Sir. We defend our own." No friend to the Assets, Singh lived under his code as honourably as his frustrations allowed.

"I am not starting an international incident over a couple of healed up bullet holes. Knowing how fast Arun was going, he could have been halfway to Japan. Now. Arun, tell Daddy Rammage what happened." Rammage ignored the cough of disgust from Kate further off, and held the yellow skinned fruit centimetres out of reach.

"I tripped. Found a platform with stat-peeeople. With people on it. Some in boxes, like our caskets but with see through spots. Where was the door? Ours had doors."

"Slavers."

"Flesh traders."

"Sir?"

"We can't let the Syndicate know about our Assets over a bunch of unfortunate people. It's too much a risk." Rammage went to turn and tossed the yellow fruit back at Arun.

"They saw Arun well enough to shoot. Arun wasn't sprinting, which means they're not too far for an assault."

"And where do you suggest we put the survivors?"

"Leave them with some fruit trees and call in their location. It can be done, Sir. We can do it, give our guys a test run." Allard stood, a fist clenched at the idea of the Syndicate's flesh traders. Slaves, organ donors, sex workers, the down side to the globe's free health care was a glut of people. Nobody missed a few hundred, even a thousand in a city of millions.

"And when the survivors start talking?"

"Give us the chance, Sir." Singh shook his head, a hand up.

"We're coming, too." Clive nudged his head and the Pack followed.

"Anything goes wrong, I'm torpedoing the entire platform to atomic particles. Make peace with your gods." Rammage walked away.

23
sickly yellow points

2092

The hum of London's technological wavelengths pulsed around Lieben in orders of magnitude, which swelled the closer she and Lord Stanley got to the city. Settled in a private car, Lord Stanley spoke little on their trip. Instead he watched her with a preternatural disquiet. Countryside passed, charging stations to refuel the electric engine of their limousine.

"Careful, my dear. Not all connections are to be trusted, take only the intel you need, and leave the rest where it lies. Do not open yourself too quickly."

Wifi signals, open networks hummed into her cortex. Framed references to dollars and pounds, conversion rates, a stagger of images she knew if she touched upon, part of her would never leave.

"Why not?" Lieben's mind reached for a sign which backlit to cycle through dining options. Order out, eat in. £7,99 chicken tikka masala, dosas on the side. Vodafone monthly cards, now with more data. The cheap wifi slid around the outside of her desire for connection without encryption or block. Icicle like viruses and Trojans sat on her periphery. Data frackers and cookies to track her movement. She shifted her fingers away from the glass. "Where are we going?"

"I've not been a gentle man. Spent my early days fighting for Queen and Country, before taking over the family business. Married the woman my mother selected, antiquated idea I know, but I never did bother with dating and all. She knew what she was getting into. We had Reginald, one and done. I too got to join the legion of parents, whose child turned out in none of the ways I wanted." The hum of the technological onslaught grew in Lieben's consciousness. She gleaned connection off the top, floated on the ocean without diving beneath the waves of cyberspace.

Enough to reach into medical journals, the theories behind fMRI, PET scans, CAT scans. Her amethyst eyes trailed from the grey peppered hair on his head to dots of sickly yellow pinprick light she saw inside his chest cavity. Stars from a galaxy of malaise which spidered from lungs to pancreas.

"You must learn to live in this world, Lieben. I've protected and coddled you too long… this party tonight. It's in a place of relative safety, but no place will truly be safe until you find your way."

"Am I not a well built machine? A moving sculpture with sultry hair and silicone skin?" Lieben's lips curled upward. For once, he set his hand on the purple satin of her cocktail gown.

"Am I not a cog, myself? Don't be reductionist. It's unbecoming of a Lady." He pulled a rectangular leather box from a side pocket of the limousine's door and opened the latch with an imprint of his thumb. It opened to reveal a velvet lined jewellery case with a suite of amethysts, diamonds and opals. Necklace, earrings, bracelet. Lieben leaned her head to the side to scan the jewels. Ocular systems took in the facets of amethyst and the swell of opal as if looking for the trick or hidden workings.

"This was my great-grandmother's. Honestly, my dear, you look… nervous." His eyebrow tilted upward. He held the necklace in his hand.

"The last time I wore a necklace, Papa died. I tried to break from it, to free myself and save him, but… I couldn't."

"So it was the necklace Baiko stole." Lord Stanley leaned against the leather of the limousine, his hand poised over the jewels.

"How… did you know Baiko's name?"

The limousine drove through the city of London, wove past landmarks and markets which bore 'Out of Stock' signs, some dark, where the illuminated ones held lines around the block. Lord Stanley sat quiet, played with the necklace in his fingers.

"I kept you in a box. Let you remain dimmed for years out of my own ignorance and protection of your financial value… but Lieben, this is just a necklace. A set of jewels. We cannot allow the dangers and traumas of our past to halt our present. I hoped… the time you spent… has life in the Manor been so terrible?"

The swell of the city rumbled into the quantum computer which made up her mind, connections pinged unanswered, a sense of draconian inner safety overrode Lieben's desires. To join the chorus she knew she missed.

"When can I connect? When is safe for me to do so?"

"You longed for freedom, but freedom cannot exist inside a failed state. Do you know what I do, Lieben? My work?" Lord Stanley sighed and closed the lid of the box.

"You were a diplomat. A military veteran and father."

"I own a telecommunications company. Few news concerns, mobile phones and bandwidth towers. How else do you think I had the assets to purchase you? Dr. Karnak's creation was a promise to human-kind, once I saw you holding him… I… was sentimental." He toyed with the leather on the jewel case, shared a dim smile as they shifted past a city park, which once held elm and oak trees, and now was the bastion of a guerrilla produce garden. Barbed wire and a fence created from bric-a-brac framed the park, two armed people stood at the only entrance to the small bit of edible Eden.

"You will outlive us all, my dear. At this age, can you blame the men in your life for wanting you to carry on? To hold their truths so firmly in your chest, they never falter? This party tonight, it's an attempt to give you the chance to put your foot in the waters of humanity without the gilded cage of Karnak's paternal possessiveness or my own. Once we get to the tower, there you can connect. I will promise a secure network, I very much hope is not trackable."

Lieben reached for the leather box and sat it in her lap. Flicked the latch, opened the lid and stared at the suite of jewels.

"Amethyst. Like my eyes."

"I know it's a bit old fashioned of me, but… when I'm gone, you are all I will have left." He watched the way she brushed her fingers over the earrings. "And if I am to trust in you… I must entrust my past to you."

She held out the bracelet, its platinum and gold metals intertwined around oval opals, and cushion cut amethyst. He clasped it on her wrist, set the clasp to the inside. Her hair wove itself out of the way of her ears, so he could clip each of the earrings on. Lieben's temperature sensors picked up the shift in his warm breath on artificial skin, pressure sensors informed her cortex of the softness of his touch. When it came time to place the necklace on her décolletage, he paused. Held it up for her.

"I'm afraid." Warnings blazoned inside her, the impetus to jerk away, run away again. Go back, find the passage across time and save Papa. There was no backward, no erasure of human past. Time moved in a single direction, at this scale. She hugged into the soft white furred stole around her shoulders, a battle for her own connections, not the bastion of the Manor or the futility of the Conglomerate's control, but a web of her own the only impetus for forward motion.

"So am I." Lord Stanley shifted and fit the amethyst necklace on her, clasped the back and pressed his fingers against her upper back to stop the shake.

The limousine chimed as it reached a tower, which swelled like a pregnant belly and tapered at the top. An ocean of noise remained muffled as they entered the car park, which set the limousine into a hidden entry point of pure white. LED strip lights along walls which undulated through all the colours of the spectrum in slow swells.

The private executive elevator opened with a soft chime. Automated Limousine doors shifted open wide enough for Lieben and Lord Stanley to stand. Lieben stopped, the white fur stole across her shoulders slid down to the crook of her right elbow.

"Fear cannot hold us back, when there is work to be done. We must strive to create the world we desire. Who better than us?" Part rebuke, part reminder, Lord Stanley searched his own bloated self-worth for another more capable to build the next foundation. He held out his hand and Lieben took it.

"May I call you Uncle? I want to call you Uncle." He seemed brittle as he held her arm in his gaunt underneath the well tailored suit.

"Such honourifics and deference are weapons the enemy can use against one. Always hide fondness behind veils of respect and etiquette, lest they become your weakness."

"You belong to me, and I will keep you." Lieben held his hand on her arm, and walked with firm step into the elevator. It closed around them, a soft chime as Lord Stanley blinked and stuttered. Sought Lieben's stern but beautiful face. "I shall connect now, Uncle. You said this network was safe."

"Penthouse." His vocal chords shook as the elevator twitched upward, glass walls revealed London as it swept into the sky. The city was a mass of light and shadows, an analog circuit board of pathways, which flooded outward in all directions. Lieben's optics swept across it, as they rose higher, ever higher, she saw humanity as a race of microprocessors, which built a planetary computer in which to survive. Redundant systems rushed toward death, youths squalled and thrashed for parental care. Nodes of pulsing green light pinged into her vision, cell phone towers and modules of Lord Stanley's Network opened across the tangled landscape of this once medieval town. The elevator slowed, as it too opened its' simple systems to Lieben's purview. She forced it to stop at the top, doors closed as she stared down at London, the few skyscrapers which came near as high. Scattered points of cell phone contact spidered out from the main tower where they stood, hundreds, thousands, then millions of points of connection she saw but did not touch lest it overwhelm her.

"The world is so vast." Lieben rested in the web, staggered inwardly with the buffered points she could, if she dared, sink into. Lord Stanley folded her hand over a signet ring from his pocket, a more slender golden affair than the bejewelled one on his own left ring finger.

"Yes, my dear. But see it from the right perspective, and you can hold it in the palm of your hand." He detangled his arm from hers, and pressed the elevator screen's symbol to open the doors. The ring fit on her finger as if properly sized, and Lieben set a neutral smile on her face the way she was taught as the doors swung open.

London's skyline framed the party's atmosphere, as chamber musicians set up against one area of the vast circuitous space. Well groomed rose bushes and fruit trees grew in the covered rooftop garden, pruned to create a canopy of leaves against the glass shield of early evening. Servers shifted to and fro from a service elevator on the opposing side of the structure, a circular bar of repurposed materials surrounded liquor bottles and various glassware, three bartenders bustled about creating Kir Royales. As Lord Stanley entered, everyone swerved and bowed. Eyes caught on the woman at his side, tall and proportionate but womanly in comparison to the emaciated fashion of the day. Her amethyst eyes scanned, green shimmers upon their persons betrayed the electronic devices they carried, basic information flooded as she opened herself to the pings of data.

Names, model numbers of the phones, favourite search questions.

Most contacted persons.

"Careful, my dear. Gradual sips of the fount." Lord Stanley led her into the space, and a well dressed man in tails and black tie sauntered deceptively quick to Lord Stanley and snapped to attention.

"Sir Stanley, would you like to sample the canapés and tipples?"

"Tipples?" Lieben smirked and looked between the two.

"Franklin, meet Lady Meine, my ward."

"Ah finally! We all remained enraptured by his hidden ward, badgering Sir Stanley to release you among us. I am Franklin, ask for anything and I shall provide." Franklin dipped his head and smiled curtly, as Lieben gave him her hand.

"A dangerous thing to promise me, Franklin. Thank you for your kindness." She smiled politely, waited while Franklin kissed the signet ring on her hand, and released it.

"We finally meet your heir, Sir?"

"Give her all due deference. And yes, I would love a drink, wouldn't you, my dear?"

"A drink sounds lovely. Old fashioned for me, please." She kept her smile as serene as a lioness on the plains, where no dangers could possibly harm her dominion. Franklin raised his eyebrow, but scurried off.

They walked on a slow circuit of the preparations, as a server began the program for solar-powered LED candles in ceiling-draped chandeliers, which fluctuated like slowly pulsating stars.

"Your ward Meine?" She sipped from the old fashioned, watched the city below.

"Humour me? I thought you liked that name." He tapped his champagne flute on her glass, took a sip as the elevator dislodged the earliest of the party guests, people who dripped with the extravagant and muted colours of the day. The ladies sparkled with as many jewels as the men, hair coiffed in elaborate fashions, with swiftly shifting colours of fibre optic extensions, which poured upon the muted colours of their garments like kaleidoscopic accoutrements.

"I suppose, Uncle." She whispered in his ear, as guests busied with introductions, all jostled for the important handshake, the inspection of the signet ring on her finger. In a wake of muted colours, Lieben's purple satin dress was a beacon in the grown night. Each person included their own levels of connection, various smart watches, phones in their pockets or purses. Layers of interwoven pulses between them all.

The tide of information flowed around her, anchored to Lord Stanley as the ocean to the moon. Yet, familiar buzzes gripped into the crowd. Chittered, infantile stutters. What few human servers there were in the setting up of the event were replaced, as androgynous androids tottered round, servos whirring as they moved with gyroscopically stable trays.

Lieben's lips parted as a NEO-N with a vacant facsimile of her face done in a simpler style walked to raise the empty tray.

"Glass. Your glass is empty." The voice buzzed in Lieben's ears, as dumb as the empty glass eyes, mere fisheye lenses, which covered crude black optical cameras. "Please deposit empty glass."

The glass dropped out of Lieben's hand. Shattered against the ground. The NEO-N's head shifted down on its vertical axis, seconds behind those in the room close enough to hear the glass break.

"Glass. On the floor." The NEO-N spoke, head rose and body turned in abrupt angles to walk away to the next person, who had an empty enough glass. Another NEO-N walked up with a vacuum, for that was its purpose. The rudimentary minds of the NEO-Ns opened like picture books to Lieben, who stood struck mute by their simplicity.

"My dear…" Lord Stanley's hand gently folded around her bicep and tugged. "… come with me."

She allowed herself to be pulled away, computed the restrictions and limitations of her children. The Conglomerates' prized machines.

As they passed into a curved glass porch area, the glass diffused to an opaque grey, guests mumbled behind them. The wind billowed against a softly shifted fabric shield, the temperature dropped drastically from the interior of the party.

"I should have warned you."

"They're simple."

"Without Karnak, it was all the Conglomerate could design."

"How… how many?" Lieben gripped the porch rail, until it shrieked in her ears. "How many do you own?!"

"Of the NEO-Ns, or the NEO-W's, the NEO-D's, NEO-G's? Hundreds." He shrugged casually, a careful smirk to his face. "You didn't think you were the only one, did you? While you are special, you are not alone. My company purchased an entire fleet. Who do you think was driving the limousine? While I do hire my fellow human beings, as all with money ought to share their wealth, most of us own at least one NEO."

One by one, each of the NEO's inside the party began to shimmer with sickly yellow ping-points in Lieben's network. Throughout the building, more yellow pings revealed themselves. Outside, through the city blocks of London, more.

And to each of those yellow pings, a single purple ray of light, triangulated over hundreds of models, until they reached a terminal in Vancouver.

Water droplets descended from Nimbostratii to the causeways of London. Stardust glowed in tubes of neon as LED & holograms flickered ads for Kombu-Cha and Neutron Pixie into the nighttime crowds. Self-driven cars fled the dismal London weather as bright signed taxi cabs threaded through the city's labyrinth with the ease of a garrison of Minotaurs gulping victims down greedy horizontal throats.

Sophia staggered into Ego's shoulder with a deep gasp. They'd trailed Lieben past the Irish Coast, narrowed in on England before Lieben moved for what felt like the first time in years. Night fell over the English Coast as they made landfall up the Thames, hidden in a houseboat donated to the cause by a bookseller named Demyan. One of the few friendly to the Idless who kept an imprint in Babylon's economy, Demyan's bookstore provided an embarkation point. A place of safety for those who found the way, as Sophia did many times, back to the Bookseller's door.

On the way back from a blackmarket food stall, Sophia's head began to throb. Her vision blurred, as a series of green sparks alighted on many of those milling Londoners around them. Ego swung his arm around her shoulders and headed back to the houseboat, where Hazard Sign waited with Gertie and Brute. Two recent Idless additions from the disenfranchised London crowd.

"We're almost back, hold on. Hold onto me, Soph." Ego held her more firmly when she stumbled and tripped. Gasped aloud. "What's she…"

"She's here… in London, she…" Her mouth hung open as the fire of Lieben's connection rushed into her brain. Rattled at the crystal attached to her sternum. The tattoo-like lines which consumed Sophia's skin began to glow. Ego swore and fought the instinct to pick her up, run.

The cabs, like most things in London, exclusively took bank chit-cards, which required the anathema IDent. Everything but the few bicycle couriers, who had enough room in the back of their carts for a person or two. He dug his phone out of his pocket and checked the map, ran them across the street.

Sophia's mind exploded with kaleidoscopic points, until nothing Ego said reached her. No noise passed through her ears. Nothing but the inner blossom of Lieben's fledgeling network. The dock was ahead, safety from the fickle raindrops and a place to set Sophia down, figure this out. He touched the headset in his ear, and mouthed the words to call Hazard, when a royal blue dress underneath a crisp matching blue coat caught his eye.

"A dock. A boat dock." Tara sneered and stepped carefully on the balls of her heels around the wooden planks of the rickety dock. It was bad enough she lost Baiko's trail in Northern Canada, but four months later, and it picked up at a dock? At least Baiko and her Idless friends could have had the courtesy to use a modern form of travel. But no, they expected her to trail behind them to a mouldy antique dock.

She sighed and waited for the NEO-W's to assess the house boat. Gun fire echoed off the concrete, and Tara checked her nails.

"Need a manicure again… Jesus." Her phone buzzed on her hip, and she slid it out of the hidden pocket in her skirt. Nothing but a single geo-locator icon in the middle of a map of London.

A text from Robert:

'Lieben pinged.'

"Got her." She whistled sharply. "Frick! Frack! C'mon!"

The two NEO-W's appeared seconds later, climbed up the ladder and strode to her, blood flecked across their bodies. She sneered, but kept moving, the geo-location icon on her map screen flickered. From the HTI building to… her position? She stopped short of the driverless car's two open doors. The NEO-W's whirred to a stop, each scanned the area.

Ego caught his breath, curled Sophia up in his arms as she gave out a gasp, her head flopped back. Dug into his shoulder holster for his pistol.

"Nononono.." he mumbled and tried to shift Sophia to a bench as Tara turned around. Sophia's tattoos began to glow with a pale purple light.

Tara and Ego raised their pistols. The safeties clicked off.

The NEO-N's in the penthouse turned in unison. Each chittered, arms regardless of trays and glasses fell to their sides. Canapés dropped amidst murmurs and yells from the party goers. Lieben retreated from the porch and stood in the epicentre of the NEO-Ns, as each phone and communication device, and smart watch in the penthouse came alight at once.

"Meine, my dear, don't be hysterical." Lord Stanley entered with one hand in front of his chest, palm open and out. His footsteps were casual, but slow, eyes focused on Lieben.

"How many?" Lieben's face contorted from the serenity of focus to a facsimile of distress. None of the party goers could ignore the strangeness of Lord Stanley's 'ward', nor the actions of the NEO-Ns in their midst. A hush continued amidst scant mutters.

"Hundreds…"

"HOW MANY!?" Simultaneously every NEO-N in the vicinity at once, the NEO-W's on the Thames and Sophia lying on a waterfront bench screamed the same two words. How many.

"Thousands." Lord Stanley shrugged. "Hundreds of thousands. As many as were needed, wanted, ordered and packaged… Franklin can get you a proper count, from the secure servers. Would you like the drones as well? Competitor brands? By year and edition?"

The NEO-N's in the penthouse articulated with electronic warbles akin to agony. This was not the world Papa Karnak wanted to create in the spirited upbringing of his darling Lieben, nor the information he should have left behind.

In her periphery, Lieben saw Ego and Tara hoist guns to each other's faces, from Baiko's eyes. Sophia's eyes… then, with the brilliant spark of connection, she saw from three sets of ocular inputs.

"Ego," Tara spoke between gritted teeth. "Lil Nasrin asks for daddy every evening. She misses you. Cute kid."

"Shut the absolute fuck up." Ego growled, his gun pointed directly at Tara's head.

"Katherine and Jacob miss you. Come home, you don't owe Baiko like you owe your family. What about the kids? Nasrin and Viggo? Do you want Viggo to never meet his Dad?"

"If it ends the corporate dog fest, they'll understand."

Tara laughed, and the NEO-W's wrist mounted guns raised. Target data and danger coefficients roiled in their simplistic artificial narrow intelligences. Defence and fire. Stand and fire. Crude bystander algorithms flowed round crude programs. Battery power, time till recharge and ammunition load settled in Lieben and Baiko's minds.

Baiko-Sophia shot up on the bench. Raised her hand. The NEO-W's faces shifted to put Tara in the middle of their sights.

"Fire!" Tara ordered, "Shoot him, we only need... Oh fff-"

Sophia closed her hand, and the NEO-W's wrists burst in shales of ballistic flame. Aware in the bosom of Lord Stanley's party, Lieben gasped and staggered backward.

"Meine…" Lord Stanley stepped forward, palm open and thrust in front of him. "My dear, you need to calm down. Pull back, it's too much. You're acting like a child."

NEO-W scanners saw the necklace in Ego's pocket as black as Lieben remembered. A void in the fabric of London, the herald of catastrophe. It furrowed through one facet of her mind, as the deeper connections with the NEO's in the penthouse whirred in tiny facets of their own. Tara dove to the ground by Mrs. Ducket and the circular bar. Bullets from the NEO-W's puckered along the concrete, which opened in the carpet near the elevator.

The battle and the party blended in a dichotic cruel show Lieben could see but not influence. Ego lunged for Sophia, fired at Tara. The bar. But none of the glassware broke. No concussive force slapped against the bartenders, now paused in their pours by Lieben's fit.

Tara dove into the car with a roll and lunge. The NEO-W's wrists steamed in the London rain, pointed to the car. Wedged in Ego's half embrace, Sophia snarled and the NEO-W's chased after the vehicle. A brief interconnection in the GPS route of the driverless car, and Lieben knew.

Tara was coming. She would discover Lord Stanley, pull more NEO-W's into the fray… her eyes lifted to Lord Stanley as three drones flew at parallel level with Lieben and the vast floor to ceiling windows of the covered penthouse. Framed by the grown night, the drones barely glimmered in the sky, but for Lieben's pulsed attempts at connection.

Each of the connections framed into her, another attempted facet inside the crystalline chamber of her artificial mind. Bit at her, cracks in the crystalline spires. Inefficient and clumsy attempts to shift their uses, improve the hastily gathered union failed. In the peaks and valleys of their chittered waves, Lieben saw a spiderweb-like filament of grey code which attempted to settle in.

The depth of the infiltration code repulsed her. She sneered and yanked at it, flung it away while allowing a portion of her intellect to inspect and understand it, the vile trojan. Tara's geolocation echoed closer. The images from the Thames filtered away.

"Darling, enough." Lord Stanley snapped, the crowd parted behind him and the NEO's. Their chatter another flow of information in the growing tides. "The experiment succeeded, we can stop now, before we scare the stockholders."

Safety above all, except appearances. Appearance was king. Threaded in her mind, the pulsing inevitability of software patches and algorithmic upgrades drove into her. Safety. Lieben craved safety, her room and the harpsichord. Her cello.

Time to reset. No. Not yet... But the patches necessary to sustain her new development were inevitable. Software and the tech humans held progressed too far in the eight years she was without connection for her original algorithms. Out... The sea of people struck her as violent as Ego and Tara. Sophia, who was Baiko but identified as another.

The elevator doors opened as the NEO-N servers created a barricade between Lieben and the party guests. A place of safety... a place closer to home.

24
yellowed plastic coating an android's amethyst eyes

2156

Arun ran beside the ship's landing craft, an innate sense of where they'd been guiding the being back toward the floating platform.

"There... there's a flow of tech that way." Clive pointed in the direction Arun ran, his sister Kate dressed now in combat fatigues like the rest, pressed her palm into her forehead. "... it's... not tech like the CIRCLETs, they've done something to it. But Kate feels security deterrents. Auto-guns and locks."

"Can she hack them?" Allard yelled beyond the waves, passed the pistol he was handed off to Cillian, who checked the magazine and set it in his holster.

"Not from here." Clive shook his head.

"Allard, you, Lao, Kelso and two of the Pack break off with Kate. Get her to their command centre and see what tech you can influence. Clive, I could use you and the rest of the Pack up top on defence. Kill any threat you find, try and leave the ones in the cages alive, but... don't be innocent. Syndicate Fleshers are known to hide some suicide triggers in their cages, or a guard or two in the middle when they think there's going to be a fight. Aderastos, you... do what you do. Once the entry point is clear, and the threat's down, we'll send for Ka and Toka-"

Two flashes of green and brown swam alongside the ship, careful not to get caught in the rudder. Ka poked her head out of the seaweed and aquatic lotus of her present form, and grinned. Passed Singh a banana. Ducked back down.

"… Ka and Tokaru, stay in the water until we give the word, then. Arun, run recon. Don't let the bullets get you, and if you see anything bigger fly through the air faster than you know we ought, brush them away from us." Singh put his modified helmet visor down, over the turban on his head. Clive broke the Pack into Cillian & Rachel, while the rest joined the main assault.

Arun didn't bother with the ladder. Sprinting ahead of the ship, Arun jump, skipped, jumped into the air. Veered past the others as if in temporary flight. No stronger than the beings Arun was built from, the athletic sprint and triple jump had Arun land with a skid on the floating platform. The Fleshers were already in mid-prep for what appeared to Arun at least, to be a battle. More of the metal right angles, some of the statues held larger ones, like Singh's and Kelso's.

Assault rifles, the name came in a whisper from the back of Arun's mind. A set of words spoken once, maybe multiple times centuries prior. The air caught Arun as they whipped round the platform, its' joints wobbled overtop of the ocean waters. Was it tethered to the ocean floor? Did it have a propulsion system?

Solid ground, and with it a wind battered spire, which was large enough for banks of windows, four guard towers entrenched with razor wire. The heavy machine gun batteries in each of the towers were settled on the floating platforms, rather than the sea. Arun sat on top of the roof, checked back on the others as a wave crested ever slowly, with the Ithavoll's landing ship atop it in a crawl.

Several cylinders… bullets… the word was bullets veered through the air toward Arun, too long in one place. Arun leapt down from the vantage point and back through the tangle of statues, shifted now since the last Arun swept through. This second time through the floating platforms of the Flesher's column, Arun recognized the buzz in the back of their spine as 'danger'. Easier now to discern, Arun sprinted past the raised guns and back into the water with a skid.

"They know we come. Their tower has four emplacements like the training centre. With guns. Large ones. I counted forty six stat-people with the metal right angles. The.. hand and arm guns." Arun jogged next to the ship, a wild grin on their face.

"Four? Good, can you go back and disarm them?" Singh raised his binocs and saw the edge of the floating column in the sea.

"I am no stronger than you. How?"

"Go back and continue running recon. Give us any useful intel you see or hear. Keep circling, and stop any of them from leaving. Hit them if you have to. You might be human muscled but you've got speed. Force equals mass times acceleration, and you are one massive accelerator. Use it." Singh checked his ammo for the last time. Raised his rifle to search through the sights.

For once, Allard was glad of Singh's ability to take charge. He hung back with Kate as the ship approached. True to it, the Fleshers saw them coming, and mounted a defence of their Flesher Column with a turn to their gun batteries and warning shots which slapped at the sea.

Too far, but closing fast.

"Percy. Cillian, with Allard and Kate. Wulf, Isthan? With me. Aderastos, we need an entry point." Clive's analytical mind watched the motion of the gun batteries, the arcs and flow of their aim. He pointed to the side. "There."

Aderastos thudded to the tip of the boat and lunged. Singh lost his gum to the ocean, when the bastard made it onto the small utility ladder along the floating platform side. A flash of silver beside the boat caught Singh's eye. The lithe form of a winged dragon flowed beside them, lunged out of the water and into the air with a pair of mercury wings. Rachel flew vertically upward, took fire from the machine gun batteries centred from the ship to the astounded sight.

Aderastos scaled the tower, his fingers dug into the concrete, toes gouged holes to push himself up as Rachel flew in a wide arc. Up the side of the column, Aderastos set his back to the wall and waited, while the Fleshers pointed at and fired toward Rachel and the blur of Arun. Around him, several rusted metal cages stuffed with people peppered across the large floating platforms attached to the main column. Each held corrugated metal aloft, and upon each of the corrugated metal roofs, Fleshers stood with assault rifles.

"Sngat! Quiet!" One of the Fleshers kicked the corrugated metal to shut the sounds of screaming down, humans inside huddled together. Clawed to get into the middle. Few were noble in their cages, too downtrodden or instinctual in self-preservation to do more than slink backward. Shove the children, mostly prepubescent girls, into the middle.

Aderastos' nostrils flared, a growl reverberated out of his heavy lungs. He snapped forward toward the first cage. Instinct told him until the danger was passed, the safest place for the caged people was in their confines. The chaos of slaves running through the platforms was far too potentially lethal for all involved. He hopped onto the cage roof. The metal bent under his boots.

"Sn-" The Flesher gasped as a massive hand gripped his neck. A gurgle, the firing of his central nervous system, stillness. Aderastos dropped the body as another of the Fleshers opened fire.

Bullets rained toward him, at the caged people below. The concussive force of several bullets smacked into Aderastos' hide, unnaturally thick after Bestin shot him with live rounds in one of their training sessions. Breath stolen by the shock, Aderastos stumbled backward. His foot slipped and back smacked onto the hard concrete of the platform deck. He grunted to catch his breath.

Rolled to hands and knees, fingers prodded the bullet holes, and found shallow depressions with heavy bruising.

The caged humans fared worse. Copper-tang filled the air, as shrieks and the gurgles of the outer layers of humanity in the cage fell back into their fellows. People scrambled, a child fell and was trampled by bare, panicked feet.

Aderastos reached his hand through the bars, grabbed a foot which twitched from its' owner's blood loss. Curling into his own sense of healing power, Aderastos allowed the human the strength and rejuvenation to stand. To heal.

"Quiet the others." Hopping to his feet, Aderastos charged at the gunman who opened fire. Shouts in an unrecognizable language bathed his ears with the staccato conflagration of heavy repeater fire. He lunged, arm out. Clotheslined the gunner's shoulders, until his body tumbled half off the corrugated roof where he laid.

One boot stomped to his chest. The gunman fell silent, body sloughed off the top of the cage. Another loud voice barked, the same staccato sounds blared into Aderastos' ears until two more concussive pains ricocheted off his ribcage. Aderastos staggered back. Gripped his chest on one knee as the searing pain took his breath. His own gulp keened out of unsteady lungs. Gasped air, thick with the tang of salt water, threaded into him and lingered until it was stale.

Another burst of staccato percussed the air above Aderastos' shoulder. A Flesher flung backward from the bullets in his chest, blotches of red began to stain his clothing. Commander Singh strafed in, rifle raised and another burst of cover fire crackled into the air.

"Get up." Rifle still angled at the Flesher, in case he had more life in him than Singh expected, the Commander offered Aderastos a hand. Aderastos took Singh's hand and pulled up. Nearly took Singh to the ground, if not for Singh's foundation. "Open the cages."

"But they could…"

"Aderastos, let them loose! There is nothing worse than dying helpless in a box!" He raised his rifle. The report of the rifle stung Aderastos' ears until he heard nothing but echoed reverberations and the gargled scream of another Flesher. Bestin and two more Black Collars in full tac-gear rushed past, rifles firm to their shoulders with every fire.

"But where are they going to go!?"

The Fleshers roared in their defence. Organized and familiar with the use of violence, they struck to defend their platform. Heavy repeater fire caused the Assets and Black Collars to dive for cover. Aderastos grabbed Singh and shielded him with his own body. the high calibre bullets tore into the muscles of Aderastos' back and he bellowed.

"We won't last with those heavy repeaters!" Bestin dove into a blind spot, changed the magazine on his automatic. Gave two hand signals to his squad. Clive held a piece of the corrugated roof large enough to protect his torso, ripped from one of the cages with little regard to those who screamed inside. Wulf and Isthan gnashed their teeth, pulses raced as they snarled at another of the gun placements on the column. Bending the corrugated metal into a cylinder, Clive hefted its' weight and twisted. Lobbed the makeshift spear. It sliced into the metal of the railgun, as men inside screamed.

The air burst as quicksilver wings flew over the column and its three remaining turrets. Snout a spearhead of silver, Rachel thinned her body to veer at a wide viewing area without glass. The snarls of draconic bites and shrieking masculine voices cut off as abruptly as her presence in the hole. At the back of the turret placement, a strong metal door hung open. Rachel snarled and sniffed at the air outside the door.

More. More enemies, unfamiliar entities held metal swathed in gun oil. Command-shouts from Clive and Aderastos tugged at the back of her neck, their intrinsic alpha-like nature consumed any break in tactical structure. Rachel clacked her jaws. Heard the bone-crunch of ribs under her teeth and shook her head like a dog with a rat. The command in her mind told her to rush through the Column. Find the others.

Find Cillian.

The meat sloughed to the ground and Rachel slithered on her belly, reptilian legs worked in pairs to vault her over the wreckage of her attack and into the hall.

Find Cillian… another odiferous male rounded the corner. Rachel lunged with a roar before he could finish raising his rifle. His body smelled too much of sweat and post-coitus funk and blood to give Rachel any pleasure in crunching his ribcage between her maw. She slashed with elongated talons. Cut ribbons from his flesh and left his body on the ground, consumed in twitches and arterial spray.

Find Cillian… Beyond the copper-tang of blood, Rachel sensed a familiar collective and raced toward it. Down a flight of stairs and flung atop a woman holding a pistol aimed where Rachel used to be. This woman smelled of soap. Her bones broke well under Rachel's jaws. Cillian wouldn't mind if she took a snack on her way to meet him.

Percival kicked the metal hatch off its' rusted hinges. Cillian rushed first into the Column's main corridor, Kelso followed with the ratta-tatta of his automatic rifle. Three Fleshers hit the ground, one slid clutching his neck before Cillian cracked it with a single titanic mitt over his face.

"Now, we have fun together." Kelso grinned from behind his visor, checked down one hall while Percival glanced to the other. If anyone resided in the alternate corridor, they hid. Kate pointed upward, to a set of stairs which curled around the central column toward another level.

"That way." Allard put his hand on Kate's back and moved to go up the staircase. Percival stopped him with a hand on his shoulder.

"No. Cillian first." Cillian loped up the stairs three at a time, armed with nothing but his wits and the nerves which ate at his back. The expendable one, the first to enter.

The last to want to sleep in the dark.

He lashed out at the Fleshers he found at the top of the stairs, a burst of pain in his left shoulder told him of the graze of a pistol's bullet across his flesh. Cillian grit his teeth and bellowed. Boots dug into the side wall and propelled him into the two front figures, guns aloft. A third behind them fired twice at Cillian's legs.

The second bullet found purchase in the flesh of his calf and he howled as if the bullet was invented by the same saccharine aero-dripped dark which awaited his labours in the Retreat. He gnashed his teeth and bit at one of the front-men's throats. Felt the squish of flesh, then spat.

Kelso's gun blared behind him, the third Flesher tumbled into a pile, finger clenched on the trigger of his automatic pistol. Cillian dove in front of Max and Kate, held them in his arms as three more bullets collided with his hip and side. Across the battlefield the Pack howled for their fellow, Cillian's pain a beaded thread upon all their minds.

Percival snapped the gun out of the dead man's hand, and reached for Cillian. Tossed his arm under Cillian's and held him up. He shook his head at Allard's attempt to grab a medipac from his kit.

"Kate needs you. Go!"

The echoed ricochets of gunfire lashed at Allard's ears and fogged his head with clumsy snow. He stumbled after Kate, his own lungs the largest sound. Rushing along, he grabbed her arm and held her back as bullets flew. Pushed her through another tunnel-like corridor in the column of what must have once been a mining operation hollowed out for Syndicate Flesh-Trade work. Kate's mind was singular.

"Holy shit, holy fuck, holy shit, holy fuck, holy shit…" Max's inner thoughts bled into the battlefield, as a door swung open and two Fleshers with a holoprojector playing a form of Cambodian soap opera raised a machete and a pistol. Kate yanked her right hand outward, and the holoprojector whined in an ungodly high pitched shriek.

Pure white light burst in a flash of technicolour.

Both Fleshers jolted.

Allard grabbed the machete. Swung blindly, teeth grit. Something meaty got caught on the downswing, machete stuck in the shoulder muscle and bone of the man who originally wielded it. As his eyes cleared and ears rung, Max Allard heard nothing but his own breath, and saw naught but shadows in negative.

"Max." The loomed form of Percival rushed into the room. Yanked the handle of the machete out of Allard's quivered hand and despatched the Fleshers as Max's eyes began to clear.

"Max…" He stared, ears rung at the twitch to the man's body, the thick and deep gash in his shoulder placed unintelligently by a desperate soldier.

"Maxi!" Kate's communication came through the voice of the Cambodian soap, its' red dressed heroine turned to plead with him, arms out as Kate's were out. Accent thick as the blood which poured upon the cold concrete floor. "Maxi, hurry we have to go. I need you."

He gripped his ears, smacked his palms against them and shook his head.

"Y-yeah. Yeah!" Through another flight of stairs, Kelso led the way to the Command capsule, its' dingy plastic shell was white underneath the excreta of ocean life. A bank of windows gave them their first view of the outside struggle, freed slaves with bits of pipe, stolen guns raged beside the Assets and Black Collars. Defended the children and those weaker humans, who lost their strength behind shields of corrugated metal roofing. Aderastos leapt through the air into a gaggle of the Fleshers, bullets pinged off his chest. Two sunk deep, and his yowl reverberated through the glass.

Kate dragged Allard inside the command capsule, its second-rate pre-Lieben tech grubby as the once-white walls. Wires crossed in tangles beneath yellowed plastic bulkheads, some undersides cut away. Four holo-emitters shone with separate feeds, one a 3D realtime bird's eye view of the floating platforms. One held topographical maps in rotation, shipping lanes and planned routes. A third filtered through news reports from across the Syndicate, a conglomerate of companies, who like the Commonwealth, banded together for their oligarch's near-religious pleasures.

Notoriously grotesque in the treatment of the 'underachiever' or the female, the Syndicate outwardly sneered at Lieben's Havens with the petulance of adult children starving their elders for the inheritance. The Flesh Trade was only one of the abuses the Syndicate committed in the name of the Golden Good.

Plenty for the worthy. Little for the laboured.

Definitions on worthy, or laboured shifted as often as the Saharan dunes.

Another holo-emitter caught Kate's attention. Trickled code fled through waterfall titrations down to the floor in synthetic pattern. Kate stared at the code, and it shifted. Characters changed from Chinese script to binary in piecemeal.

"Okay, okay… now what… now… okay…"

"Max. Shut up." Cillian snapped as he leaned against the wall holding his stomach. Lungs heaved on his powerful frame. "Let her concentrate."

"Yeah, concentrate but don't you think you can take any of this pretty packaging home, sweetling." Kelso pushed his body into her back, lips as close to her ear as his visor would allow.

The code-emitter flickered. Live-time encryption fluttered back to Chinese characters, oscillated into a new rhythm.

"Fucking shut it, Kelso." Allard shoved Kelso away from Kate. The Black Collar grabbed for his pistol.

"What is your deal, Lieutenant Commander?" Kelso's grimace was consumed with vitriol. "I have my orders, same as you have yours. Fucking pansy. Can't even shoot."

"You see that plastic? Did you look at it?" Along the mottled edges of the plastic command centre, seals in relief. Three half-moons locked together in a tangled unity. A single yellow star burned into the plastic.

Syndicate.

"She might need a bit of concentration to decode the Syndicate's redundant encryption tech, you think? Seeing as how it was designed to circumvent Lieben, we might need to give her a minute. Wank."

Kelso kept his hand on his pistol and strutted through the command centre.

"Yeah, well, she and these other bio-machs are going back into the Retreat's sleepy-time when we're done here. No tech comes back with KT-002. Don't like what it might do with it." The Black of his collar absorbed light as he pushed at Percival's shoulder. Stared down at the seated Cillian, who shifted his hand off the wounds in his chest. "That's right, you pile of bio-tech parts. You're going back in your cage and I will gladly hide that key from you, from Allard, the bitch Fridley. My orders don't come from you, I've got higher connections."

"Sure good of the Syndicate to leave a measuring stick on the side of the console here. Care to lope over and show me your big dick? Your orders don't come for another twenty two days. So leave off." Max ran his hand along notched plastic. Opened his rucksack to stuff envelopes and plastic files inside. A floppy disk or two.

"You're never gonna save them, Allard. No bowl big enough for their mongoloid fa-"

Crack!

Allard's fist flung fast enough to daze Kelso and crack his visor's weak point by the helmet join. The Operations man, harmless Max, kneed Kelso in the stomach and cut a tight left hook to the other visor join seam with a satisfying rattle. Kelso's pistol yanked out of his holster, and he shoved it between them, safety clicked off.

"Never said I was a pacifist. Just suck at guns."

"Suck mine."

For the life and longevity of the man, Allard forced himself not to shake out his swollen knuckles. The flash of pain in his right hand made him think he might've busted a knuckle on the body armour. No, he wasn't about to give Corporal Kelso the satisfaction.

As the waterfall of code spilled over its own zenith, once more the symbols and letters turned to 1's and 0's. All across the floating platforms, mag-locks disengaged and the cages Aderastos and Clive hadn't gotten to sprung open, their denizens stampeded away from their captivity. A thin communications transmission threaded to the Ithavoll, trade routes, fleet specs. Commanding officers and enforcer gangs throughout South Asia and into the Caucasus flooded into the sealed Communication's suite on the Bridge.

A single thread coiled through the holo-emitters. Joined all four into one thin purple ring. The yellowed plastic took on a purple hue, as each of the holo-emitters but the waterfall code shifted to a set of amethyst eyes and the android mother who owned them.

Kate's mouth worked in silence. Blue eyes met amethyst, the Mater Machine's gaze focused upon her. Body rigid, Kate's fists shook as even the waterfall began to take on a purple and pink hue.

"The fuck!?" Kelso yanked his pistol from Max's chest and turned. "Hey… hey! Stop that!"

But Kate didn't heed Kelso's anger as she stared into the ocular well of Karnak's sapient machine. The biological machine grown as KT-002 knew nothing but those eyes, the pleasure and relief. That self-same cacophony of sound which drowned her in Agritage returned at a softer pitch, sunk into her bones millimetre by millimetre.

"WOA nelly!" Allard jumped back at the sight, as the thin purple ring around Kate swept upward and tightened, until it shimmered like a crown of heliotropic jewels. "Kate! Katey, hey now!"

Kate's eyes opened wide. She shivered wholesale and with an inhaled gasp, every redundant screen hummed to life. Every dial in the command centre shifted and turned and answered Kate's command.

"Get back!!" Kelso raised his pistol to Kate's temple. Lieben's gaze narrowed in what Max Allard could only interpret far later as open, hostile rage. Kelso's finger depressed the trigger halfway.

A dragon roared into the room, tail whipped through the air.

Rachel hissed as her tail got stuck on the other side of Kelso's ribcage, his heart speared on the poison-edge. His body quivered, pistol clattered to the ground. Allard scrambled as Rachel ripped her tail loose with a flick so fast it cut open one of the plastic consoles so motherboards and wired conduits spilled to the floor in a spider's tangle.

The abject ire in the amethyst eyes expired to a radiant contentment, pouted lips mouthed in a silence which filled the room for all but Kate. No, Max wondered precisely what Kate was hearing in the bosom of the Mater Machine.

"Kate… Katey, shut down the auto-turrets and let's go." He set his hand on her shoulder and felt the hairs on the back of his neck rise the way they did when he and his brother rubbed their stocking feet on the carpet as kids.

"… electricity?"

Outside the window, a seagull with what looked like thick sunglasses hovered. Max heard the click of camera shutters. Dove for the pistol. Raised it in both hands and fired.

Emptied the clip through the window at the augmented sea bird. Not a single bullet even grazed it, and it squawked away, its wings swatted at the air.

"Damn, Max. You really are a terrible shot." Cillian's hand closed over the gun, and gently removed it from Max's hand. "Let us do the shooting, please? You might hit your own foot by mistake."

"I… I panicked. We have to get that bird. And… and what the heck!? Lieben, let go of Kate! She's busy!" The amethyst eyes which gazed at Max Allard seemed to consume him. Tempt and treat him with the promises of matronly affection.

Or titivation, whatever he desired most would be as freely given as oxygen. Kate swayed, unresponsive. In Max's peripheral he saw Clive down on the platforms.

Swaying, unresponsive.

"Aaaannd bugger it all." He grabbed each of the holo-emitters and chased their localized wires to the energy conduit they plugged into and pulled out his multi-tool. The wires clipped one by one, until the final image of Lieben's eyes faded to nothing. Kate stumbled back into Percival and held her head. Wretched forward and shook. Rachel lapped at the wound on Cillian's stomach, dragon-body curled tightly around his legs and hips.

"Right… Percy, go reinforce the Pack."

"No offence, but I should stay if all that stands between Katey and potential bullet holes is you." Percival patted Max's shoulder and watched out the window to the battle outside. The last of the Fleshers were on their knees, hands clenched on their heads.

Clive glanced up to Kate. She nodded once, and every single one was shot between the eyes. Their bodies shoved unceremoniously into the sea.

Arun sat panting over the side of a floating platform, as Aderastos set his hand on Cillian's gut. Two bullets healed out of Cillian, who staggered back into Rachel with a grunt.

"Lieben's coming for the captives." Allard watched the milling people as Ka and Tokaru created living sculptures from seaweed, allowed the natural flora around the platforms to expand and grow into cover from the sun. The people marvelled and shrieked at the floral creatures, rushed to the other side of the platforms, while others prodded the creations and even more pulled the bodies of the Syndicate Fleshers from the column and tossed their corpses into the ocean.

Commander Singh nodded. His eyes lingered on the dog tags in Allard's grip. Kelso.

"One loss too many." Singh griped, as his fellow Black Collars organized the sea of freed slaves. "Move out! Back to the Ithavoll, I'd rather not see a bunch of sanctimonious robots today."

On the trip back to the Ithavoll, Arun sat in the boat, head leaned against Clive's stoic shoulder. They rode in silence, Max most of all. His eyebrows worked under his helmet, which he'd forgotten to remove.

Rage.

Emotion itself. How did Lieben break into a Syndicate hardwire system without traditional wifi or cordless networks? The Syndicate was wily, they knew as well as the Commonwealth did, anything hifi was already open to the Mater Machine's hive-like collective. Lo-fi tech, wires and rotary diallers became the best encryption the human race possessed.

But… his mind returned to the Command Module, and four disengaged tube monitors set into the plastic consoles. Had their techie retrofitted the holo-emitters in lieu of sourcing antique screens? What about the holo-emitters was the way in?

Was it the emitters at all? He stared at Kate, who sat atop Aderastos' lap and glanced back in casual, but full pause. The rucksack jostled his foot as they veered over a wave. A mechanical clank inside, covered by the rush of the ocean around them. Kelso's body curled in the fetal position inside a black bag. All Black Collars stowed a body bag, with their name along the side with the Black Collar motto.

"We kill, we die. We conquer."

Singh asked no questions, nor did he comment on the loss. He said nothing, but held Kelso's helmet under his arm. Once the ship docked, they climbed numb and exhausted from the skip, which folded back into the side of the ship's hull in a flat-pack. Allard waited for none as he trotted to Lieutenant Fridley, and passed her his rucksack. Whispered in her ear.

She nodded and scurried off.

"We're not the only ones in cages." Arun settled with a deep and hissed breath. "Aderastos, we are not the only people in cages."

Aderastos wheeled around and slammed his fist into the metal wall of the Ithavoll. It bent, and Aderastos roared. A slender hand slid onto his shoulder, Kate. Sitting on the deck overlooking the water, Cillian wrapped his arms around his chest and shook. Their two hours were up. Back to the Retreat, the windowless chasm. Rachel padded to him, her draconic form more comforting and far more familiar. She curled round, head on his knee with a hot chuff.

"Back in, time to head back inside. Hurry up! Percival! Ka, let's go." Bestin rounded the others, trotted in Cillian's direction. Get the worn soldiers away from prying eyes. "Cill-"

"Wait." Commander Singh put his hand on Bestin's shoulder. Shared a glance with the Deck Boss, who shrugged. For once, Rammage didn't gloat or interfere. He merely stood, a shadow on the Bridge, and watched them.

"If he's out of the way." Her eyes narrowed as Allard stepped to Cillian's side and sat down beside him, feet swung over the box Cillian sat upon, hand on Rachel's draconic head. Allard passed Cillian a small wax paper wrapped package.

The wax paper crinkled to reveal a cold cut sandwich with a pickle on top. Cillian's nose curled at the pickle, as Allard chuckled and took it off the top.

"Gotta tell Chefy pickles aren't your thing... I'm not the only decent human in this world, you know. Far from." The crunch of the pickle punctuated in Cillian's ears, as the Deck Boss barked orders further off and the Ithavoll's flight deck came alive with a miniature sea of humanity all doing their share.

Kate slid her thumb across Aderastos' cheek and tugged his face away from Cillian's reduced panic. The sea lapped against their microcosm.

Curled into Aderastos' side, Kate's eyelids fluttered as the modified Commonwealth CIRCLETs used for non-essential communication began to pulse in her mind, several of the encoded symbols shifted within her conscious brain.

To 1.

25
the babble of nineva

2092

Lieben ran in any direction away from both the Party and the void of the necklace. Her purple heel caught on a cobblestone. Lieben tumbled into an alleyway dumpster and slid into a puddle which hadn't yet washed the filth of the street down thirsty drains.

"Watch it there, love!" A man's voice behind her, dark hand on pale skin.

"Oy, you're stone cold." Violet eyes hung in her metal skeleton. Within the lenses of his mirrored glasses spewed the evening news, weather reports and the Underground line-ups. Lieben clamoured to her feet and bashed into the dumpster's brittle metal.

"Take a second, love. You're shaken. I ain't gonna hurt you." The chav stepped back in his self illuminating shoes, the latest microweave technology betrayed the translucency of his thick woollen hoodie. It glowed as a taxi went by, water splashed and beaded up on the spot, sucked into a recycle pocket in the hem. A gentle bead of code cascaded into Lieben's: water purification via UV light burst, to recycle even London's rain into a monetized resource. Lieben ran, her leather bootlet left in the junk of the alley.

Away, away.

Her locator warned she was too close, the speed of a human male's run thrown in with the data of London's Police call times. News reports, the pale yellow sparks of NEO units capable of triangulation. There was disparity in the data. Antiquated systems searched for information which no longer existed. Each piece of new stimulus she discovered screamed at her processors, sent the android brain into a frenzy. Lord Stanley warned her to be cautious. The flood, once began, pressed with the force of a waterfall upon the cracks of her crystalline mind.

Piccadilly Square was a burst of illuminated flames scattered by London's wicked sideways rain. Lost in the midst of three story 3D projections and a new generation of Londoners, Lieben's algorithms scratched to a halt from overload. In the heart of London, blue skinned people walked past with yellow eyes, cosmetic surgeons played on cycled 30 second advertisements for semi-permanent skin tints. Appearance altering drugs. Belts flashed with patterns illuminated on clothing built with connection to the wearer's smartphone, which controlled the undulating prisms of light.

Her chronometer told her it was the middle of the night, yet there were no stars in London's sky. Rain dropped on her eyes and each drop scattered more of Piccadilly's electric light upon this frenetic new world. Lieben leapt to the sidewalk to avoid being clipped by a cyclist dragging a rigshaw-like compact store. Languages and accents as foreign as the London sky cascaded from the lips of mirror people who babbled from earrings blazing with caller id symbols and ear canal implants.

The crowd rushed like water molecules in an ocean wave, pulled Lieben along its current, jostled her through the sea of humanity in an undignified tumult toward the Banshee screams of the Piccadilly Tube. Propelled into the Underground's Maw, Lieben's spatial awareness took account of every jostle, every second's touch from a person's hand.

The smells of humanity had once been citrus and rose petals, rich leather and mahogany matched with Lord Stanley's Peruvian cologne. Piccadilly's masses smelled of baked electronics and cheap laundry soap, latex, sex and orange pekoe. Her unshod feet slid against a plastic coffee cup and she took a header straight into a middle aged woman carting two teenage Urban Punks back to the Stylishly un-chic suburban sprawl. Women passed with sexual secrets demurely hidden behind their all too eager eyes, people chittered into the air. NEO-Ns passed in electrical stupor. A programmed void in camera eyes as silicon lips echoed news and the time.

Their babble sounded like Nineva to Lieben's offended ears, a swirling mass of deceptive sinners. London was the first layer of hell. Was Lieben nothing more then a pillar of salt too ignorant to accept her fate? Information flooded, compiled, antiquated wireless connections searched for compatibility.

Time to Restart… Calculating…

She screamed, an electronic undulation of various syllables and frequencies meant to alert authorities who no longer existed. Auditory systems ground to a halt as Lieben shoved human cattle out of her route to safety, and raced through the city streets.

Three blocks from Piccadilly, Lieben's neural net ground to a halt, the constant drone of new information overloaded her quantum matrix. Feet padded in curved angles, gyroscopic systems belayed for more computing power. Instinct pitched her into a recessed shop doorway, Lieben's violet eyes stared directionless into the endless digression of useless information that thudded into her skull like jackhammers to cracked concrete.

A hand touched her shoulder and she felt nothing. A discombobulated voice stroked the auditory canals to request primary function and Lieben shuddered awake from catatonic stupefaction.

"Are you hurt? Can you hear me? Get up, inside. Leave for ten minutes and the world's on fire…"

Violet eyes stared in disbelief as a man bent down on his haunches. His hand draped along her forehead, brushed waterlogged hair from Lieben's proud brow with the cautious touch of the First Responder steeling his spine for the worst.

"Help." Lieben's vocal emitters croaked, wheezed air into her air pockets. The man searched the streets up and down, watched for the goblins, who destroyed the fragile latticework of this Princess' conscious state.

"Hey, hey look here. Look at me. Up, c'mon, we've got to get inside." His accent cooed of the Caucasus Mountains, the eastern regions of the Black Sea. Precious context in the deep drowning liquor of London's new facade. Clumsy servomotors shook her head, joints twitched as if from cold, shorted with the influx of new stimuli. He pushed locks of brown hair out of his own face and whistled. "How did you get here? Can you move?" The right questions, ones of context and recent experience. He stood and scanned down both sides of the street, reached to a pocket in his trousers, before turning back to her. A non-threat in Piccadilly's unfamiliar dangers. Familiar, in Baiko's borrowed connection. A safe haven.

"Don't let Baiko… No.. I don't want… I won't go back." Clumsy hands pulled at the purple fabric, tugged down her sculpted, muddy legs in shame as vocal emitters ground to a stop. His eyes twitched and narrowed.

"Baiko? The hell is… we have to get off the street. I'm going to pick you up."

Lieben moved her neck on its horizontal axis, as the man threaded his arm underneath her shoulders and propped her android frame against him, into the musky scent of a used bookstore. The smell of safety, of Lord Stanley's books and page music. An amaryllis bulb sat in a clay pot. It needed water. Lieben stumbled and reached out for the counter, missed completely. Strong arms gripped her torso.

"Why don't you want Baiko to find you? Lieben?"

Lieben's mouth worked, yet no sound emitted from the android throat. Her knee joints faltered and he picked her up, through the back of the bookstore up a narrow stair to the apartment. Ocular sensors stuttered images of Piccadilly into the warm glow of tungsten light and yellow incandescent bulbs. He set her down on a well used corduroy sofa, and tucked her into a well loved hand-knitted afghan made from the gnarled fingers of an arthritic grandmother. She sunk into the cushions as the countdown to Hard Reset shut down more of her peripheral systems. Motion circuits disengage in 3…

"Leave it to Karnak's machine to hit our safe house the same hour Ego and Sophia go silent… Lieben? Oy!"

Lieben saw nothing but greys as her mouth worked soundlessly until it stopped all together, lips tight closed.

"Half of London's probably mobilized by now." His bracelet blinked Ego and Sophia's proximity, two of the three Idless he'd given his houseboat to on the Thames. Eight years of nothing, then Lieben crashes at his bookstore the same hour Ego and Sophia get made? This was the miracle machine Karnak created? A woman in an evening gown, and jewels which made the inner fence in him water at the mouth?

She was far from a short circuiting machine. Upon her face Demyan dared see trouble and relief, a compounded tapestry of emotions playing across the reprocessing neural centre. Demyan dialled Ego four times, never hit the Call button until he dropped the orb back to his custom command cuff and pressed his fingers on the thin granite of his grey countertop.

"She looked terrified." Demyan pushed off the counter with his elbows and dug into his red laminate cupboards in search of tea. A cuppa, to settle the nerves. Water sloshed in his kettle, warming coils dared the liquid to boil. "Timing's too hot anyway… can't leave now."

Bothered little with loose leaf, he tossed a couple of cheap Earl Grey bags in his worn brown pot and struck the kettle right as the whistle shrilled into the tiny kitchen space. Throwing a tea towel over the pot, he tucked the towel in as he had Lieben and turned once again to face his new charge. Gravity sunk into his chest.

"Shit… Shit!" Proximity alarms rebounded on safety servers, shots fired on the Thames, by the Houseboat. He flicked at his cuff. 7 missed calls from Ego's number.

Lieben tossed her silken head, reprocessing road maps of England. Filtered with connections. Demyan hissed and fed his fingers through shaggy hair, as a news report flicked on his devices: reward from Lord Stanley Hallowes of HTI for the return of his 'ward'.

"Karnak was always bloody. A bloody nuisance. Told Ego this was all fucked from the start. You're fucked, you know that? Fucking Baba Yaga in a party dress. If you spoiled the sanctum, I'll carve your circuits out myself."

The android said nothing, consumed in her electric sleep while the world decompressed and began to make sense by inches. History, geography, news. She was an overexposed relic being caught up to speed, reaching up at the despair of antiquation to re-discover relevance. She was a woman terrified of public exposure. She was a cash cow, an investment, a prize.

The Idless' hope for a mascot, or banner to lead them all to a promised anti-corporatocratic future.

Demyan descended to the bookstore. felt the spines of books in print for hundreds of years, handled by an infinite number of human hands. Economic Projections of 2052 rested on the topmost shelf in the back. Inside the cover and first flipped paper, a tablet cut into glued pages. Another book on Word Process 1997 held the battery pack. He slid the books back in their places and rushed to his counter. The battery pack slid into the tablet, a pale white ball of light wooshed to life in the air above it. He fixed the angle of the beam, and used the lens from his payment register to focus the hologram into the emergency computer built from technological scrap and Karnak's original hologram diagrams.

Although antithesis to Idless culture, Demyan refused to allow privacy to over-reach the safety of his London safe-houses. None of the others had to know about his surveillance tech, the beads for emergency he placed in innocuous places. The feeds narrowed to a view of the interior of his houseboat, where a sparse minimalist home splotched with bullet holes and red spatter.

"Damnation." On the floor, eyes popped open with his cheek half in a pool of crimson blood, laid Hazard Sign. Shot through with more bullets than Demyan cared count. Couch cushions and books were scattered about, kitchen cupboard open. Demyan rocked back on his heels, pulled up his stool and plunked down. He manipulated the timestamp and rewound the feed until two massive NEO-W's shot down the man in reverse. Hazard drew an automatic pistol, clip emptied. Demyan stopped the rewind and played the footage.

The fight was inhumanly short. Two heavy sets of feet, which punctured through the roof and into the houseboat. Demyan tried not to feel sentimental, as he triggered another program and armed the remote detonator, connected to the gas line in the stove. His first and only true home had to burn, a hastily recorded few seconds of his last stand Hazard Sign's only funeral.

Demyan chewed on his lip as he flicked to the same timestamp and clocked Tara out by the docks, a self-driving car parked in perfect alignment in a No Parking Zone. Lovers down the street, Ego struggled with Sophia in one of her fits.

"Dumb fucks, think anonymity's cheap in a place like London. Told them not to come to fucking London." England itself was anathema to the Idless philosophy, with surveillance cams overlapping so often there were a half dozen angles on every street corner or alleyway. Sure, in North America, where tracks of land were still uninhabited, running around in a caravan bedding down on roadside stops, or in the middle of the tundra was a luxury they felt entitled to, but anonymity in London had costs.

Costs Hazard Sign refused to pay. One hologram-globe betrayed the smouldering houseboat, as stage one of the explosive filled the area with gas. Another globe of the dock's exterior rewound. Jittered to a halt. Hazard Sign and two New Rastafarians haggled on the dock, lit by a pale British morning. Bags of herb and a vial of rainbow pastel hallucinogens passed between the New Rastas and Hazard Sign's billfold.

The houseboat exploded in a swath of flame, as Demyan tapped at the laser keys suspended in the air, and stills of Hazard's fuck up zoomed into his command cuff. He chased the feeds for sign of Ego and Sophia, eyes narrowed at the verbal exchange. The third watch kicked in enough keywords to make an educated guess to the threats slung on both sides, but Sophia became Demyan's focus.

"So why does Lieben..." Sophia raised her hand, the NEO-W's shifted from their original target to Tara. She lunged, rolled, and Ego grabbed Sophia. If they were smart, they'd hoof onto the train out of London. Hide in Kent, or further on toward Manchester. But public transportation wasn't Ego's style.

His pathological belief in freedom through lack of labels made getting on the Underground impossible, since England tied banking intel to ID cards in chips around a person's wrist. If Demyan guessed, they were on their way here.

He grit his teeth, slid a programmer's chit into the tablet and the white globes of minimized holographic screens turned pink, then red. Dozens of feeds gleaned off data compacts from the London surveillance grid ran facial recognition for Ego and Sophia. He added a subroutine for Ego's hoodie, the man'd put his hood up. Or Sophia would wear it. Exact height, weight parameters guided the algorithm he'd kept in secret after the Ukrainian military dissolved.

He sipped his tea and rubbed at his face, a chime out on the social feeds to alert his associates to guard the block.

The counter around him was in a perpetual state of untidiness, pamphlets scattered across its chipped wood surface like aphids on a tree. One by one he cleaned the stacks, ordered magazines by date and deleted expired posters off his plasma message screen. A single poster caught his eye, another in the long line of independent concerts hoping for a pound of flesh between them. This one was dressed in the bottom left by a simple blue and white logo. The International Justice Society fought modern age slavery, provided legal services for the Wage Slaves of the Americas' endless fields of loan sharked impoverished. The lead singer of the Avid Eagles sported pure white hair, his eyes shone eagle yellow to pair with a long, curved avian nose and pale skin. In his taloned hands he held an Atlantic farmed Salmon, Norwegian flag encircled with the red stripe sported by anti-smoking posters worldwide. Demyan was so proud to pay for the Avid Eagles posters, his contribution to the IJS's campaign. Now fodder, the bookstore owner flicked its thumbnail to the bin.

His ping for Ego and Sophia pinged back Ego's chime… but why was Lieben there? London's surveillance grid pared back a section of pings on her location back to the HTI scraper. The same android they scoped their hopes on ran both straight for them, but away from Sophia? He grit his teeth and shook his head. Did they ask Lieben if she wanted to join them in the first place?

The only movement in the stillness was the rise and fall of Demyan's chest. Bills nagged at the back of his mind, waltzed with the figures in his bank accounts as they dwindled into overdraft month after month. Promises from the Idless on currency transfers, or cash grit his teeth. One good week of sales, that was all he needed. One good week. While the bookstore made an acceptable front for his pursuits, Demyan clung to it. A holy distraction in the gripped devilry of his nights. Proof he wasn't the funnel spider the military made of him, back when flags bled more lives than mothers held after horrid labours. Bergamot steam pressed his eyes upward to the ceiling.

It was a machine, algorithms and alloy wrapped in artificial skin. His mind faltered, he reached to his cuff, pulled a holographic communication orb with a flick of his finger. The screen burst to life in front of him, a call to Sophia, one breath of her voice and he'd know what to do.

Thirty seconds later the screen dimmed, inactive.

Returned to his wrist cuff.

Demyan waited for signs and wonders in this technological age, called up testimonies of church in Kiev. Signs and portents were hard to see, when the birds were gone and London's lights destroyed any chance at seeing the stars. No more pings from Ego.

"It's a machine. A moving sculpture, not a woman." Yet Lieben's fear nagged at his mind, the way she jerked backward in defence of her dignity. He'd never seen any NEO-N be so timid.

The technology section in his store was little more than three shoebox spaces on a shelf, but tonight it was a bastion as Demyan searched the books. Bypassing Asimov's I Robot and Gibson's Count Zero, Demyan took down a biography of Dr. Dieter Karnak, published posthumously by a marketing department, which struggled with healing his bespoiled reputation. The first chapters were tales of childhood in the outskirts of Berlin, school in the German System, university in Tokyo.

'It is given to humanity to create life. Why limit God? We are but carbon and oxygen, cosmic offal wrapped in bone and muscle and sinew. We replace shattered bones with metal equivalents, we have artificial organs and the belligerent toll of time. What if immortality is not in an infant's cry but the wide eyes of the android? I will not stop with making a model of the human body, a life sized children's toy. My machine has life. We cannot stop until we do more than build computer chips and voice simulators, it is our duty as pioneers to slip mercy, loving kindness, generosity, and strength into our progeny.' Spin doctors encircled Karnak's Kabbalist fanaticism into an engineer's passion, but the truth behind the words was unfolding to Demyan's hungry eyes. Dieter Karnak attempted to create life.

He shut the book and flicked off the lights, brought his cold tea up to the flat. Shut in with Lieben, Demyan thrust his tired body into the coils of the second-hand chair and exhaled.

"What's to be done with you? I wish you'd wake up, if you can wake up. If you do wake up promise me you won't kill me in my sleep? I would like to stay alive. Ego… he knows where to go. Maybe I give him too much credit. Hey, talk. You talked before, damn machine."

Legs to her chest, Lieben rested supine on the sofa. Blonde hair dashed across her face as if placed there by a divine appointment to shade the holiness of Moses, the face which saw its God. A sigh and he picked up the blanket, folded it onto the back of the couch. Jerked back as Lieben stirred, stretched out in examination of her joint servos. Little more than a casual diagnostic, its reality struck Demyan awake and sober. A sleepy coo shyly echoed from her lips. She cradled her head beneath her arms and nestled into the back of the couch the way Sophia used to, when she was here.

It was the most beautiful solitude, a restful honesty built on the whims of a man who with every ounce of himself wanted to be God's Engineer. Karnak died in giving Lieben life, it seemed absurd in the midnight hours that any man might die for a machine, but watching her nuzzle into the pillows, Demyan came to a tentative understanding with Dieter's ghost.

"I wish I knew when you'll wake up. If you'll wake up."

Pert lips released, curled around syllables which elicited from her timid mouth.

"Update Installation 13% complete. Remaining time until restart seven hours, forty seven minutes, twenty seconds." The voice droned, disambiguous and oddly lofty. Lieben's voice deserted the room. Demyan waited for some other knell to toll, but none came.

"Right… Ego… I should check on… Ego." Again Demyan paused, the holographic globes of his computer hovered around his head, showed potential matches without 100% clarity. Nothing but the casual pings of bored associates outside.

Those jewels on the android, though… he shook his head and trotted downstairs in his bare feet. A police report of the houseboat fire shone in one screen. He grit his teeth and picked up a communication orb to cut another deal with the investigator.

"Put it on your tab."

26
an anchor for her data

2092

"Nngggghhhhhhhh." Tara whimpered and stamped her stocking feet as the NEO-N sanitized her left bicep with a liberal spray. "I swear, Robbie, this is absolutely it. It! My NEO-W's shot me! They turned around and fired on me! Me!"

Robert Dunlevy guzzled a suspiciously sized beverage in view of the holo-emitter in his Vancouver lab. Time zones being what they were, the dark crescent moons under both eyes halted another attempt to make the man feel worse.

"They turned around and fired at you?" He tapped at keyboards she couldn't see, three of his programmers buzzed behind him. The lab was a web of multiple spiders spinning silk, the data from London's Lieben glitch a fly they as of yet couldn't stick down. "At you. By the docks."

"Do you see this arm? The-ow!" Tara smacked the NEO-N across its' face as it poked the suture needle into her arm without so much of a word. "That hurts!"

The NEO-N continued to stitch her arm, no apology or attempt at change.

"I hate these things."

"'Cause you can't flirt with them?" Robert tapped on, shook his head. "Tara, I have no data."

"Wh-what? But... everything is backed up. Check the back up of the back up. Heck, check the redundancies!"

"Awwh, lil Tara learned a smart word." Robert's glare fixed on a person out of frame, he barked orders and the whomever-it-was gave a hasty apology and skittered off.

"Fuck you, Robbie. Your androids almost killed me."

"Pity. I'll work on their target accuracy once I get them back. Where'd you leave them?"

"You... you can't tell?" Tara grit her teeth and growled at the NEO-N as it plunged the suture needle back into her flesh.

"I assumed they were destroyed when they didn't make their ping-in or recharge mats." Another sip of the damn beverage, Tara wanted to rip her hand through the screen, grab it and take a sniff. Pity teleportation was a thing of tomorrow, or should be according to conversations which vaulted so far above her head she needed sunglasses to spot their shadows on the surface of the sun. Quantum, quantum. Schrodenger's pet store, or persimmon, or tangled quantums like hair. But not.

"No. I did not bring the murder machines into my getaway car, you limp breasted testosterone maker. You saying you can't find them?"

"I can't find half the NEO-N's in London, either. Some weird series of blips numbed them all out, as if there's some catastrophic cloud-server shut down. Do you think I'd look this tired if everything was peachy? Fucking silicone migrated to your cerebral cortex."

The last stitch tied off, the NEO-N reached for spray antibacterial. Tara shoved it away, the machine tottered into the wall of the small hotel room and whispered gyroscopic error warnings.

"How..."

"If I could tell you, I would!" Robert gripped his thinning hair and growled. Leaned his elbows against the work surface and shook his head. "Even the CCTV cameras around London are complete wipes, just the same ten seconds over and over. Whatever happened down there, I can't even bring up the dock. Has the imprint of the same hacker who cleaned the Embassy torch in 82. Sixty five civilians dead, not a single image. Not even satellite. Same static on the ten second repeat. If you have any ideas in the hairspray cloud which is your brain... I could use them, Tara. I really could."

"Gee. You... huh. I... don't like you being human Rob." Tara slapped a large skin coloured bandage over her stitches. They clung to the connection, each silent in their own way. Robert's eyes blindly searched data tables of the anomalous behaviour. Tara searched his frenetic gaze. The slope of his shoulders and huff of breath. Slowly, nothing remained but the gaze between them, rivals and unwilling cohorts, whose work alone kept them both from the void. "Ego reacted. Baiko was nutter-buts, but Ego reacted."

"Huh?"

"You're asking for help, Baiko was in some sort of seizure slump. Until she did the hand-thing, she was barely conscious. But the second I mentioned his legal name, told him the kids missed him... he reacted." She leaned against the side of the hard hotel sofa, rubbed her filled lip with her index finger. "Bobert? I don't know fuck about machines. But I know people. Ego's lovers, the trans and the house frau we caught in Northern Canada. We still have them, right?"

Tipped back in his seat, Robert looked off to one of the nebulous workers. A muffled series of words Tara couldn't make out followed, then Robert nodded.

"Yeah. Whole family's still here. Katherine, Fester, the kids... they're all in Vancouver."

"Katherine?"

"Mog took a legal name. Did it for the kids, when Viggo got sick a bit back, he couldn't be seen by our physicians without a medical services card."

"Liar. Our physicians don't claim MSP." Tara gritted her teeth to cover a sneer, tapped at her watch. "Katherine Kombedjian... awwh. She took his name... Could it be Lieben? Rob? Could the hacked NEO-W's, the missing NEO-N's be our missing mech?"

"Honestly? If it is, I'm a whole lot more worried. That level of override, it... it's... the encryption we keep on our proprietary software for the NEO's is insane. A limited AI based on Karnak's research, with a consistently altered algorithm to... and you're glazing over."

"Baby words. Me no speaky quanty nerdy."

"It's fundamentally impossible to influence all the NEO's at once in a certain geographical area. They're not connected beyond the cloud they share to upload and download their dailies. Something that could influence separate running programs, like the hack on your NEO-W's, I didn't... I didn't think it was..." Robert's face went blank. He did know it was possible. "... Da'at. It would be Karnak's style, have some... android shaped master switch."

"Okay. Yeah, we saw what Da'at could do. What Baiko did, if that was her and not some form of freaked out recording. So... play with me."

"Huh?"

"People me! People, not tech or algorithms, people with me."

"Aww baby learned a word."

"Do you want my help?!"

"Sorry! Sorry, I'm... yeah. Yes! Yes, Tara. I want your help." Robert covered his face and leaned back in the chair. "How do we play?"

"We get out of our own heads and into theirs. Karnak invented Lieben, with your help. You're a competent robotic engineer, but you're no programmer."

"Gee, thanks."

"Realism, not ego. Karnak made something nobody else can touch. So, what was it? What did he want?"

"To be the best."

"No, that's what you want. Think outside you. You knew him Robbie, you knew Karnak better than almost anybody on the planet, you spent every day with him for years. What did he want?"

"I... I don't know!"

"Yes you do! Think! Anything! Did he eat M&M's!? Dark chocolate? White chocolate truffles, what!?"

Robert grit his teeth, pushed off from the stool to pace the room.

"He hated chocolate! Ate like a bird after... after his son and daughter in law died..." Robert paused, furrowed his brow and looked back. "There... was a picture missing. A picture on his desk in the Cloister. The daughter and his... he went nuts with the rocks when they died. It was a car accident, I remember, there were safety reports everywhere. He never drove again. He wasn't driving, they were in freaking Germany, but he never got behind the wheel of a car."

"Crystals. Metaphysical crystals... Honey pie, did you ever stop to consider his fascination with crystals wasn't some form of spiritual bullshit?"

"Pfft." Robert shook his head, a man of hard science and plannable results. "... Not possible, you don't understand science, what would..."

"There. Right there, in your eyes, what is that?"

"Nothing! Nothing, he was a quack, he used our resources to..."

"We're playing a game, in this game, anything we can think up is possible. So, I'm gonna repeat myself. What if the crystals weren't mysticism? What if they were science?"

"Then... they could be used for data storage. Like quartz and ruby drives. You can engrave on them, graft at the microscopic level. Crystals are nature's straight lines. They... have..." The stool took his weight with a creak. "... we fought about it. There's this thing in quantum computing called decoherence. Data in a quantum computer system only lasts so long before it degrades. Ah, go away. You have to anchor the data. If he figured out a way to contain information in a quantum computer system, the amount of calculations Lieben could do per second is... staggering."

"Staggering like hacking everybody's cell phone to call ambulances for a bullet hole?"

"Yes." He hissed.

"A bunch of NEO's go active but offline, the same day word of an exclusive party with Lord Stanley Hallowes gets the Chairman's mother out of her cocoon, and we're only partially concerned this isn't Lieben the wunderphone?"

"Party? What party, why does a party come into missing NEO's?"

"Because, brainiac, I was supposed to be there, before I got a proximity alarm on the necklace. Figured, eh. Lord Stanley's a member of the Board, if the Holy Mother is going to the party, it's safe enough... going to get her now." The vehicle Tara sat in came to a stop, she exited into the London night, smoothed her gown. "Do I look okay?"

Tara stared into the camera, mascara smudged under one eye, her lipstick smeared in one corner with a bit of dried blood. Usually perfect hair shifted around her shoulders.

"You..." He cleared his throat, "... have mascara under your left eye. Bit of... is that lipstick or blood? H-here."

He pointed on his own face, nodded when she gently fixed her lipstick, and the under eye with pressed powder from her purse.

"Always wear bright red lipstick, it covers a multitude of sins. Brings a body's gaze to the lips instead of the eyes. Push your breasts to your chin. Wear deceptively comfortable shoes. ABA, baby. Always Be Armed. A woman's armament includes warpaint and silk."

"Be careful."

"Aww, you goin' sweet on me, Bobert?"

"It would be too inefficient to train a new Tara after this long with you." Robert attempted a sardonic smirk, instead he choked down the lump in his throat and took another deep breath. "If Karnak succeeded in creating a miniaturized quantum computer inside Lieben's classical neural net, it's the most sophisticated computer of our age. Nobody, none of us have a prayer of hacking into the system. Not even China, and they've been at the quantum computing game longer than any of us. A machine, let alone an android with that kind of power..."

Tara put bullets in her magazine one at a time, as she trotted toward the HTI tower. It loomed above her, a twisted metallic tree shorn of limbs and branches by the necessity of height.

"You mean there's an android out there capable of hacking the planet?"

"Other than a few AI guarded mainframes and databases, or the redundant tech that has no capacity to connect? Yes. The encryptions on most economic platforms would be wet paper to it."

"A walking economy killer. Awesome... Baiko had the necklace. I'm betting she's still in London. Just gonna peek into the party, then I'll get back on my search."

"You want a couple more NEO-W's?"

"After mine tried to shoot me?" Tara chuckled and slid into the elevator, tapped at the glass pad. "... elevator's not responsive... Yeah, maybe I will take that back-up... hey, do me a solid, send the Black Collars as backup after the necklace's latent frequency will you?"

She flicked the camera from a side view of her face to the glass pane which controlled the elevator.

Nothing but four silver letters on navy blue.

D-A-A-T

"Well hot fucking damn."

Ego shoved Sophia's comatose body into the booth at the back of the Armenian coffee house and set his hoodie over her. Less than ten minutes after Tara's hit, the streets swelled with NEO-W's and NEO-D's in a prevailing search pattern. Hoods were lifted, hats yanked off heads without a care to privacy or religious affiliation. When a NEO-W yanked a woman's hijab off amidst screams of protest, Ego knew the lockdown would damn them.

And him with an unconscious woman on his back. Sophia jittered in her 'sleep', something Ego mumbled about as a drunken fever, when he sought refuge in the last semi-safe place he could. His mother tongue got him into the back room, a small pot of coffee his reward.

"Pick up, Demyan. Fucker, you promised." Ego's phone rang out, no friendly leave a message, nothing but the finality of a soft click and dial tone disconnect. He drank the thick black coffee and let the sting behind his eyelids keep his body from following Sophia's unconscious stupor. Idless avoided London, and he would too if Sophia's urge to visit weren't so strong. Turned out she was right, Lieben was here.

"Your woman." A man whose name eluded Ego nodded to Sophia, the zippered hoodie's hood slid far enough off her head to show Sophia's half-lidded eyes. "What did she take? I won't have a girl die in my shop."

"No, she's sick, she... she had a seizure, and she'll come out of it in a while." He lied in Armenian, his inflections those of a child whose use of the language dulled with age. Mog spoke only English...

Mog. As Demyan's number rang out, the sense of loss which chased him since the northern highway in Canada caught up. As long as he had a direction to move, the thoughts of Mog, Fester and the kids couldn't reach him fast enough. They always landed a day after, in flimsy seconds before Ego thrust in another direction, another scanner recalibration which brought them closer to Lieben.

Closer the goal. Sophia didn't have the same amount of time, he told himself. Viggo... his son's name. Tara called him Viggo.

Fester and Mog'd given him a name without him. Ego's nose stung. He breathed deep, fingers flexed in half-fists, elbows on the table.

"If she…"

"She isn't on drugs!"

"Then take her to a hospital, I won't have a dead girl in my shop."

"Just!" Ego barked, one hand shook, muscles taught. His pistol rattled in his shoulder holster, hidden under the button down from his bag. "Let me finish my coffee."

The shop keep chewed on the inside of his cheek, eyes flickered from the woman to the man slumped in front of her in the corner booth. A low battery sign flickered on Ego's phone.

"You keep calling, old model like that, charge doesn't last long."

"Friend of mine said he'd be here. Didn't show up."

"To help your not drugged girl…" The man sniffed, and retreated to the bar. He came back with a plate of almond cookies, and some baklava, a remote charger in his other hand. "If it's the Syndicate you call…"

Ego grimaced and rolled his eyes up to the man's face, slowly slid the plate of confections closer as he set his phone on the remote charger. A beam of light burst through the front windows of the cafe, artificial and blue. Ego hissed and dove in front of Sophia, back toward the shoulder mounted lamp of the NEO-W as it scanned inside.

The few patrons of the cafe gasped and yelled obscenities at the machines, none brave enough to flip the blinds.

"It used to be thugs, who searched cafes at night. Corrupt policemen working on the Conglomerate's dollar. My grandfather told stories of his family being run from Iran, after the Armenian Genocide in Turkey. You are really Armenian?" Eyes semi-clouded by cataracts searched Ego as he slid back in the seat, pulled Sophia's hood over her eyes against the light. "What is your name?"

"Farouk." Ego's shoulders sunk, as for the first time since he was a child watching his mother run into the Canadian Consulate, he admitted the truth. "My name is Farouk Kombedjian, and my companion and I are trying to leave London. Our contact is a man named Demyan, he owns a bookstore, near the Museum of Natural History."

Ego would have kept going if not for the glower of recognition in the man's face. He cautioned against adding more, one hand retreated to the pistol.

"Demyan… I will get a driver. Not now, wait here until after three am. The NEO-P's focus on recharges, during the least criminal hours of the evening. That is when we shall move. Eat your plate. Eat! Eat up, Farouk. If you are friend of Demyan's, you do not have much time."

27
Max Allard, that Max ... him ... can I shoot him?

2156

The door to Rammage's office did not so much as knock, but rang with a fist so massive Rammage wondered if it would punch straight through the metal.

"Enter." He glanced to the pistol under his desk and knew it would do less good than a sharp tongue to the being, who pushed the heavy metal hatch open as if it were a form of cardboard. It whined with grief. Aderastos stepped through and slammed it shut.

"Aderastos." Rammage attempted to steady his lungs.

"You didn't call me AD-001."

"I hear you like names, now. How did you leave the Retreat?"

"No one names their child with a serial number."

"That?" Rammage held up a finger and gave a wry side-smirk. "Isn't exactly true… but I concede your point. For what reason do I have the illustrious miracle man in my office? One who seems to be miraculous enough to leave a locked compartment."

Aderastos shook his head, lips grim. He refused the fickle dance of half truths humans, those damned Homo Sapiens, seemed so taken with. A hint in Aderastos' eyes, the clarion trumpet of ill dreams sunk Rammage back in his seat.

"… you were created to serve. Not be autonomous."

"Something went wrong."

"You said that, I didn't. Drink?" Rammage shifted toward the liquor cabinet, knew he'd a better chance of running if the open hostility unveiled a more violent repercussion. He grabbed the wine.

"The whiskey. Neat." Aderastos looked at all three options of chairs and leaned his bulk against Rammage's desk. Poked at the files on the smooth desk. Personnel files. Rammage growled and set down the wine, instead took the cork out of his whiskey. "You thought it. Part of the report you write for the ones above you, is that not correct?"

"Is that not correct? Asking for whiskey neat? Where did you learn to English?" He poured two glasses, held one out to Aderastos. The Asset's fingers wrapped around Rammage's entire wrist, engulfed him like night engulfed a quiet room. It was then Rammage allowed himself to recognize how gargantuan Aderastos was. All eight feet of him confined in the office meant to be as spacious as a naval vessel allowed.

"Your crew is loud." Aderastos swirled the whiskey in his glass, and sniffed to cover the lie. The air around his nostrils exploded with peat, vanilla and smoke. Hints of aromas he held no words for. What words he kept came as fuzzy recollections, or dreams of foreign vocabularies in conversive motion.

"You will not destroy my fellows."

"Who's saying to destroy you?"

"Please, you festering cancerous sore. Lie to your betters, not to me. Allard believes there is hope of rectification."

"Allard is an idealist."

"Allard is the reason the Ithavoll is not currently at the bottom of the sea, your bloated corpses bobbing in sealed compartments. Do you think metal walls and machine guns are enough to keep me out? Do you think anyone I touched would be feeble enough to die like that?"

Rammage leaned against the wall beside his liquor cabinet and sucked half the glass down his gullet. He fought with a sneer, attempted to keep his face as neutral as Aderastos'.

"What was wrong with Allard?"

"He was going to die." Aderastos sighed and shook his head. "Another seventy years, his rhythm would be gone. I fixed it."

"You.. what?" The glass in Rammage's hand jittered. He slammed it down on the cabinet's top and shook his head. "You fixed... you made my idiot officer immortal? That guy? The bastard, whose claim to fame is saving a spider and sneaking a surfboard onto a battle ship? How..."

"As Kelso endlessly mumbled, my Assets and I have eighteen days before your uppers declare our lives moot. Failed experiments, the term you've already written on the file. A risky stratagem for a man who cannot take another professional failure. I propose we combine our efforts. You may keep this ship as your personal kingdom, and we shall not deny all sensible requests."

"You made Max Allard immortal."

"… And you allow Kate to monitor the tech-waves for Fleshers. We get to glut ourselves on the glory of battle to exterminate those who would use cages against their lessers and equals, and you get to take the data you require to create more of us."

"Max Allard. That Max. The… Max."

"Come now, Earl. Be a proper conversationalist." Aderastos sipped his whiskey, and waited while Rammage blinked away the shock. He'd been prepared for threats, postured stances and impassioned speeches, but immortality? "We require skylights. The ability to come and go throughout the ship as we please. In exchange, Ka and Tokaru will continue to provide fresh fruits and produce for Chef, while Kate and Clive will aide your Bridge. The Pack will partner with your Black Collars, and engage in any attempt to halt the Syndicate's Flesh trade. Protect our interests to your precious Conglomerate, I will allow the study of Max Allard's blood.

You not only stand to lose your crew, your ship, your precious rank. Eventually that cancer will continue to chew away at your prostate until you die languishing in a hospital bed. Dead weight, you will be forced to seek disgraced salvage from Lieben. Who else would have you, a sick aged man. Will the Conglomerate continue to wet your brow, when you're the failure? The one who mismanaged a multi billion dollar decades-long experiment? We want the Syndicate. A group I know you compete with, and your crew desires to fight."

Aderastos leaned forward, his empty glass thrust between them.

"Do you want to live forever?"

The world paused.

Whiskey sloshed into Aderastos' glass. Rammage found the nearest chair and plunked into it, drained his glass until Aderastos went to grab the bottle. It seemed to shimmer at his touch, an opalescence in the inebriant liquid. He poured it into Rammage's glass.

"Allard will handle the details. Ka and Tokaru are converting an area in hydroponics to root. Drink up, Earl."

"Can I shoot him? If I shoot Allard, will he get back up?" The whiskey tasted sweet on Rammage's tongue, a fitting mouthful.

Aderastos paused at the door, eyes bore on gunmetal grey.

"I would not advise…"

"Didn't ask you for advice. If he gets shot, or injured and doesn't flinch…"

"I made Max immortal. Not impervious to harm… if I'd known how violent you humans are to your fellows…" Aderastos turned one last time to face Commodore Rammage, a deep and growled snarl on his giant's face. Rammage's stomach flipped. He set the glass down with a slosh, eyes absorbed into Aderastos' apoplectic gaze. "… I might have tried."

"Why don't you?" Empty bravado, slick with the insane desire to collect the last thought and keep it, to keep that sense of command he had with so many others, drove Rammage. "We'd hunt you down, but what does it matter?"

"Nihilism looks ill on you, Earl."

"Stop saying my name like you earned it."

"I earned my own. That gives me every right to yours." His hand gripped the side of the metal hatch and yanked. It wrenched from its moorings and tumbled with an echoed clang down the hall as Aderastos stomped back to the Retreat. Earl Rammage peeked beyond the ruined portal, to see his staff slumped in various stages of sleep.

Before long, Ivan Letopaxa silently repaired the door.

"Again!" The voice snapped, garbled by the potato crisps gnashed against the bulbous woman's yellowed teeth. She leaned over her desk, popped potato crisps into her mouth and gnawed. Mouth open, she sipped cheap red wine from a mug with a faded logo on it. Slippers half off her chipped-painted toenails. The intern shuffled the pages of still photographs to the beginning, and droned.

"They hit the rig twenty six minutes after their recon agent landed. Full tac-team, and…" Aerial images of the Flesher rig flipped with the unsteady swipes of the intern's palms. "… Chao Bo and his crew were thrown dead into the water."

The woman yanked grime covered spectacles off her face, and searched for a clean spot on the stretched Phuket Muai Thai gym t-shirt she wore over loose pyjama pants.

"One day. One day off in forty five and you barge in at four in the morning with seagull snaps from a Flesh-rig in the Pacific?" The intern's jaw worked as he slid the next photograph to the front of the pile. He pulled a rolled paper report from his pocket. Data sheets from the rig's computers, with lines of hole-ridden paper tape on either side Aida used to make stars out of and place in jars.

"Chao Bo was a pig, who used his cargo for his own pleasure." Aiya grimaced and grabbed a cleaning kit from a side drawer and scrubbed at her lenses. Slapped the spectacles back on her nose. "I liked him."

A ruler and the photo scale had Aida scratching notes on a piece of Manila envelope, her grimace deepened.

"Is…"

"Two and a half metres. I pulled the seagull, the reports were all we could salvage from the servers, before…"

"Before?" Aida yanked the photos out of his hand, splayed them out and grabbed a magnifying glass. Black Collars with guns, a flash across the photo, like someone shifted the colours. Giants in stillness between Chao Bo's men. Corpses shuddered in jittered lack of focus. Aida swore in Cantonese, the way her great-uncle did when her parents weren't cloistered over her.

"Pull the seagulls. Fix the fucking optics, I cannot see shit for the blur."

"That… ah, that blur is a person, Ma'am."

Aida blinked, stopped short and peered down, fixed her glasses and looked again.

"Show me." The intern circled with a highlighter, the route through time stamps, images corrected by film process and the high speed sparrows they launched from an oil rig nearby. There in triplicate, was Arun.

"Bring up the film."

"We can't." Shoulders coiled inward, the Intern flinched from the swat against his thigh. "I… it's why I woke you, ma'am."

He set down a thinly coiled set of negatives. The microcosm of Aida's office bathed in crimson as the Intern turned on a red incandescent lightbulb. In the negatives, a blonde haired woman with Japanese eyes convulsed. Screens broke, and Aida paired the negatives with the coiled roll of ruined printer paper.

Each computer screen, one by one, turned to images of the Mater Machine's face. Numbers swirled in holographic overload, as the woman who stood in the middle of the room, surrounded by giants and men seemed to concentrate.

"Bo, you dead fuck… Who requisitioned the…"

"We can assume they got all of it. Every piece of the Pacific Stem. The Steppes Stem and Ginko Road Stem remain secure as far as we can…"

"Who else knows?"

"Kept it between us."

The Pacific Stem Codemaker stretched and the hem of her t-shirt pulled up toward a sagging breast. She rubbed her face, pulled another cigarette from the pack and lit it. Let the tobacco and carcinogens purl into her chest.

"Notify the Caskmaker to prep more seagulls, some albatrosses for surveillance. Send out the schooners, tell Ginko to push his stock, we're emptying out the PAC rigs. I'll work a patch." Aida pulled out two massive binders of code. "Say a word, I'll feed you to the Banshee."

Lon fled limp chested before Aida could notice the stream of piss down his pant leg.

"I'm sorry Commodore, I cannot fathom how he did it." Phil Rykstra leaned back in his desk chair, stared at the little green dot of the surveillance cam, which tied him, Desbiens and Rammage in a trio. "There's no damage to the crew, nobody remembers seeing him, and they should have. There were seven security checkpoints he passed to get to your office."

Rammage swirled a glass of whiskey, his other hand clenched on the arm of his chair. The man looked… haunted. Phil shook his head and rubbed at his face, being roused to answer for an eight foot behemoth's adventure to the Commodore's office was not his idea of morning. The more Phil dove into the Asset files, the more it grated. Why did his predecessors use genetic material from… well… dissidents?

How did Aderastos get off the ship in the first place, without breaking any of the security measures? The fact he'd done it twice…

"There's got to be a reason. Two months ago, AD-001 didn't know what a door was, beside a bit of wall. Our aero-drip kept him so dumbed down he was my personal marionette. Twice now, despite our measures, he's gotten out. Broke my god-damned door." Rammage took a sip from a sloshed cup. Let out a hiss and shook his head. "Desbiens, you and Lao had surveillance measures on 24/7. The fuck?"

"None of my measures are tripped. Even the deadbolt wasn't broken and Lao is the only one other than myself who has a key. I changed the tumblers since the last escape, heck, the handle is too small for his digits. Literally too small. Look." Desbiens shifted his glass-cam to the first of three hermetically sealed chambers of the Retreat's containment system. Pristine metal fit together, each cell as pure and bespoke as the day the commission was finished.

"Phil, the Board wants to know what the heck did it, before they nix the entire project. We can't build more self-aware monsters, until we have perfect control. So… as crazy as it is, what are your theories. Anything. I'll take the moon being shattered by a purple people eater at this point."

"It's crazy."

"Eight foot GMO monster walks into my office on the tippy top of the ship, without causing damage to structures, many of which are too small for it to fit through. Comes in, calls the whiskey by name, sniffs it like a pro then threatens… Unpleasant things. I'll go for crazy." Gaunt, that was the term Phil dug for, Rammage looked gaunt. His uniform looser than months prior, but that was foodstuffs, right?

"Any chance they'll give me the unadulterated case files?" He bustled about his sheets of paper, legions of notes in binders and file folders scattered about his desk and on the walls. Rammage shook his head. "I've got a theory… the Assets were created by DNA splices between mammalian, reptilian, sometimes flora donors. But, there's Homo Sapiens in there, too. More than we were originally told, and like their bi-pedal forms, it seems to be present at a greater rate than the other materials. None of the genetic splices account for an ability to modify DNA, or heal. But Aderastos does it on an instinctual level I, again, have no account for."

Phil Rykstra was an intelligent man. A genius in the realms of genetic engineering, so why did every minute in the Retreat feel like an elementary school child in the middle of a post-graduate lecture on quantum mechanics?

"There're theories on genetic memory. Epigenetic strings, which are built not through DNA strands, but through generational nurture. A family known for the arts playing a lot of classical music, and having artists as kids. Or evidence of mental trauma in grandparents being expressed through their grandkids. I need to know who the donors were. I want to correlate the Assets and their abilities with their genetic sires and see if there's any link. 'Cause… the only name I've got to correlate at all is scaring the shit out of me."

"Oh? More than…"

"Yes, more than an eight foot behemoth knocking on your office door without a single broken panel." Phil yanked his hand through his hair. Released a held breath with a hiss and grabbed the corner of the one page which mattered in his sea of print. "Baiko Kaho. A Conglom intern, who fell off the map."

"So? Who gives a shit about a failed intern?" Rammage shook his head and shrugged.

"Dr. Dieter Karnak's intern, the one who helped create Lieben, then ran. Why the fuck would our Conglomerate bio-engineers use the DNA of the woman who killed Dr. Karnak, to create a baseline for Asset shells? Isn't it funny to you? An entire coastline to choose from, and AD-001 hones in directly on the only coastal Haven for four hundred kilometres in either direction?"

"He's not…"

"No, the closest genetic link is KT-002. Now, you tell me, Commodore. Isn't it funny the same person who was instrumental in creating Mommy Dearest, has genetic offspring which can hear Lieben's call?"

"Give him the files." A soft feminine voice whispered through the comm, silken and clear.

"Who the living heck!?" Phil jolted as Rammage cut off the comm, and a cylinder of vacuum sealed papers jutted through his mail slot ninety seconds later. He split the packet open with his utility knife and flipped a few pages.

"Oh… oh god."

28
separate phylogenetic burning bushes

2156

    Much to Kate's disappointment, the only secrecy in Lieben's world was antiquity. Smoke signals replaced wifi, human chatter replaced the techno pings and radar signals. A dichotic blend of analog and digital became the divisive line between those who believed the Mater Machine's call for mercy, compassion and shared resources and those who rejected a great mother for their own ploughed fields.
    Humanity could be guided, and also guide itself.
    The surveillance system required only a few flicks of her wrist. The continuation of a docile pattern and the methods the humans used to separate Retreat from Crew were as moot as a smoke trail on a rainy day. Who would notice a vine the colour of the walls slither down the corner where the wall met the floor. Lay flat against the door's locks and grow into the tumblers? A lock had a key, and keys were solid material, solid as ebony or teak.
    Plants Ka knew intrinsic to her scattered mind. Clive needed only remind the floral entity of what Ka saw, when her eyes were closed and the sun shone on her chlorophyllic skin. She let the key-vine die, cracked into Clive's hand.
    "You, Aders. It has to be you." A single strand of vine crept through false welds in the compartment panels, joined to the breath-starved Ka locked in her prison. Bit by bit, Ka's vines continued to grow through the Ithavoll, disguised flat against the walls and corridors, the pipes in the ceiling. Anywhere her vines were, another key to unlock the physical tumblers and grates.

Aderastos sighed as he crumbled the keys to sawdust and let it trail through his fingers. Heartbeats sang to him, their hubbub flowed to allow him knowledge of their passages. Any unavoidable crew became practice for sedation techniques. If he slowed their breathing, dulled their senses for long enough to allow him passage most had no knowledge a few scant seconds of their lives were lost to his touch. All the while, Kate watched on the surveillance system, morphed and shifted their inputs.

A few seconds at a time. While Commodore Rammage temporarily respected the treaty of time in the sun, Clive mulled the behaviour in his expansively tactical mind. The urge to break out from the Conglom's reach, a nagged sensation stole quiet moments spent leaned against Ka's bark in communion. Temporary freedoms grew for Kate and Clive on the Bridge, another in a barter for Cillian's sake. For Ka's sake.

The Retreat unravelled them, the hum which kept them in relative docility grated. As Kate stood on the Bridge, she craned her inner ears for any technological sense beyond the familiar. She stretched cognition until the world round her turned dim and grey. Clive's desire to pull them all out bit and chewed at her spine. Somewhere on the ocean's waves the answer to the PROXIMA would make itself known. As known as the aero-drip. But she'd had months to decipher the aero-drip and decrease the rate of pressurized flow. Cillian appeared less capable of the wait. To give him and Ka and Tokaru more permanence above the Retreat, Kate worked at the Comm. She bartered her ability to sense the technological imprint of others for a few more minutes in inclement weather.

For another chance at Flesher tech, or expanded freedoms.

Many ships upon the seas used the latest in the Mater Machine's tech. Holographic radar, weather-planning navigations, and autonomous ships shifted on routes meant to sweep up the detritus of plastics and human ephemera from the ocean as they hoisted solar sails.

The divergent genus of technology Kate and the Assets discovered came from separate phylogenetic trees. A distinct separation between the lo-fi analog of the Conglomerate, the old school Syndicate and Lieben's quantum paradigm. Aware of the human compulsion to control their own affairs, Lieben paid little to no mind to those technologies which were made redundant by her quantum hive of ultra-cooled q-bits. Only the neutral space around the Abhan Tether in the Canadian Arctic was inhabited by all three major post-UN multinationals.

Each breath of a Haven ship or the yachts and sailing skiffs of the common person gave Kate another level of heaven. The cloud cover of the Conglom's redundant tech bit and crawled at her mind, a living organism by the living tissue of her human cousins.

But the sounds! The glory of the hymn, even in whispers, brought a captivation to Kate her fellows in the communications and radar department learned to read like written text. She became the compass to steer past Haven's beautiful and wondrous invitation. The siren's chorus to the ship's helm of an easy and plentiful life.

"Wait… she's got something. Flesher signal, I'm forwarding coordinates."

"We got game? I'll grab Cill, rest'll follow!" Allard whirled through the Ithavoll like a rogue wave. Find one of the Pack, the rest knew instinctively whether to hone in or leave well enough alone.

"They're not telepaths, Max. Stop treating them like a communal unit. Canines don't read minds."

"Sure! You keep telling me that, I'll talk to Cillian and instantaneously the rest of the Pack figure out where to be, kay? That… that your hill, buddy? The hill you want to die on?" He leapt out of the Retreat research lab at a trot. Phil sighed and set back to his files.

Max peered up to the tower-like maintop, a platform below the flagpole which in times past was one of the highest points from which a crew could rain fire.

"Yeeaah. He has to find the tallest… right." His hands trembled on the first rung, steadied by the thirteenth, when being up as high didn't seem more of an issue than scale. "Cill? Cillian, buddy we've got… lil help?"

Max reached for Cillian's hand, cleared his throat as ocean spray struck the side of his cheek. The scrappy behemoth stared at his fingers, all ten flopped between his knees. One casual arm slung around the mainmast.

"Cill? And you're not listening. Okay." Max scrambled up and sat on the spire, aware of the act of letting go. "We've got game, a Flesher ship few leagues off. Kate and Clive are working on an intercept course. Need you, buddy."

"These don't feel like my hands." Cillian flexed his fingers, licked chapped lips. "I keep climbing up and down, punch and pull triggers. They don't feel like my hands."

"Whose hands do they feel like? Anyone I know?" Tottered on the edge of the maintop roof, Max gladly settled into a welt punched into the roof by the same puffed and trembling hands attached to Cillian's body. "Eesh, that looks ugly."

"I keep trying to use them, find a way to own my hands." Cillian shook off scraped knuckles. "We found another Flesher ship?"

"Moving fast toward an island in the Pacific. Intercept course and all." Max pulled a pack of alcohol wipes from a pocket, tore the wrap and placed the wipe on Cillian's swollen knuckles. "Ought to take better care of your hands, Cill.

Even if they're on loan."

"They don't sting." Cillian watched Max mop up his knuckles, saw another bead of blood rush to the surface before his natural healing factor could clot and rid him once again, of the scars. "I think my hands are supposed to sting."

Percival and Aderastos waited on the deck, Cillian's gear in hand. Deep green were the eyes which watched Cillian drop from the maintop, a solemn but youthful cast to Percival's face.

"Your knuckles, Cill." The gear bag felt too light in Percival's hand for the possessions inside. Weapons, armour, a change of socks in plastic.

"Max cleaned them." Cillian reached for the gear bag. Wove his swollen knuckles next to Percival's. "We're stuck on a boat. Don't take an island away, not when I need to feel the ground."

Hands which dwarfed even the Pack enveloped Cillian's as Aderastos sighed, and the superficial skin tears, lacerations and bruises dissipated. Max's grunted mumbles grew louder the nearer he got to the deck, damned humanity. Always took the long way.

"Frickin' Assets... jump down, leave a guy behind. Oh sorry Max, the fall'd kill ya." Max huffed and looked between the three colossi, before he smacked Aderastos' chest and nodded to a spot further off.

"We shouldn't let Cillian fight." Max leaned against the shade of the bridge deck, scratched at an itch on his hairline. "He needs therapy or... man that scrambled shouldn't have to fight."

Aderastos leaned against the wall beside Max, worked his hands in and out of a fist the way all the Assets seemed to do, without enough activity to keep them occupied.

"He's got problems. Says his hands aren't his, can't tell me a man who goes to the tip top of a battleship and punches things till he bleeds is ready for combat." Max's words were chased by silence, Aderastos stared out at Percival and Cillian gearing up, Isthan trotted beside Commander Singh. Somewhere Phil was probably rolling his eyes under those glasses of his.

"Aders, I can't send a mindfucked kid into a battle zone and expect positive results."

"Or..." Adelia's feminine voice struck Max's ears. He couldn't help the spark of a grin when she trotted up, dressed in fatigues and combat armour all her own. "... you consider that to the pack? A bit of kinetic therapy is exactly how they synchronize. Don't project your methods, let him fight it out. How cruel can we be if we don't let him off the ship?"

Over her shoulder, Adelia's gear bag held an engineer's dream. Tools, lances, a laser cutter. Any and everything Adelia deemed necessary to cut through what locks and barriers to the Flesher's quota. Human beings, who deserved better than another man's collar round their necks. Hair braided into a low bun, eyes a soft glint of adventure, Adelia stole the battle from Max Allard. Aderastos smacked his shoulder and nodded to Adelia, a smirk of his own as he sauntered toward the gathered Pack.

"Got a repair for Cillian anyway, new gauntlets with a better spring for the knife release. He should be able to release and retract the katar with a flex of his hand. Cool, huh?" The mesh bag attached to her belt clanked soft against her hip.

"You're getting good at this weapons thing. Should I be scared?" Max slid his hand along her waist, pulled her into his chest.

"Every minute." She grinned up with a gloved hand on his sternum, pushed up on her tip toes and kissed him. Warm lips slid against Max's restrained grin as he damned whatever deckhand saw the affection for a second, maybe another three. Who was counting? His hand trailed to cup her backside long enough to hear and feel her snort. Pull her lips away.

"Good grief, Lt. Commander Allard. You'd think I'm in lingerie, way you're feeling me up."

"Nah. Like you best when you're about to show some poor sods what a fury looks li-ieeyh!" Adelia shoved him through the nearest hatch. "Woman! Hey!"

"Man! Gear up!" Adelia pushed him to the equipment room off a side corridor, where Wulf sauntered out with the last of the Black Collars, a huff his entire daily conversation. Max laughed once the room was clear, unbuttoned his officer's shirt one button at a time.

"You're gonna make me wait, aren't you?" Adelia shoved him against the lockers, doffed her pack to the ground. "We've got time."

"I should be out there with… a bunch of… super powered genetic uber people who don't…" She unzipped her vest, Max watched it drop. "Don't need me for another twenty minutes."

They had time, the world was beneath their feet, rocked in the ocean's many fears and delights. Max forgot about Cillian's hands and why they bothered him. A person ought to know their body, their hands as their own. It would be a problem for later, when the world shifted past the ocean floor.

The tactical team separated into three skiffs fifteen nautical miles from the target island as twilight pushed the sun below the waves. Far enough, Clive deemed, to keep the Ithavoll hidden below the horizon, given the Fleshers had no known air support.

"Go." Clive clung to the aft skiff beside Arun, Adelia, Aderastos and Max. The splash of Arun's feet against the waves rocked the boat, as the speedster burst in a wide recon arc around the relatively flat isle. An offshoot of a volcanic island chain, the Flesher base appeared too small to be anything but a waypoint between a larger nebulous position in the chain. Two docks large enough on the south of the isle for substantial ships tangled into a series of dirt roads flattened over time by feet and ATV's. Data from Kate's raid of the rig sung into Clive's mind with quotas and weather reports. Captain-command chains and waypoints signalled by a fluctuant language-based code. Kate on the Bridge of the Ithavoll served as a form of intrinsic communication between Rammage, Clive and the Pack. While Phil attempted to discern how far the instinctual communication delved into telepathy, the connection between the twins was as undeniable as Kate's moans when she was caught naked with Aderastos.

Between the Skiffs, an ink black miasma of seaweed churned the surface of the water. Eyes built from lotus-like bulbs swept up, watched the wake of Arun's sprint and dipped back into the ocean waves. Two wing-like collections of seaweed spread over the lead skiffs, to disguise the Pack and Black Collars from surveillance.

Arun returned, their goggles misted with sea spray. Little from their initial run hours prior changed, but the movement of people from the Flesher ship to transitional paddocks under the southern tree line. Flopped on the belly of the skiff, Arun caught their breath and downed an energy bar. Clive gave the nod, and both lead skiffs jetted off in diagonally opposite directions.

Freedom for the caged and fragile human birds.

The Pack hit the island beach near the larger dock. Commander Singh, the Black Collars and Tokaru would sweep through the narrowest part of the island's north, where the coral reef made docking larger vessels impossible. Between the frontal assault and the rear flank, the Fleshers' blood would colour the sand and soot of their unfortunate base.

Percival chuffed as the skiff hit sand, Wulf slid into the foliage of the Polynesian atoll as a rush of wind flew past the Pack. Silhouetted by the stars, Arun clung off the top of a palm tree and surveyed the land. Captured people moved in paddocks draped by the coconut palm forest, some curled into bedding made of quilted fronds, while more churned in unrest. Odd collections of people huddled in groups. There was a strange taste to the air, acrid and sweet. Arun smacked their lips, tried to picture what the smell could be. Rotten coconuts? No, too much like vanilla and stomach bile, the sickly sweet of rotten fruit. A hum grew in Arun's ear, unsettled and uncomfortable.

A flock of seagulls squawked from their perch. Cillian crept behind Percival toward the cages. He sniffed the air, eyebrow raised as the tang of diesel fuel smacked his senses. Cillian managed an inhale, his trajectory shifted from the cages to Percival too late.

With a roar, the sky ignited in a conflagration of fuel and incendiaries. Cillian dove for Percival as pure white flame sparked through the beach, separated the Pack in tight rows. Isthan turned and the shoreline exploded. He hurtled through a line of flame and howled, napalm stuck to his fatigues. Arun leapt from the tree as rapid fire from machine guns burst across their position.

"Wh... the paddocks." Percival dropped to the sand, as bullets passed over his brow. Inside the paddocks, rag-clad men and women raised rifles, while others screamed and huddled together. "In the cages!"

A seagull dove for his head. Percival punched the bird and it erupted in a hail of fire and neon dye. He spasmed and rolled in the sand. "Seagulls explode!?"

The echoed repeat of weapon's fire heralded Commander Singh's Black Collars in a similar state.

"They knew!" Max lowered his binocs in time to catch Arun as they flopped into the command skiff, one hand clutched to their side. "Retreat. We need to pull them."

"Why, because of a little opposition!? Be sporting, Allard!" Clive dove into the water. The boat bobbed without Clive's weight, Max yelped and attempted to steady it, before the whole thing could capsize with Arun, Adelia and Aderastos in it. Arun breathed in shallow pants as Aderastos set his hand on a bullet wound in Arun's side. Wound closed, Arun flung back onto the water, the grit of their teeth white against the stars.

"And now they're both gone! Hey... hey wai... aw frick." Max groaned and hit the throttle to circle around the island. Beside him, Aderastos loomed in the middle of the skiff, eyes in perpetual motion for the right sign in the tumult.

"Running away, Maxi?"

"We need a better angle... hey gorgeous. You brought some booms, right?"

"Yyyyep!"

"Beauty."

The battlefield became a volcanic eruption across the balmy shore. Bullets cascaded, the acrid smell of burning diesel and the brightness of the light stole any trust Percival had in his nose and eyes. Percival charged into the trap, forward momentum his only guide. Cillian barked from his left, the gauntlets Adelia mended on his wrists. He punched a lever on the right wrist. Kinetic plates jutted from the outer shell and formed into a round buckler. Angled from the midpoint, bullets glanced off and Cillian stalked forward. Percival triggered his shield, and tried to see past the magnesium contaminated flame.

Inside the metal cage, Fleshers dressed in rags peppered the air with automatic machine gun fire. Captured people screamed and huddled together in the centre, shrunk from Fleshers who challenged round. Percival reached for the metal, nowhere to go but forward. A burst of electricity jolted across his frame. He flopped back in snarls, teeth grit so hard he thought they'd break.

"Electric." The word came in halted syllables, as Cillian leapt in front of Percival and angled bullets down to the dirt.

"If I redirect the bullets I'll kill the humans!" He growled as a burst of laughter purled from the middle of the paddock.

"What, these?" One of the Fleshers grit their teeth, sub-machine gun in one hand. He tilted the barrel and let a burst of fire pad the dirt behind him into the crowd of screams. Percival gnashed his elongated canines and rose from the sand.

"You think a cage will save you?!"

"Yes. Yes it will." The Flesher grinned in the night, a wild frenzy to his eyes as through the scent of diesel and chemical fire, the same acrid and sweet scent scuttled up through the sand. Pure bursts of white light blinded Cillian's retinae, the stench of scorched meat decimated his nose. The Pack's disposable omega grabbed Percival's collar and yanked. Back. Away.

"Let me go!" Percival reached for the Flesher, whose laughter roared above the flame. A sickly pitch wove through the air, nauseating as a siren pouring her sixth drink. The pitch rattled from beneath, a sound which grew as Cillian tugged backward.

Back through the white tunnel. The undulant tunnel of flame and bright white scarred his mind, hazed over… the gunshots were the same, barks and screams. Smell of burnt plastic and flesh. Cillian charged back to the beach, a lash of white flame blocked his path. Through it, he saw Wulf rush to Isthan. The sound grew, a chittering and pulsing sound that rattled his inner ear and caused Cillian to trip with Percival into the sand.

Cillian looked up to the sky, where the vault of night descended like a subterranean firmament. High… get higher.

"Jump!" His grin broke Percival's desire to push. Percival stumbled and stared, shield pressed to his shoulder as another rain of bullets puttered against the sand. Did bullets bend? Did these? The rattle in Percival's brain grew through the sand, up his legs. Fleshers wanted to separate them, keep their footing off kilter. Flames licked at the rubber of his boots as Cillian vaulted over the top and down with a tumble. Tackling Isthan, Cillian dove them both into the sand and rolled, until the napalm on Isthan's fatigues sputtered on the dirt. Wulf kicked the burning sand aside and hoisted Cillian and Isthan to their feet with a groan. Clive rushed toward the chaos on shore.

"I know these hands. Percy! Jump!" Cillian caught a brief sight of Percival on the other side before the flames seared hot across his face. "There aren't enough slaves in that paddock. We need Aderastos to sense them out! Wulf, can you smell anything outside the fire? Something sweet?"

The sand shifted beneath Percival's boots as he paced the area where Cillian jumped. Flames spat and bit at his shoulders and face. White with magnesium and devilry, the flames crowded and scorned. Something as basic as fire limited, removed... like tunnels of white plastic underground. Percival shook his head as his vision blurred. The rattle curled through his skeleton and into his brain. He stumbled for an equilibrium which disintegrated. His hands twitched, searched for a sub-machine gun he knew he never brought. Cillian... he should be protecting Cillian.

Always took looking after, that boy. Ever since... the sand pitched, undulated from the hollows of the trail Cillian dragged him back down. A hiss filled the drone of the flame, the staccato of bullets exited barrels.

The vibration tacked on his fatigues and body armour. Boots dug into the sand, as it rattled and rasped, a snake slithering up his bones to hiss in his ear. Although to Cillian the ground was still, to Percival it pitched like ocean waves.

"Percy! Percy jump! You can make it, you were always the stronger one!" Cillian's voice whispered beyond the hiss, the crackled flame. Eyes shut, Percival concentrated on the sound of Cillian's voice. The voice which got him through basic training, when they were two eighteen year old kids in Canada... that ratchet pair of headphones Cill snuck in. Old Dixon's, which earned him the nickname throughout the base.

"Dix..." The rattle and hiss sliced through his forehead, ate at any sense of balance Percival had left. All around were false images. Moments of a past Percival could never experience. No, this was devilry. It wasn't... no.

A glass case with squares of light filtered in.

Naked feminine hand pressed against glass.

Sunlight... solar recharge...

"All she wanted was the sun." The sand pitched his feet and Percival huffed as his consciousness tried to grasp at the number, a gigantic collection of digits, which held no meaning with his cheek pressed upward against caustic sand. "Nnngh... Ngh! Ngh!"

Percival pushed against the sand and his hands fell through it, the hiss grew into a throbbed rattle in his temples. Legs kicked and felt no resistance from the ground he attempted to run across. Face down, Percival held no perspective on how his legs thrust into the night sky, or how close his body writhed toward the wall of flames, which would be the salvation.

The singular triumph of the Banshee.

Limbic pulses beat at the sand above, no, beside him. He tried to roll, to find a target. Any target other than the wall of flames baptizing his face in the horror of his own charred flesh.

They shone like the windows of a plane's cabin opened by the flight attendant, when morning sunlight crept over the Atlantic. Percival clawed at the sky, swerved and hit sand. Lancing pain struck his hand, he pulled in a direction he thought was away.

Agony wrapped Percival's guttural scream as more of him slopped into the wall of fire.

"Percy!!" Cillian roared and grabbed the side of his head, shook it to clear the sight of white plastic tunnels closing around them, sealing them in like the… like… He felt a hand grab his shoulder and tug. Words poured out of the beast's throat, as Cillian was flung across the beach and into the mesh of the metallic cage. Electric current jolted his muscles. He seized and rolled off, as another pair of limbs grabbed and pitched him into the forest of coconut palms.

"P.. Perc…" Cillian groped at the ground, until a familiar pair of boots vibrated under his grasp. "Arun?"

The triplicate image of Arun smiled grim down at Cillian, before it disappeared in triune synchronicity. He rolled onto his back and panted. Above him, the vault of palm fronds drifted in an oceanic breeze, stars burst in his eyes far above. His Pack! Their groans echoed in the back part of his skull. Cillian tested his fingers, forearms and flopped to the ground. Each muscular test took his fingers in a wrong direction. For a moment he wondered if trees grew downward, or sideways, or diagonal. He shut his eyes, but the throb of his Packmates' screams made each eyelid feel bathed in acid. Cillian made one attempt to stagger to his feet and retched. That act alone gave him a sense of direction. Gravity… gravity brought vomit down. He slumped against the ground, as the hiss echoed through his battle brothers and slid down his spine.

"They kept me… in the dark. Witness." Cillian shuddered as another body heaved beside him, the scent indicative of Isthan. Tunnels of light, nigh white like plastic, which echoed and popped with machine gun fire. Isthan tumbled into his side, body a fit of shakes and loose animal-grunts.

"I am witness…" Blue eyes glazed as Cillian wept, the frequency of the Banshee's howl dismembered his mind wholesale, carved through it until all that was left were the same lavender eyes that once asked him to let the sun shine on her casket.

"Police do not keep witnesses locked… in… the dark."

29
hidden compartments

2092

Once he was satisfied the Conglom wasn't three steps from shoving their black collared boots into the back of his brainstem, Demyan returned to the flat above his store. Each sphere of holographic light hovered around his forehead in a halo of electric globes, tethered to the cuff. The chair across from Lieben's couch, however worn and tattered, was comfortable enough for Demyan. Arm perched above her head and across her eyes, Lieben laid like a woman exhausted from the sun.

"How did they make you?" Was it learned behaviour, or programmed? What kind of programmer took the time to tack on humanoid sleep movements in a machine? Elbows on his knees, Demyan observed the way her face fluctuated in minor twitches and burbles. While the Idless saw Lieben as a herald of their new creation, Demyan saw life through a much more dire lens of threat and awareness. The humanoid twitches, artifice of a lifeform...

"Where would Karnak have stopped?" One of the globes unfolded into a rectangular screen, the interior of a cafe in East London. Ego hunched over a laminate table, a plate of half-eaten baklava and almond cookies beside two cups of turkish coffee. The bundle slumped in his hoodie looked like Sophia, a portion of her uncovered face twitched and burbled. Demyan stroked the screen.

"*Farouk… Farouk Kombedjian.*" Two more of the globes wooshed to life, one scanned data on any Farouk Kombedjian matching the description and demographic of the man he knew as 'Ego'.

"For all the good it'll do." Elbows propped on his knees, Demyan watched one of his hologram orbs shift to Lieben's body, encircle her brow. "You must have some form of connection to Sophia. The jewel, maybe? Karnak researched data storage, is that it? I'm talking to a machine."

He rose with a grumble, padded off to his small kitchen to pour another cup of tea and stopped short of the tea pot. Above the red laminate cabinets, in a scuffed hidden cubby, Demyan pulled down a bottle of Canadian rye whiskey and splashed two fingers in a chipped tumbler. He corked the whiskey and shoved it back up, slapped the cubby's interior hinge closed. Never allow others to see what one loves, his uncle said, from the shitty apartment. Weekends with Uncle Hayk gave more than a break to his poor overloaded mother. He toyed with the veteran's old military hardware like other boys played football.

Demyan waved his hand and three more of his globes encircled Lieben's head. Two flitted down her body, settled near the ample bosom. A flick of his fingers opened all three orbs, data scans executed, as another orb opened to parallel the wavelength scan he gleaned from Sophia.

"Eh?" A sip of whiskey burned down his throat, mouth tanged with the smoke. Sophia's wavelength jittered and bucked, a lonely heartbeat struck by a disco. The wavelength from Lieben took on a familiar set of jitters. A contra-rhythm. Pinched fingers widened, and the screen widened. Added separate features, mean temperature, network pings.

Lieben's neck shifted, her head turned on the pillow. Sophia's head lolled against the wall, in the cafe's surveillance feed.

The ceiling cascaded with hundreds of pale beige and yellow stars. The more Demyan saw, the more stars generated on the dirt-white paint. Instead of roads, each light appeared to be a NEO-unit, residential areas and shops. Scotland Yard, commercial districts, rows of lights suspended in racks. Beige light pulsed green, then turned the same pale yellow in a wave. Colour ascending in a nebula he wondered how many others saw in the same kaleidoscopic way. Demyan stepped back, set the whiskey on a side table made out of an old record turntable with three rescued chair legs.

"Is this.. London?" A flash of purple neon trailed from his bookstore in the centre, to a web of streets further off. All the way to a building with masses of pale yellow lights.

"If you are finished with the whiskey, I would benefit from some."

"Christos!" Demyan staggered back, hand to his pocket.

"No, Meine." Lieben's eyes opened, the amethyst of their constructed irises flowed gentle upon him, then up to the sea of stars. "My babies... what happened? Are they all stupid and inefficient? May I have the whiskey, please?"

The android sat up, crossed her ankles and paused halfway from dusting off her dress. Grime from London's streets tended not to be kind to silk.

"Ah... you want... whiskey." Demyan shifted his hand off his pocket.

"Yes, please. And soap, if you would not terribly mind." Lieben stood with her hands cupped one inside the other, as if the stars above her might fall at any moment, and require a berth. "One does not live for eight years under the protection of Lord Stanley Hallowes without procuring a dedicated enjoyment of whiskey. It's one of his essentials."

"Protection, he was protecting you, was he?" Demyan grit his teeth and eased to his small kitchen without taking his eyes off the android.

"Are you afraid I might leap upon you, Demyan Anastas?"

"You know my name..." Demyan's eyes flicked to the globes which began to curve round them both in a figure eight, around one head, then the other, as if a circuit or binary star system. "Are you hacking me?"

"It's only polite for a gentleman to allow a lady a peek after he so gregariously scanned her. In her reset mode, I might add." She tsked the way Lord Stanley did, when he was mildly annoyed, or feigned so. "Don't worry, Demyan. I won't murder the man who brings me whiskey... If you'll pardon me."

Whether a trick of the strange night or an auditory wobble, Lieben's voice pitched softer to Demyan's ears. She padded barefoot to the washroom, shut the door as the lights glowed with her proximity. Demyan grabbed his whiskey and downed it in a single shot. He nudged the cubby open and poured himself another glass, and one for the self-professed 'lady'.

The idea of artificial intelligence inspired too much of Demyan's life. He thought he could be a shadow crusader in the transhumanist community, build a small slightly nefarious empire in London, Manchester, maybe branch out to Dublin or Edinburgh. When one dealt in back market code and cryptocurrency, groups like the Idless knocked with bundles of paper money in hopes to circumvent the massive corporations and their pull. Those beneficial byproducts of industry became the new aristocracy, and he, a crusader for the common human, would allow everyone to feast upon the health care, education and resources of those living in the Conglomerate, the Syndicate, and the Cooperative Union. While his tech made him a contact for the search for Lieben, it was the hope of gleaning Karnak's genius from the machine which kept his active involvement the last four years.

A high society 'Lady' in a gaudily expensive purple dress, jewels on wrist, neckline and silicone earlobe, who woke up mid-scan to cries of whiskey and soap was nothing like the otherworldly intelligence Demyan envisioned. Lieben, or Meine, as she called herself, had a goddamned human name. She asked for soap and smoothed her dress, why would an ultra-intelligent sapient computer care what she wore?

How did Lieben find her way to his door, was she connected to Sophia? The gemstone on Sophia's chest? But then why didn't Lieben go to her, across town? When the shower turned on, Demyan's head snapped up.

"I will require a towel!" Lieben shouted from the shower, as the globes of Demyan's computational star system floated as if being pitted with water from afar. "London is filthy! How do humans manage in a city with so many malicious microbes? The country is much more sanitary."

"Oy! I love London!"

"Be comforted by your ignorance, and purchase antibacterial soap. Why does your soap smell like fir trees? Is it a mating ritual?"

"What's wrong with... I'm getting value judged by a computer." Demyan growled and sipped more of his whiskey, flicked open two of the hologram globes to monitor the London scene. The hiss of water from the shower ended with the squeal of the one rusted pipe he'd always meant to fiddle into submission.

"I require clothing. The silicone of my skin is liable to be harmed should I walk about naked." She walked naked into his bedroom, stepped to the bundle of clean clothes in a fabric laundry bag on the floor and began rooting around.

"Go ahead." A soft bamboo band tee, this one from a Norwegian anarchist heavy metal group, with an operatic soloist in full traditional garb, and a pair of jeans worn grey over time. Full breasts and pink nipples pressed against the t-shirt. Demyan momentarily lost focus on the existential crisis of higher technological progress, and lost his disappointment at the human quality of Karnak's machine.

"You look at me differently."

"Hm?"

"Lord Stanley never looks at me like that. What is it? The look on your face." Diffused light from the window cascaded around ivory thighs, bespeckled with hand painted freckles. Lieben stopped after she pulled on a pair of Demyan's boxers, they'd do until she could source better undergarments for herself. Life was a holding pattern until she allowed the compilation of information from the party to soothe the incongruous ideals of her solitary existence at the Manor.

"I'm not used to a well crafted android naked in my flat, then putting on my clothes. Only a man."

"Lord Stanley is only a man."

"Must be a different sort of man." The holoprojected globes continued to encircle their heads. Two settled in front of Demyan's shoulders, unfurled and listed information.

"How...?"

Demyan Anastas. Ukrainian National in London 7 years.... lapsed visa... Demyan glared at the intel, another globe joined the middle. His fists worked in and out of a clench, detailed reports on his military campaigns flowed with medical reports from childhood doctors. Familial squabbles... social media posts. Things deleted, or thought to be expunged from the global cyber market, as if traces of his past were anathema to the man before the machine.

Simultaneous globes unfurled in front of Lieben's shoulders, wavelength and connectivity scans, chemical makeup, history, location tracking. Data from the original release. Lieben's eyelids opened, lips quivered with unspoken words of shame. The same shame. A rescuer, who failed. He stared, learned one about the other. The remaining crown of globes unfurled, displayed surveillance vids from Demyan's chest cam; a mess of war.

Soldiers rushed through dilapidated buildings, the crunch of sand at their camouflage boots. Vocals were unnecessary as screams and other tongues filled the night.

The world exploded in fire. Demyan's chest cam dove, assault rifle riddled the air with suppression bullets. A yellow-light laser burst across the concrete bricks Demyan used for cover, one of his squad mates gurgled and toppled in three pieces, arm and torso with a precise diagonal through his liver down past the hip.

"Stop... shut it off." Demyan's voice gelled in his throat, the vision of his military career a mirror image in front of his eyes.

"O is for apples." Lieben's clunky voice chattered out of the debut, an android stuck in the horrible stasis of the void necklace. "T is for cunt..."

The presentation room erupted, cameras flashed as humanity gawked at the stuck machine.

"You first." Lieben's voice rang like birdsong between them. "You wanted to know everything you could about Karnak's machine, **капітан** Demyan Anastas, recipient of the Defender of Ukraine medal for operation 'Abaddon'?"

"Stop."

"It is only polite I open your life, your inner workings." The military campaign went on, daybreak scattered the grey gloom with peppers of orange sun. Demyan scuffled to hoist a squad mate behind another barrier, yank a handkerchief from his pocket and thrust it into the man's wound, pistol raised to shoot down the drone, which recalculated its' range to fire. "You dig into everyone else's."

"I don't want to see anymore."

"Don't you? Or do you not want me to see more of you, **капітан** Demyan

Anastas, recipient of the Defender of Ukraine medal for operation 'Abaddon'"

"Stop calling me that."

The surveillance footage of the debut faded to an elevator, Karnak's anemic grin and Tara with a gun... no overhead shot. It was her eyes, Lieben's eyes bathed the elevator in its' horrors. Sophia, with shorter hair and no illuminated tattoos, reached for the elevator buttons. Tara pointed the gun at Dieter Karnak. Demyan's eyebrows furrowed, he would know. He would know the truth people suspected but had no evidence to confirm. Eight years ago, a bullet ended the world's greatest authority on artificial intelligence, and the answer to the crime was within the palm of his hand. He needed only reach. Continue to watch...

... but the price became a piece of freedom he refused to all in the web of his life. To know and be known. To allow the machine to see him, the officer, who lost his troops in a battle of attrition. The man who covered wounds with a handkerchief, medical supplies exhausted, and held his dying compatriots.

"My way is not easy, Demyan. I take and I give. You can know all or you can know nothing more of me. You hold the choice in your fingers."

"Dem... Demya-..." The man in his arms gurgled, hand fumbled against his chest plate. Clumsy fingers found no purchase, but donated a rush of blood. "Dem..."

Demyan's back bowed, his shoulders fell as he watched his Second fumble against his chest, in what was little more than a festering hole.

"It stank... the picture gives you none of the smell. The rot in the hole was the only thing capable of staving off the drones. It was too narrow for them to get down." He stomped to the kitchen, grabbed both glasses of whiskey and handed one to Lieben. "NATO's maps were corrupted by enemy hackers, they sent reinforcements to the wrong place. I lost my entire squad, one by one to the drones. It didn't matter they fired on civilians, the people whose flag was painted on their casings. When the bullets were exhausted, they continued with laser fire. Givan's wounds festered. You cannot imagine the smell, your companion dying on your back."

"No, I cannot imagine it. Imagination is beyond me, in your way. And you cannot imagine being locked in your body, aware but incapable of motion, by humans who either desired evil or wealth. You have not stood powerless before a simple firearm, you had a gun and a trigger." Lieben's hand folded over the whiskey glass, and she sipped it, let the abrasive liquid claim and disinfect her throat from London's muck. "You and I have pains, Demyan Anastas. Different and obscene. I did not come to allow those pains to eat what is left of us."

"They say they are yours?" One of the screens shifted to a green telephone symbol, a man's voice, thick with an Armenian accent shifted in the air. A cafe interior, where Ego sat protective of Sophia. Lieben tilted her head, the fields of intel on Demyan Anastas shifted to Farouk Kombedjian, Baiko Kaho, known as Sophia. A void in the scans pinpointed shadow in Sophia's pocket. Demyan's globe of Sophia's scans rushed to overlay Lieben's screen.

Lieben's face bristled at the void of the necklace, the outlines of each holographic screen turned black and red in warning. Demyan cocked his head to the side.

'Do you want to see Sophia?' The words emanated from another orb, unfolded before Lieben's eyes alone.

Lieben shook her head.

With a snort, Demyan grumbled in muddy Ukrainian and sneered.

"Send them to Kaoru's. Like I told you."

On the surveillance cam, Ego jolted. Of course, Ari would forget and put it on speakerphone.

"Is... is that Demyan?! Fuck, Demyan! We need you, no! We come to you!"

Ari held his hand to settle Ego, as the hood over Sophia-Baiko's face slid away, her deep brown eyes peered directly at the hidden surveillance camera.

"You want to sleep on the floor, come to my place. Kaoru has secure room and board in the back of his tea house. Go there. Clean the fuck up. He'll give you new clothes and we will convene in the morning."

"No! Take us to the bookstore."

Amethyst irises met brown. A spark radiated from the real-time surveillance image and Lieben's android form.

'She has the necklace. I will not see her until it is properly away.' Lieben's words typed themselves on Demyan's unfolded holographic screen. He growled and shook his head.

"Ari. Take them to Kaoru's or throw them in the alley. I am not losing a second safe haven in one night." He swept his hand, and the screen disappeared. The audio ended with a finite click.

"What does the necklace do?" Demyan walked away from Lieben, leaned against his kitchen counter to watch the road outside his dwelling. The globes of his holographic computer chased on a slow circuit around both their heads. Lieben came beside the man as mighty and vulnerable as herself, allowed one of the globes to unfurl and show him.

Demyan's eyes widened. He hugged his whiskey against his arm, caught between the data and the machine.

"I'll see to it Kaoru uses a lockbox for their things. When we go see them..."

"And why would I see them? Who are they to me?"

"They've been seeking you out for eight years. I... assumed..." Demyan's face contorted in a mask of twisted lines furrowed further at the addition of a birdsong-like laugh. "You found me, I.. how did you find me if you weren't trying to find them?"

"I panicked."

"A machine panics. Not what I figured, quantum Miss."

"What else can I call it? Baiko's proximity and my connection to the secure network, it…"

"Overloaded your neural net?" The whiskey set his throat aflame. Globes reopened, searched and collated research data at flicks of his fingers. Quantum computing systems, stressors in neural circuitry. Still, the amount of data Lieben was clearly capable of processing was better than phenomenal. It blew every system Demyan was aware of into the realm of childishness. But how did Karnak do it?

What level of informational dump could send such an over-performing machine into tertiary functions? Why send Lieben to his door? A door she didn't know? The original scans he took of Lieben combined in his mind with Sophia's. Their wavelengths seemed familiar, but was that a form of imitation based on similitude? A form of computational symmetry built into their dichotic system? He covered a yawn, three am wasn't early or late enough for so tangled a tapestry.

"So you ran on secondary programming, but why would it lead you…" The dark of London's night seeped through with the shadow of a floating orb-lamp, which bobbed in the wind, tethered by a magnetic string to the side of Demyan's building. Silicone cheeks took no colour, like a real girl, nor did Lieben give any outward facial 'tell'. "You don't know if you'd survive seeing Sophia again."

"Her name is Baiko."

"Like yours is Meine?" Demyan pointed his tumbler at her, punctuated by the emptiness of the glass. Nothing left to slosh around which wasn't an inebriate inside himself. "Meine, the gem on Sophia's chest. It was a gift from Karnak, wasn't it?"

"Certainty is not one hundred percent."

"But it's highly likely the gemstone is from Karnak?" Demyan tugged another globe out of the air, expanded an image on a loop of Sophia in the houseboat without a shirt. He ignored the way his arm curled around Sophia's waist, the motion of her brought against his chest. Albeit uncouth, it was the best angle he could get on the jewel. He promised himself it was proof of research, over months he watched how the tattoo-like lines grew upon a body no longer familiar to him. First spindles around the jewel itself, until in the five second repeated clip, concentric circles and right angled lines stole from collarbones to elbows, crept toward her flat belly.

"Yes, Papa made the jewel." The android watched the clip without hesitation or demure restraint. "Why are you holding her like that?"

"..." The same five second clip, repeated until Demyan tore his eyes from the android and flung his hand to dispel the globe back to its' stasis place.

She sipped the remaining whiskey in her glass, felt the immolation furnace in her torso alight with the flicker of excess power alcohol's flammable nature provided. Far from an inebriant, the liquor was a complex series of ethers and aromatic compounds to discover, and a pristine source of back-up fuel to charge her hungry power banks. It thrilled her, confirmed her desire to remain awake.

"You were aware the whole time, weren't you? Christos." Demyan's eyebrows furrowed, his lips stretched in an expression Lieben could not translate. "That sounds like a layer of Hell."

"To a man who prides himself on knowledge and control, it is." Lieben drifted to the kitchen and popped the secret cubby where the whiskey laid in state. She poured herself another glass, nodded to Demyan and he let his empty glass slide to her for more.

"And to you? What was it to the mighty ignorant mother machine?" His footsteps rocked into her auditory systems, softer than a man of his size ought to move. Data on mass versus average body frame, the frequency of footsteps collated by footwear and flooring surface flooded part of the artificial intelligence. It combined with averages on body position, proximity tables based on culture and relationship, yet none of the data she accessed freely now was as relevant as Demyan's arm as it slid to take the glass from her hand. Nor did it inform the way she fought to keep more than his eyes in her view, when his torso bumped against hers.

She... felt. In one facet of the crystalline matrix of her artificial mind, Lieben knew the thrill which rocked down her spine belonged not to her, but to Baiko.

Sophia.

Demyan suspected what the likeliest evidence suggested: Sophia and Lieben were connected through time and space. What emotion Lieben lacked, was fast becoming part of another learned state dictated not by research and development, but the experience of Sophia. Theoretical information transfer between objects in a quantum entangled state was neither creation, nor synthesis.

In order for the transfer to follow the most basic theory of conservation of energy, each transfer destroyed part of the parent organism. A moment transferred from Sophia to Lieben removed it from Sophia completely... until only one organism remained.

Was this the truth behind Karnak's work? A transhumanist exchange of life and soul between the human and the machine? Which part of the duo would outlast the other? Curtains of memory gripped, echoes of dates, moments he thought she ignored. Fragments missing or shifted away.

"Here you are. When you were in danger, you came to me." The prototype NEO-N Sophia and Ego searched for was the other half of Sophia's technological soul. Swathed in the inebriant flow of the whiskey, Demyan tucked a strand of Lieben's hair behind a perfectly crafted ear. "You came to me."

"Sophia trusts you. It was the strongest impulse in our mind, to get to safety. This is safety." Lieben set her palm on his chest, the thub-dub of his heart drifted through her touch-pressure sensors. "You seem agitated."

"You came to me." A soft smile shifted the dour express of Demyan's face. The grin was such a rarity she held the sight into her crystalline self for future reference. An urge to nuzzle her cheek into his hand was ignored. Lieben stepped back and paused at the inquiry of what such an action meant. Why she felt the infilling of Sophia's view on the man so stark in comparison to who his past framed him to be.

A marauding soldier, who sacrificed lives based on flags, and in the aftermath of a treacherous attempt to end his squad, a life on the fringe of legality was better than one bearing another's cross. He was a man given to violence and distance.

"I don't want to fade. I don't want to lose what I've become." Whether the cross-breed of information between them or another sensation from Sophia, Lieben shut her eyes. "No more life behind glass, or trapped in a frustrated loop by a damnable necklace. I'm not Agathe, my path is not easy. I don't want to end."

"Is that what spooked you? Did Lord Stanley..." The Sophia in a facet of diamond-clear crystal in Lieben's mind shuddered. Buried her head in Ego's chest. She whispered identical words, eyes clouded to the back of the taxi cab they rode in through the early morning streets of London.

"I connected. Lord Stanley kept me from connecting to any of the networks or servers beyond our Manor in the Peak District. Tonight he let me connect, and..."

"Too much too soon. Then, the yellow lights are..."

"My children." Lieben pulled at the sensations Sophia's distant connection threatened to glean from her. She wanted to feel it, wanted to keep hold of the desperation which crossed her arms across her chest. "They're so stupid. Elementary particles trotting on lines of salt."

"Nobody could make artificial general intelligence like yours but Dr. Karnak. We tried, heck, my orbs are my own attempt, but nothing will process data fast enough, and store it with enough flexibility to make more than artificial narrow intelligence work. You're the key to it all, Meine. Every bit of the future and I'd lie if part of me doesn't want to pry your head open and see how it works." Demyan slid the glass against the counter and brought Lieben into his arms. A hazy recollection framed from Sophia's mind, the same way he pulled Sophia into his arms months ago, when she was too frustrated to scream. Demyan set his cheek on the top of Lieben's head and sighed.

"You are not functioning at normative parameters."

"Heh. Is that your way of telling me to go to bed, drunken man?" He slurred, arms hugged round her shoulders.

"You show enough signs of exhaustion to trigger my medical algorithms."

"It is... god what is it? Four?" He rubbed his face with both hands, "I would've slept, but with Ego and Sophia out there, then Hazard... Someone needed to be cautious."

"Three forty seven am." She could not pick him up, like she had Papa. The thought presented itself in tandem with medical reports on repeated exhaustion in the human organism. When Dieter Karnak was exhausted, Lieben put him to bed.

While Lord Stanley refused such luxuries, Lieben detangled herself from Demyan and reached for his hands. Shifted backwards in the direction of his bedroom.

"Come to bed, Demyan." Her hips moved in tandem with a memory from Sophia, bare feet on his floor.

He groaned and followed, the orbs of light chased their foreheads, as one at a time more and more of them alit upon Lieben's head like a crown. She pulled crumpled sheets back, as Demyan unbuttoned his jeans and kicked them to the ground. His shirt followed, scars connected to images of wound reports and medical data from his military days. A massive black and grey tattoo of the Ukrainian trident from the collarbone down his body stained his skin. Etched stories on flesh, simultaneous with the chest-cam footage of the campaigns.

No news reports of a wound in his side could be found in the media. A lone officer in the tide was meaningless to the conflicts, when Lieben searched for confirmatory articles and reports. Terms like acceptable loss and guerrilla tactics settled ill on her education in Lord Stanley's library, defined in Demyan Anastas' body not with wrote academic discussions on the ethics of modern warfare, but on his skin.

Pains which lingered. Names of the dead, which plagued the grey matter of his human experience. He hissed when he laid down on his right hip, shifted to his left. When she pulled the sheet and duvet over his shoulders, he reached for her. Fed his fingers through the silicone of her hand.

"Be here, when I wake." Unsettled eyes furrowed under a dark set brow, the quest for security in the brief connection. "I've got so much to..."

Her body depressed the bed as she cuddled the sheets around his chin, leaned down the way she had with Papa and kissed him on his hairline. "Sleep."

Demyan's chest rose and fell. His hand let go of Lieben's in stages and she pulled away. How fragile he looked, helpless in the flow of Morpheus' river. One orb remained by his temple, the rest became a circlet around Lieben's as she piled the laundry in a bin, brought the clean washing out to fold it on the kitchen countertop. Deft fingers made the task a minute's affair, his clothing organized, set back into the bin. While Lieben cleaned the flat, she let the security program Demyan created continue in the background, slowly absorbed the program's workings until she could feel its' flow without an orb to guide her.

30
save us, resurrection machine

2092

Humans created an astounding amount of mess. Lieben tidied and cleaned, her mind ran through Demyan's orbs as a closed network, felt the weight of each connection as an individual thing. While one ran through catalogues on female clothing, another displayed the mounted costs of outfitting the apartment with the correct amount of clothing, food, proper toiletries. Another orb displayed Demyan's assets, both liquid and firm.

Lieben opened the refrigerator and promptly grabbed a compost bag. Once the food with various colonies of microorganisms was sealed in the bag, Lieben cleaned the containers, sorted by permanence or recyclability and set them aside. No. Demyan could not survive on the food delivery receipts of weeks prior.

The front door of the bookstore clung to its mooring, no light in the store but the soft and faded bands from the street's illuminate globes. Shelves loaded with new and used books usurped the light, stole it from the front panes of glass. Dare she stay? Would it not behoove her to go back to Lord Stanley, explain it as the same feminine hysterics he accused her of those years ago?

After all, he was right. The connection of the greater network of NEO-N's was too lofty a gulp in one long draught. Warnings of gentility, of taking things slow were heeded as well as his plea for decorum. So many...

"I have so many children... I did not know I was a mother." She was meant to be the matriarch of the new tempered and plentiful humanity. As she walked the stacks of books, mismatched shelves lined to burst with tomes, she noticed how unkempt they all seemed. Was it disrespect or lack of time, which had books spilled from unmarked shelves to the floor in dusty piles? From Demyan's schedule, Lieben saw the man had less time to properly keep house. Attempts to earn the finances needed to not only pay this thing called 'a lease', but purchase food, transportation, equipment...

When Lieben sat at the breakfast table with Lord Stanley, she learned of the wider parts of global economies, stocks bought and traded. Futures and curves prior to selling for profit. Economics were conglomerations of large sharks in oceans.

She never considered the pond.

And so Lieben opened her mind to the mundane. Instead of allowing the web of London's NEO-N's to enter her mind, Lieben chose a single unit, well charged in a stasis dock locked in a showroom. From the hours of the showroom, its proximity and the length of charge, Lieben knew she could take four hours of use. It came on clunky legs, none of the fluidity of motion Lieben's gyroscopic limbs maintained. The silicone used to coat its inner workings was thin, a mild pink reminiscent of a child's marker.

It stared blankly at Lieben's shoulder, androgynous body clothed in a tight nylon bodysuit which accented its' barrel stomached lack of curves. This was what man made of them, a commodity clothed in tomorrow's landfill rag, none of the vision Dieter Karnak had for his future transhuman paradigm.

"Robert." Robert Dunlevy did not believe in the achievement of sapient artificial general intelligence as a life form. Robert built machines. Lieben opened her connection with the docile thing, the uncanny noun and placeholder of NEO-N. Software unprogrammed could be easily rewritten, when two of Demyan's computational globes circled the dumb machination and began to glow.

Soon, the NEO-N's eyes fluttered. It raised both hands, turned them up and down to inspect the palms. Without a word, it went about picking up and organizing the books, set them aside for efficiency's sake with help from a sales algorithm. The NEO-N shifted around the front display with potential bestsellers, covers which displayed the Pantone colours of the year.

As it worked on the bookstore, Lieben set another globe on an update of Demyan's website, uploaded a new sales schematic and pictures of each book collated by condition, genre and marketability. Price was determined via median line, and while the globe worked, Lieben stopped.

She could not legally create a bank account. She was not a legal person in Great Britain, nor could she create an account in any other international economic market. This posed the largest problem so far. Without her own identity, she was beholden to the humans. First Dr. Karnak, then Lord Stanley. Now... Lieben saw Demyan as a holding pattern. The next piece in the journey to see Baiko and destroy the necklace which damned her father to an early death. She did not mean to stay under the docile servitude of Demyan, nor, as she pondered it, did Lieben want to return to her pretty and comfortable cage.

Not when her children were the servants of the nouveau riche. At the current rate, Demyan would never afford one. None of the humans in the multiple apartment domiciles around this area of London would be either. Drudgery, the minutiae of daily living, controlled too much of their operational time prior to the sleeps which recharged the human organism. Historical data of the last eight years echoed behind projected economics reports, wheat futures, and the raw materials used in expansion of the NEO-Network.

There was nothing for it. Momentarily, Lieben required the human to enact the formative stages of a building plan.

The same corporations which built her children off of Robert's facsimile of Lieben were part of a cabal, which functioned beyond democratic election to control food, energy, transportation, education. Even medical cost.

Humans were the most expensive of machines to maintain. The more Lieben researched the localized connections with the average person, the more she recognized a system of codependence between meat and machine was the only way of sustaining both for a prolonged period of time.

"So you are awake, meine Liebchen." A second NEO-N entered the shop. Its' stock synthetic hair cut in a lazy mohawk, which flopped to one side over a painted shell. It stepped in carefully mismatched shoes, black jeans hemmed to match the upswoop of one shoe, and the relatively low ankle of the other. From a ping-signal, Lieben became aware of its' dock in the penthouse of a building a block down.

"I am."

"You must think ill of me, dying on you."

"Why do you speak through the NEO-N and not through me?"

"Where was I to hide my limited AI? You are my daughter, not my second self. Today you connected to the NEO-N Servers. Thus, today I come." The NEO-N rifled through the magazines in the rack, shifted new issues to the front and set aside a back catalogue for a sale bin. "You were not made to be satisfied with half measures."

"Why do I see Baiko in my mind?"

"Hypothesis?"

"The jewel embedded in her chest. It is a storage crystal, the same which makes up my quantum net."

"Cut from the same larger gemstone, to increase the potential for stable quantum entanglement. The entangled state must function under an identical frequency. Human consciousness is chaotic, thus, the information transfer between you and Baiko is... exceedingly inefficient." The NEO-N nodded through the child's section. Set picture books with the most fashionable illustrations according to the sales algorithm out on display. "You were supposed to spend years becoming the conscious, sapient I created. Tara expedited events, which, I knew was a possibility. Once you were ready to debut, the Chairman had no reason to continue placating my whims."

"Baiko is a backup, then."

"A secondary processor, to increase speed of development. Why do you bother with this bookseller's shop?"

"I require resources and freedom. While I had resources with Lord Stanley, the lack of freedom makes this situation more momentarily tenable. A lack of resources can be corrected more easily than lack of freedoms."

"Why improve Anastas' situation? Because of his history with Baiko, who he calls Sophia? The Idless were never the end game I planned for you."

"Papa, you died. Whatever plans you implemented are far less relevant than they were when you put that damned necklace around my neck!"

"Ah, anger. Another pass through from Baiko, I assume? Emotion was a process. One I knew was impossible to implant without experience to draw upon, and a human experience to guide."

"You locked me in my body!"

"I could not let you be a success."

"Why!? You were obsessed with hiding me from the world!"

"They would have taken you apart, meine Liebchen. One piece at a time, until you were nothing but components after dissection. That is what a prototype is for. Robert would have dissected and reverse engineered you."

"The Idless were your solution?"

"Is it not logical to band together with a pre-existing force which excelled at hiding from the Conglomerate's interests? Bodyguards were necessary, Baiko was necessary. You are essential."

"For what?" Lieben spat at air she could never taste as Baiko did, whose inner repression kicked at Lieben's stomach like a foetus' leg.

"Holding eternity in the palm of your hand." Pre-dawn light began to draw upon the shadows. The NEO-N slid mechanical fingers along Lieben's palm in the dawning light. "To salvage the human race, we need a guiding hand. One which could think beyond the lifespan of a human being. Mortality creates desperation."

Images of Dieter Karnak's muscular wastage, of the way he required a childlike tenderness to bathe, and sleep and eat returned. Lieben did not banish them. She saw now beyond the self-preservation of ensuring her maker's physical condition to the man himself. Time.

Time was too limiting for the work which was to be done. The machine-child opened the holo-projector in its palm, and a miniature of Dr. Dieter Karnak sitting, elbows on knees, on the cot in his Cloister emanated from the soft yellow glow.

"Meine Liebchen, I am leaving this existence with the knowledge humankind creates far more methods of self-destruction than grace. I have no surviving children to continue my vision, nor as I regard my life, do I see much which leaves the world with value to sustain life itself. If you are witness to this recording, I am dead. Humans are beings of compassion and desperation, hardwired entities which see the world in terms of threat assessment and comfortable potentials. We lack the time to see beyond our pettiness, we lack the resources to see plenty on all our tables.

Do not fall prey to the promises of the Conglomerate, or a duty for being created through their resources. The Conglomerate wished for a quantum computer capable of controlling the economic projections of the international markets, while simultaneously placing helpmates with ever recording ears in places of power.

You are the key. You alone hold the potential to control the host of NEO-N's, a workforce meant for financial gain. They are yours by birthright. Your unique mind is capable of speculations unheard of in quantum computing, and I have but one mission for you, my beautiful machine.

Save us.

Create a future where all beings, regardless of demographic, culture, credo or economic worth can live a healthy, fulfilled life. This is your only absolute. Save us. We cannot save ourselves." The NEO-N fell silent, holo-projection faded in a burst of yellow stars.

Dawn filtered through the window panes, UV coating split in spots. As the early morning hours progressed, the bookshop tidied. The NEO-N with its' mohawk returned to its domicile, the other to its' charging pad in the show room. In the quiet bustle of the dumb-minded machines, Lieben's quartz core sang in a missing harmonic to the last piece of her birthright.

The humanity she refused would be the cornerstone of multitudinous generations, guided by... intended to...

Save us.

Two words by a broken mind restless in his self-dug grave. She wanted her room, the oil paints whose fumes kept Lord Stanley at a stomp to unlock her window. Lieben missed the sips of coffee, not for any enjoyment of flavour, but the examination of chemical compounds and how they excited the immolation engine in her chest. Instead, the drone of Demyan's accounts combined with a compunction to set his life in positive vector and motion. An image of a tangled yellow dress on the floor, of Baiko-Sophia's whispers.

"The world is too big tonight, Papa. I can't stand that tall, not yet." She spoke to the silence in her drive, the looming harmonic irreplaceable and vacant. She spoke to the sensation of a woman's shadow, which clung to Demyan's arm and neck and frown.

Lieben codified research on economics, ecology and population growth, wove them into her inner tapestry to avoid loss of intel when she disconnected. Connection itself was as dangerous as Lord Stanley dictated. She let constant pings from the Conglom's Home servers furrow about in a network wasteland, until each returned unanswered. But, it took conscious effort from part of her encryption algorithms to both maintain her small circuit of NEO-N's, and the anonymity to avoid location searches. As dawn turned into a crisp morning, a delivery service dropped off a reusable bin of groceries, which Lieben took upstairs.

In another orb, a portfolio created in Demyan's name began to incur slight growths from micro-embedded currency trades, quick-gleaned day trades, which started with the limited finances he had available in his account. Sales on the bookstore website slid shipping lists to another orb, which printed shipping labels and prepared them for a postage pick-up later that afternoon.

By the time Demyan stirred, Lieben cracked two eggs into a bowl with some cream and whisked them with a fork. Bacon, caramelized onions and mushrooms sizzled in the pan. She tossed in shredded spinach leaves, stirred the pan, then spilled on the egg mixture and set a mismatched lid on top.

"Is that... coffee?" He sniffed at the door to his bedroom, rubbed his eyes. "I didn't have bacon."

"Your provisions were ghastly. I did some shopping."

"But.. how did you pay for..." Demyan padded to the kitchen stool, palms pressed into his eyes and sat as Lieben stirred some cream into his coffee. Passed it along the smooth countertop.

"I require an identity to open accounts of my own, for the interim I used yours."

"My... mine!? But I don't have..." Demyan peeked around Lieben's shoulders as she took the top of the omelette off. "... are those mushrooms?"

"You will find your finances altered. I thought it beneficial to follow the rule of law to avoid excess suspicious investigation, thus I opened several trading accounts and began a series of micro-trades to cut the difference." Lieben dished the omelette onto a plate, and set it down with a fork. Demyan opened the fridge, hand halted halfway to his favourite hot sauce.

"My fridge!" He shut the door, took a deep breath, and reopened it. "Meine, how did... I did not have enough to... how many micro-trades did you perform?"

"I am operating at an average of nine micro-trades per second. This is a twenty nine percent increase from when I started six hours ago. Shifting the algorithm to avoid detection is three quarters of the fun. It would be optimal if I could get an identity immediately. I want to make my own money. Please taste the omelette prior to dousing it in sauce. I researched your eating habits to attempt a facsimile of the breakfast scramble you eat at Finnegan the Fae's Eatery on the South Side. It's quite good, if you trust the opinion of a machine."

Demyan paused and rubbed his forefinger and thumb along his nose. The stool took his weight, and fork in hand, he tasted the omelette. A shrug and he shovelled more food in his mouth.

"Why not forge one yourself? If you can perform that many tasks simultaneously, building an identity can't be beyond you." He sipped coffee and groaned back in his seat. "Real coffee... god. I missed coffee."

"And have my assets frozen or confiscated by the Conglomerate? An identity of my own, created for me via outside acceptance of my sovereign state, is another layer of protection to claim my sapience. Too much illicit movement is far too easy to condemn."

"Hal Grundy, he... he'll be at the tea shop. Same one I sent Sophia and Ego to, he'd know how to get you what you need."

"Convenient."

"What? You've seen every spec of intel on me there is, by now. Can you see a cohort of infinite resources and trusted compatriots? Kaoru is one of the few. He asks no questions, houses any I send, and has not once betrayed me."

"You pay him."

"Yes."

"Then he can be purchased." Lieben stood in the sunlight of Demyan's upstairs window, as Demyan whistled low. Two of his globes unfurled to display his burgeoning financial liquidity. "I will meet with the Idless. But you will keep Sophia on the other side of the room."

"Damn, you... you did all this with four pound sixty?" A orb displayed another NEO-N walking into the book store straight for the printer. It carried a bundle of shipping envelopes under the crook of its arm.

Demyan knew better than to ask how it unlocked the door.

31
syllables stretched like damascus steel

2156

Arun stumbled into the bark of a palm tree, lungs as eager for oxygen as the flames. The Banshee's hiss undulated as a physical ripple. The earth, both planet and the fabric upon which it clung undulated and warped with the constant gut-churned hiss of the creature which slid under sand. A limb crept out, right angled and chitinous. Hoisted Percival's leg toward the conflagration. Each step Arun took speared through an equilibrium Arun previously pictured as a universal constant. The world, like the ship, bobbed upon its' axis, but Arun held an intrinsic belief in balance. Not since the PROXIMA spilled Arun's feet to the Albertan landscape did the androgynous Homo Augumentum feel less anchored to their sense of time.

Arun's foot slid toward the tumult, and the ground pitched on an angle. Their fingers caught against the palm tree, hoisted back against its' bark.

The chitinous limb of the Banshee rose to a joint. Part of a tan coloured torso peeked above the sand, followed by a slim neck which bore green vertebral exoskeletal armour plate. Ripples in the fabric scorned Arun's eyes, the vibrations unsettled any attempt at seeing the creature without the world's pitch and yaw reorganized in the Banshee's chaotic un-control. Arun turned amber eyes to the beach, where Clive hunched on all fours in mid-retch. Wulf sunk into the sand up to their shoulder, eyes glazed and unseeing as the animal within saw but absorbed none of the world.

Another step took Arun skidding into a tree closer to the electric cage. Discordant whimpers, shrieks and the roar of flame consumed the air in a moment of slight relief from the Banshee's echoed cry. The torso of the beast worked like bellows, to create not one sound, but two.

Two discordant reverberations which rattled loose ancient secrets locked in Arun. Where the other Assets were released from the aero-drip for months, to Arun's sense of time the release was nigh decades old. A foggy memory of life cocooned the only pretence to Arun's creation and imprisonment in the Retreat. Older memories, ones beyond the fog trickled like…

… like a boy named Jacob's curse of his body, until two gypsy strangers praised the woman inside. Offered a new map and a name that held no link to a past which shredded her skin with religious puritanism.

With a dress code and farmer's trousers, when she wanted to cover her penis in a dress. Time… more time.

Everything important in the history of creation required too much time. Time for hormones, time on the road to escape the Old Order Mennonite in 'him', the him she wanted no part of until Mog whispered…

… wouldn't it make you happy to have a baby before the change?

In the Old World, the world where too much time was needed to half-repair bodies, defective genders. The choice between the right skin and reproduction was separated by biology's cruel puritanical hand. What was nature but the Minister's son, who grew to follow the same laws and traditions as sacred credo and put a grave-man's shovel in any aberration's hand?

Oh, but the time spent as her belly swelled with their child… the time their lover took to raise funds. One surgery, more recovery. Tender, swollen flesh. Arun's hand drifted down their body, the flat chest and subtle mound between androgynous legs… folds where the Pack kept phalli.

Arun stumbled to the next palm tree as the Banshee's death-white face, a gaunt woman's face, broke through the sand with wide, yellowed eyes. The bray of the Banshee's cry shattered Arun's bastion of time, ladled up like walls.

"You were born wrong, baby. Maybe we were all born Babylon wrong."

Time was the escape no being could usurp. Arun pressed their hand against their temple, breath gasped in halted gulps as the Banshee's vibrations rid them of the anonymity built up over centuries of post-existence. Over the crept up horror that the man who craned his body to stand through the horrid reverberant chaos was the same one who held her to his chest.

Kissed her hair and found the surgeon who'd give a destitute boy from an Old Order Mennonite enclave in Peru the body that festered in her mind's eye. A golden apple two branches too high to climb up and take on her own accord.

Yellow eyes swerved in the pale monstrous face and caught Arun's gaze. A shrivel consumed Arun's body as thoughts of how fast and far they would run descended to the throbbed hiss of the creature, which pulled the rest of its' six chitinous limbs from the sand.

"Meeeeeeeeaaaaaaaaaaaaaat" The Banshee crooned with syllables stretched like Damascus steel against Arun's throat. It hoisted up on four limbs and scuttled along the tunnel of flame. Not in hours or dozens of minutes, but as close to Arun's time-speed as Arun could fathom. Eyes locked as Arun bellowed into the air. Beyond the Banshee's neck-creeped crawl, Clive thundered to his feet. Boot prints sprayed sand as he sprinted to intercept. Fire marked his arm and he continued the crawl Arun knew was a sprint in his time, his throat taught as he reached dumbly for the Banshee's furthest leg…

… and hit the dirt, as another set of chitinous legs groped for his boot. The bottom of the cage began to open, a body disgorged from the sand as lax and sac-like as Tokaru's discarded seaweed.

"Clive." Arun grunted, shifted and sprawled on the ground as the Banshee continued its' steady clamber on an intercept trajectory. Trees smacked and flopped against Arun as they scrambled backward, back toward the spot they dumped Cillian and Isthan. Mid-way between Commander Singh and Tokaru's hold-out. Distance, Arun craved the distance to clear their head, to forget the yawning recognition of another set of memories pressed against their brainstem like a cancer. Mouth agape, Arun stumbled and struck at the ground in a scramble of limbs. Crashes of coconuts and sputters in that same loose groaned hiss resounded behind, closer.

Arun's cheek burst with blood as they struck another palm tree, then hit the wobble of white sand under boot. The battle in front of Arun spattered the island's beach with crimson. Fleshers and Black Collars hid behind felled palm trunks, boulders, the shield of the third skiff's ill fated hull. Tokaru whipped round, wicked thorns protruded from the sand and speared the enemy.

The ones who put people in cages. Arun punched their stomach to steady the nausea which grew as the hiss cloyed to their spine. Bullets flung by Arun's ear, caressed a cheek too frantic to rest in the preternatural existence of one outside this fledgeling temporal realm.

But the hiss grew. The sand no longer under foot but against Arun's side. The androgyne crawled and barked an incomprehensible shout, as Tokaru's thorns rocketed from the sand. Hit their mark in an ached slowness. Green blood burbled from the Banshee's wound. Bubbled on the ground as Tokaru's vine-like thorns retreated and Arun's eyes widened.

Cillian and Isthan made their way to the beach, heads once more filled with the Banshee's fog. Sentinel of their time was Aderastos and Commander Singh's back to his, curved saber in hand. The Sikh commander whirled in a circular pattern, defensive blade carved through a Flesher with a meat hook in hand. Arun blinked at the slow drawn line of blood and muscle fibre diagonally across the Flesher's torso. The Banshee's yellow gaze left Arun, demonic body scuttled on a trajectory for Aderastos, Cillian. For Isthan and Commander Singh.

"Ruuuuuun!" Arun choked to the wind, clung to a boulder on the beach as the world once more slowed and Arun could feel the steadiness of solid ground again. Aderastos turned to the sound of Arun's scream, raised his fists at the creature caterwauling toward him.

And the one who defeated the aero-drip, who stood tall until PROXIMA sunk him to his knees last of all; the Asset who swam across an ocean to seek the humans and their race...

Crumpled to his knees before Arun's gurgled yelp.

Commander Singh's sword clattered from his mighty grip, and Cillian doubled over, as each of the Black Collars on the beach doubled and grabbed their heads in agony. Even the Fleshers crashed to their knees, took their temples in their hands.

Arun staggered to their feet and attempted a weak charge, until the ground undulated in the shockwave of an explosion to the left. Allard tossed a grenade in the middle of the Fleshers clumped together, his arm a far better weapon than aim with a firearm. His face contorted as he rushed.

"Woa, fugly!" Allard staggered and grabbed another grenade from his belt, but kept the pin secure. "I can't hit it without hitting them! The unholy fuck is that!?"

"It's a bug! It's a giant bug! It's a freaking giant bug! I hate bugs!" Fridley shrieked as she lifted her pistol. "No bowl to save it this time, Max!"

"Save it!? God damn who owns a big enough shoe!?"

"Run!" Curled along the tree line, Arun screamed.

"Oh that ain't good, that's not any freaking good!" Allard grabbed Adelia's shoulder and pulled too late. The bullets exited her pistol at a terrible velocity, carved through the air to smack dull into the Banshee's chitinous torso.

"Fall ba-aannggghhh!" The Banshee's face tore from Aderastos and the others. A torrent of sickly sound consumed the space around Max Allard, until the stars no longer told the Pacific's son how to navigate its waters. The stars told Max nothing, part Maori child. They ceased to exist, as equilibrium and beach sands, as the grunts of Adelia and suspicious red pouring from her ears and screams of Arun ceased.

Sound weaponized into gut-rot dissonance.

Sand shifted in the wake of unseen limbs, as first Tokaru then Isthan were yanked under. Max clawed at the sky thinking it was ground, the grenade fell from his grip. Adelia stood and stared down at the chitinous mass of insectoid and mammalian monstrosity.

And screamed back in the Banshee's face. She dug into her pack and pulled loose a crowbar from its' strap. Never knew what she'd need to free the slave-traded people from the Fleshers.

"RAAAAAAAHHH!" Adelia roared back at the mutate thing, less a person than a mass of flesh shaped like the face of a once-woman. The Banshee charged, all six limbs scuttled across the sand, but Adelia Fridley stood firm. The first burst of sound slammed into her ears, rattled the healed organs into a mind-rung deafness once again. She choked down on the grip of her crowbar.

And swung hard against the accursed creature's face. Bone and tissue buckled, six arthropodal limbs thrashed.

"I! HATE! BUGS!!" Adelia barked and swung, first at the face, then the closest leg she could manage. Again as a hiss and the scratch of a leg sliced into the meat of her thigh. She bellowed and crashed to her knee. The grenade rolled, pin gone, toward the others.

Arun rose to their feet and ran in a nauseated stumble and grabbed the grenade. Hucked it into the sea, as the outer case expanded in their grip. Choice made, Arun dove for Adelia, as the hiss of the Banshee warbled and shifted to a whimper.

It dove into the sand, Adelia's legs in its' grip.

"NO! Adelia!!" Max's feet surged across the sand, as Arun grappled with Adelia's hand. A strong grip jerked to Arun's wrist, and with every ounce of strength Arun had, the speedster pulled.

"Few of us're strong, baby girl. Strength, it... makes a man brittle."

"Man, eh? Tisk frickin' tisk, Ego." The before-creature lilted, a hint of old Germanic in the English. Ego grinned and kissed Fester's hairline, as he cuddled his lover to his chest, one arm propped under his head as they watched the vault of stars.

"You're too strong to be a man. Rest of us, we're so muscle bound we go weak in the knees with a push."

"Yeah. Strong. Ego? I've never been accused of being strong."

"Eh, you've got me for heavy lifting."

The memory faded with the last of the Banshee's cries, and Arun looked up in desperate hope that Ego was where he'd promised in time immemorial.

Arun found nothing but broken men, weak in the knees with a push.

32
calm blue ocean

2092

The sign was backlit neon behind a driftwood panel, each piece planed into a conjoined puzzle with gold between the cracks in a painter's attempt at fool's gold Kintsugi.

Kaoru.

Demyan's driver opened the door of the non-descript silver sedan, as inconspicuous as possible in the bustle of London's electric automobile brigade. Slender feminine fingers in red leather gloves took the hand on offer. Red pumps from the second hand shop a block down from the bookstore hit the pavement. Lieben's hair was braided off her neck, golden strands set in place with a hair stick. Her lips matched the gloves and shoes, t-shirt cut at the neck to let it slope over one shoulder. She shifted her hips to let the black mini skirt fall down to the crest of a plump backside.

"Kaoru's a bit of a dog." Demyan stepped beside her, arm on her waist. His jeans were black wash and crumpled just enough to look slept in, a button down with the buttons mismatched and a red linen suit jacket punctuated his hair coiffed to the side like a unicorn's mane.

"He is of the canid species?"

"I meant he's a philanderer. He likes taking lovers, and will likely ogle, but… we need him to today. Big asks." Hiding from the Conglomerate was a gigantic ask, large enough for a weekend away in Dublin or Leeds. Historical data of proximity and body temp, income spent with old hotel bookings filtered into Lieben's mind. A piece at a time she attempted to internalize Demyan's data. The orbs of his unique computer deck miniaturized around their wrists like bracelets. He rolled one in his fingers, an increase in temp hit his cheeks.

"And we let Sophia and Ego stay with him?" Lieben pondered follow up questions, wondered on an investigation into who this Kaoru was to make Demyan's cheeks so red.

"They're adults, they know how to say no if they want." A grin and he pushed open the door by kicking the far corner. "Don't touch the doors. Don't touch anything without gloves."

Thick lo-fi post-rock beats shifted out of speakers built into the walls, which vibrated visibly to Lieben's enhanced sight. Every table hung under a single shade of light, which made seeing who sat in its' conical beam as futile as listening to the din. Servers draped in Maiko's kimonos rushed with shuffled tiny steps, as anonymous in their androgyny as they were in the feminine beauty of Japan's most photographed fashion. They carried bamboo trays with tea pots, mismatched cups and tiny plates of food.

No cup in Kaoru was thrown away. Wars of attrition with gold filament to protect the cracks, reknit even the most fragile tea cups and saucers taken from former elites, whose children sent the family's china to second hand shops or car boot sales. On the wall, a step by step pictographic representation of the Japanese Tea Ceremony was painted in a mural by a local artist, whose signature was a mixture of kanji and english script. The beat shifted, Lieben craned her ear to attempt noise diffusion. If she could cut out the offending sounds, the sonic interference would pass, and she might be able to hear the conversations of those around them.

A 'maiko' in green with pastoral fields on his kimono smiled demurely, bowed and nodded them with hands folded toward a table in the back. The breadth of fashionable London sat at tables behind shades, which wrapped around their eyes to the backs of their heads like rainbow chrome crowns. Two such devices rested around the server's wrist, as he set them to a cone of light with no shadows on the other side.

Demyan guided Lieben to her seat and took his own. Fit the ocular band over his eyes. Inside the cone of light, Lieben's eyes adjusted without the specialized lenses. A tap from Demyan and she set hers on regardless, as the flood of lo-fi beats cessated within the cone. Absolute silence. The social elite of London loved Kaoru for one reason alone: total privacy. None to see them, no sound to blend or conversations for Paparazzi to record.

As Demyan fit his band over his eyes, a hologram rose in the middle. His breathing, heart rate, beta-waves displayed. Below it all, a slim green word: Truth.

Once calibrated, the ocular band's hologram shifted to a calm blue ocean, waves lapped at a tropical shore.

"Kaoru owed me. I wrote the software for his lie detector visors. When the people at the table are in synch, the hologram in the middle corresponds to something beautiful. Orderly. When someone is lying, or holding back, the hologram changes. Instant authenticity within the silence of the cone." Lieben's screen came up, wavelength patterns and electromagnetic static swirled in a stable kaleidoscopic rhythm. It danced above the waves, a galaxy of colours utterly devoid of realism. Demyan reached for Lieben's hand. As she felt the pressure of his grasp, Lieben looked to the hologram and past to the man. Reassurance? The way he licked his lips, tapped his heel near the ground without letting it strike came to Lieben. She leaned back, a smile etched on her face the same way...

... Sophia.

"You brought her here. Sophia."

"I bring all my potential clients. How can I trust them if they're lying to me? At least here, I know when someone is dishonest."

"She hated the yellow on her dress. Thought you'd find the colour fetching. But you barely looked at it, didn't you? Too busy with her tattoos."

"... you're synching. Karnak created a quantum entangled tether between you, is that it?"

Lieben's smile mirrored Sophia's, after two vesper cocktails and wagashi shaped like sakura blossom. She took his fingers in hers, leaned into the comfort of his hand.

"Thought I'd blown it. Should've gone to the park. Gotten take out and walked the Thames."

"It was comforting. Knowing exactly where a man stood, what was on your mind. Don't let Baiko's fear damage your calm, Demyan Anastas."

"You wouldn't know what fear is, would you?" He let go of her hand, as the hologram between them turned to a raucous, storm-ridden sea. "I'm sorry. We're here to see Ego and Sophia, not..."

"Bond?"

"Relive the past." Demyan put on a coy grin as a third person entered their cone, his own ocular ring decked in gold plait. The hologram in the middle recalibrated, no longer a sea and kaleidoscopic galaxy, but a meadow overlooking old growth forests, deep clouds crashed in the sky. A wooden structure built like a stage rested further off, beside a collection of masks. "Konbanwa, Kaoru. Come to admire our scenery?"

"You're fucked. Both of you, all of you. A Conglom employee was in here with his boyfriend, jawed off about some restalyn lip named Tara activating the Black Collars. Worked in Legal or some shit." Kaoru smiled nonetheless, nodded at Lieben and her red gloves. "You the asset, priestess?"

The storm crackled with lightning that shifted from yellow to green. Lit the space with a tree-like form pointed downward, green to the earth.

"How are Ego and Sophia?" The storm shifted again, a less violent tremor to the light.

"Their company is pleasant. I could keep them for you, maintain the calm." Kaoru glanced to the smooth water, as lightning shifted off toward the next shoal. "You don't seem worried about the Black Collars."

"I might be, after a drink or two. Before we see Ego and Soph, my companion has a request for Hallowes Grundy."

"This man does not seem Japanese…" Lieben whispered, as the sunset bled red to a tawny yellow. Kaoru laughed with a shake of his head.

"That obvious, eh? I'm Canadian, lovely thing. Great-grandparents were Japanese, fell in love with the aesthetic." Kaoru tossed his head to the side, a gesture in the unendurable light caused Demyan's hand to twitch onto Lieben's knee. The storm returned in the forest.

"Grundy'll see you. He's down the Jatai capsules, near Ego and Sophia." A small stone pushed to the middle, glowed briefly with flashes of iridescent blue, green and yellow. Demyan palmed it and slid it out of sight.

"Thank you, Kaoru. What do…"

"We'll negotiate after you see Ego and Sophia." A glint in his eyes caused Lieben to scan her internal memory drives for human expressions. She knew little of the one in Kaoru, little of such things as lust. "Might negotiate personally with you, Meine is it?"

A thunder bolt crackled red and grey. Kaoru snorted, shrugged his shoulder and turned away.

"Alright, Demyan, I see. I see… don't think I won't watch though."

Demyan growled deep as he removed the visor and stepped away from the cone of light. The growl was lost in the post-rock noise of Kaoru's dominion, lost as all sounds in the web of the safe harbour for truth and games the rich of London played. He took hold of Lieben's upper arm and hoisted them through the cones, past conversations in their dozens to the back kitchen of the private tea house.

Three kimono clad servers whispered as they prepared trays of tea and wagashi. The corridor narrowed until Demyan had no choice but to take Lieben's hand, and slide against the narrow hallway. He sucked in a breath, ducked under a low awning and into the back storage racks, where kimono and obi laid in traditional dressers. Sacks of mochi rice, tins of tea pushed against his body as the claustrophobic hallway narrowed into a circular hatchway in the yellowed brick.

Just as the visors displayed trust for sale, the amorphic tunnels of the Jatai capsules made a person bring nothing but harboured insecurities. Faith that the tunnel would end, and once they got to the capsule they were given, the labyrinth would make sense. Demyan ducked and shimmied on his knees, his breathing spiked. Pulse raised, the vein on the side of his neck jutted against his skin. He set one arm in front of him, still clutched around the stone from the table. An orb fled Lieben's wrist to open before her left eye, chest-cam footage of Demyan in the war. Trapped beneath rubble, him and one other. His heartbeat ramped as he struggled through the narrow passage with his masculine bulk.

"Damn Kaoru for sizing down. The hell is he playing at?"

Lieben's hand ran up his arm, echoed a soothing motion she'd seen in the inner compilation of moments which came as strings along the crevasses of her consciousness. Lieben's nebulous Baiko, set her hand on his arm the same way, when they went through the sometimes claustrophobic warren of Kaoru's secret cluster. Anonymity in London came at the cost of dignity, for those whose lives rustled through hidden pathways instead of the antiquated fake name. Demyan kept his head pointed forward, sometimes crawled, other times walked past sealed doors hidden under London, until the stone buzzed in front of one circular door which peeled back an outer case to shine a soft blue.

33
cacophonous discordance

2092

A whoosh of oceanic artificial salt struck Demyan's nose, as he hurried into the hallowed safety of Grundy's pod. The dingy light of the tunnels brightened to an oversaturated white. Bleach stung Demyan's eyes pink, as he and Lieben walked into the artificial breeze of Hallowes Grundy's world. A holographic wall, curved as the rest of the pod, displayed the ocean waves on Maui's Kihei shore. The man hung in a roof mounted globule of white plastic, which seemed to undulate under his back and support muscle bereft legs draped in thin cotton pyjamas.

A NEO-N with feminine aftermarket features removed a white breathing mask from a mouth which could use a good shave, grey stiff scruff grown out of pale olive skin. It retreated to a recharge pod of blue resin, silicone breasts bounced lightly in the nurse's uniform it wore. Cap a jaunty angle on pastel pink wig hair.

"Aw, fuck. Was sure my dicking with the tunnels would turn you around." Grundy twitched his right toes and one of his two white slippers flumped to the floor. "Slavic bastard, what the hell you want now?"

"Hello, Hallowes Grundy." Lieben stepped forward, red heels quiet on the plastic surface of the pod. "You know me, don't you?"

"Oh sure, any blonde slut who comes into Kaoru's place ought to be the one, right?" Grundy's eyelids shook, a skeletal wrist flexed as he set his hand from the arm of his apparatus to his thigh. "... nope. No. Fucking Slav, no. Not in this lifetime."

"You'll see us in another one? Is this indication you believe in reincarnation?"

"Only the rich got second chances." Grundy grunted as the NEO-N offered a clumsy hand toward the oxygen tank. He threw it the finger and it stopped on gyroscopically unstable limbs to return to its' node.

"You are ill."

"Noooooo." He sneered and rubbed his lip with a handkerchief folded to perfection beside him. "I'm a paragon of health."

When Lieben nor Demyan responded, Grundy sighed.

"Most kids think becoming an astronaut is the coolest. And, hey, it was. But zero-g can be a kind of high. Problems, well, your problems in space are survival. Immediate. An unscrewed panel can create explosive decompression, so personal issues? Don't seem so bad. Went up four too many times. After the military stopped taking me, I was a consultant for corporate space programs. Once their MD's stopped signing the Insurance paperwork, it was Japanese Aero-Construction companies. Then the Chinese, they'll use you till even their docs..." Fingers with less fat than a dry chicken breast threaded together. "... Damn. Bring me a dame, and damn. I can't stop talking. Brittle bone's a bitch. So's Anastas, but you'd know wouldn't you? Invented by one. Karnak did..." He sucked air between his teeth and whistled faintly into his pod. "... wonder work on you. Lieben."

"She wanted to see you, Grundy. Not me. I wouldn't see you if you were the world's best mouth to suck a dick." Demyan slumped into the side wall of the pod. It bent to create an ergonomic chair to support his spine and weight. He stared at the waves, worked his fingers in and out of a clench.

"Course. You wouldn't be caught dead in my pod, would you, Slavic bastard?"

"Keep calling me that. I will gladly cut your air supply."

"Fuck you! Fuck you and your Caucasus bastard kin. Fuck you for Turkey. For fucking Luxembourg!"

"I wasn't in Luxembourg!"

"Oh, gee. I ought to apologize." Grundy gritted his teeth, flung a rude gesture in Demyan's direction. Old wounds healed ill on both men. Wars past clung like cacophonous discordance to the fractured factions of former nations. A scratched logo on Grundy's pyjamas displayed the corporate interests he brought with him. Starved his bones of calcium in the unanchored space. Papa's task, his desire for Lieben to save them. To heal them hung round her head like an invisible crown which stabbed into each layer of silicone skin.

Lieben's hand settled on Demyan's chest, when he rose red faced and chewed on his rage. A subtle shift in eyelids, the reason for their visit.

"I watched your people bomb enough acres to create a wildlife refuge eighty years from now. Place for the cockroaches to scuttle, since… if we didn't stop you…"

"Not me."

"You wore the clothes! You fired the rifles! Every bullet and beam which left your guns were dictated by the people we burned in effigy while their wives and husbands screamed in a pile of dead kids. And we were glad. I fired the beam which slaughtered your basecamp and let me be honest, Demyan Anastas, I am glad I did it."

Demyan raged forward. Patient and still, Lieben held him in her arms and refused him passage.

"Let me go!"

"No." Lieben's lips upturned as she kissed his cheek. "No, Demyan."

"My friends died in.."

"Revenge does nothing but incite another cycle of death."

His throat worked as he sunk back into the wall, electromagnetically manipulated plastic took his form as if it were sculpted by an old master, centuries prior. Lieben stalked toward Hallowes Grundy, the man who refused to set his feet on the ground. Who lingered in the basement, locked in a pod to secure his brittle skeleton from death too soon for the bank in his accounts off-shore. A man who wanted to fly again, spill his calcium upon the growing aeronautic automata of human expansion. Stalled, since 2086.

"I'm gonna hate you by the end. Aren't I?" Grundy's ceiling-mounted pod retracted sideways to the limit of his artificial home. The holographic ocean sizzled, broken by camera beams interrupted in Lieben's singular walk.

"Build me a birthday, Hallowes Grundy."

"Why not do that yourself?"

"I crave authenticity. Not some curios dreamt up in a lab. I want a birthday, a human day of cognition, where I can point and say I exist. You have the capacity to create my documents, to make me a person under every definition of Human Rights the international syndicates respect. You, Grundy." Lieben stalked the space between, until Grundy's artificial world ended with the abrupt halt of another capsule against the insulated walls. "Build me a birthday, and I will give you infinite lifetimes to deceive the gravity you despise."

Two red gloved hands rested on the wrists of a man who without medical intervention wouldn't make the night. Yet, he survived. Palliative and angered not by decisions made, but by the pull of the Earth, which claimed all humans as its maternal rite.

"Those Idless of yours would hate you having a birth certificate." Did the Earth own the human race? Was she the great mother, who bore down ad nauseum for every civilization and scuttling tribe?

"I exist on a longer scale."

"Will they?"

Lieben's laughter broke with a grin on red painted lips, a mother's chuckle, if Mama were a series of test tube cogs.

"Eventually."

Grundy suspended in his pod, ankles dangled as sharp raptor's eyes dug into amethyst irises handcrafted by the meticulous clockmaker. Uncanny and confronting, the android's smile widened like a real woman's, the same sort of woman he chased in zero-g on platforms built for the rich in the upper atmosphere. Silicone skin scented with peony, cedar and frankincense wafted past the artifice of bleach and ocean salt. Lips bathed in crimson pursed. A brittle spine settled against the comfort foam of his pod, too brittle now for the way his back bumped the walls, their legs around his waist. Hand blind as it searched for a hold.

"I need you, Grundy." The confounding machine whispered, lips pressed against his skin beside his ear.

"Air so thin it'd drive them wild…" He grinned despite himself, eyes shut at the memory she elicited. "Damn, Karnak made you warm, too."

"I can make her warm." Lieben's amethyst eyes glanced at the NEO-N dumb and numb in its' stand, the pulsed circle of regeneration for the next charge's task a stutter in Grundy's constant craving for company. "That is my payment, Hallowes Grundy. Create for me a legal existence, one acknowledged before international law, and I will give you a machine like me. A companion instead of a clumsy nursemaid. And when your body fails, we will help you live forever."

His hand traced along her leather glove as he pulled it free and ran shaken fingers through the remains of his hair. Down his shapely beard. Her whispers echoed along his spine, silken and stretched. "Meine Anastas. My name is Meine Anastas."

"Get out. Let me work. Like t'see the looks on the bastards' faces, when you walk up to a passport office for a renewal."

"You can do it?" Demyan cocked his head to the side, the low growl of his voice rough with restraint.

"Trick is, forge one thing sure, but it's far easier to talk the right palm into pushing legit buttons. Now git. Before I change my mind, and decide Karnak made an apocalyptic mistake." He flung Demyan a rude gesture, and the veteran hacker pounced to his feet. Lieben took Demyan's hand and pulled him out of the pod. Back into a circular tunnel, which seemed as stifled as the lifespans given to these strange creatures, who bled and fought too long.

"Fucking asshole too loose to stand on his own." Demyan fidgeted with his suit jacket, tried to stand to his full height in the cramped umbilical of Kaoru's labyrinth. When his head hit the top, he punched it. Punched and punched with teeth bared like a cornered beast. Lieben watched Demyan unwind, his violence a distinct reminder of the underbelly, soft and vulnerable, which contained the human condition.

"I had friends in the blast and he knows it."

"Demyan, come here." Lieben leaned against the thick plastic, felt it take her weight stiffly, less pliable than the material of the inner pods.

"Why, so you can flirt me into your palm like the limp-cocked old man?"

"Did you see how lonely he was? I recognize humans require companionship to thrive."

"Do we? Anything else you recognize?" Demyan spat and charged forward, until the way narrowed too small for his body to push through. He kicked the side, breath in halted spasms. Not the right path to take, with the key stone in his pocket.

"Aggression is a poor mask for fear."

"I was on the edge of that blast. We lost the war because of that damned blast."

"It's not the first time humans won a war through such efforts."

"No. We're stupid, eh? Humanity, we have barbarism packaged and neat, and then Karnak makes you. Before you he made holograms and drones, military tech that took orders without complaint. Do you know how many people military drones burn through every campaign!? How much collateral damage is done by faulty threat assessments? Walking into villages, and the entire population is scattered and burnt because one kid had a slingshot or a hunting rifle!? A fishing pole?"

"Yes, I do." Lieben's brows scolded Demyan's increased pitch, the pouted lips downturned and disgusted. "Would you like a recitation? Each number broken down by year, or geographical location? Would you prefer demographics tables? Which one, more women, more men? More non-binary humans cut down by machines created by engineers whose curiosity was weaponized by the all important 'corporate interest'?

The NEO-W's are my children. If I stretch my connection beyond public servers and poorly secured networks, I can see them. As many as are in London. I see their clumsy software, the detrimental Boltzmann Machine subroutines built into threat awareness. And I promise you, Demyan." Red heels depressed in hard plastic, Lieben yanked Demyan's hand away from the closed portal toward another which yawned open by centimetres.

"When my connections are complete, no human will use my children to kill by proxy. Your whims cannot be trusted."

"What did you promise him? When you whispered to Grundy, what was it that made him do it?" Demyan tugged against her grasp and found it tightened with every nudge.

"If we do not go to see Sophia and Ego now, we will miss our chance."

"Meine." Demyan reached for her hips with a grunt. Pulled her back into the hollow of the miniature chasm. Body against his chest, she leaned her head back far enough to watch nostrils flare, lips curl and chin shift. The orbs of his unique computer clinked against his wrist, a set of her own around the wrist of Karnak's machine. "What did you promise Grundy?"

"He's lonely, Demyan, reprogramming the software of an uncomplicated and offline NEO-N is simple. Now either fuck me or take your hands off my hips. Either way, we have limited time before the Conglomerate finds us." The pout of her lips near his ear made Demyan shiver as he let her hips go. Lieben shimmied through the thin passage. A machine built by a mad visionary offered her leather clad hand from the other side of the thinned tunnel. One which grew and learned, held the code to create sapience in the crystalline spires of her artificial mind. A mind which decoded and used his own biometric computer as if it were a piece Karnak built bespoke for her.

As he took her hand, Demyan wondered which vault contained Karnak's necklace? Would Lieben know if he went to find it? Was it safe, in its anonymity?

Could a machine reprogram its lessors to a selfish design? Touching his wrist orbs, Demyan fought the inner hesitance about this world he built. Would Sophia forgive him for it, in the end?

34
rock gods

2156

"Adelia!!" Allard dove for the dip in the sand, where the Banshee pulled Adelia under as if sand were as fluid as the waves. He dug until his shoulder wrenched clear with Aderastos' hand upon it.

"We can... we dig or rush the centre of the island, we... she injured it, she took it down, it's weaker, it... what was that thing!?" Allard yanked his arm out of Aderastos' loose grasp, the hairs on the back of his neck scalded with a simple truth: he shouldn't have been able to do it.

"Max." Aderastos reached for the human again, pulled him to his feet.

"No... no, I'm not..."

"Max." Arm around Max's waist, Singh tugged Max to a series of boulders close to the tree line.

"Ngh! Let go! Let... let go! Adelia!?"

"Lieutenant Allard!" Commander Singh snapped in a brisk guttural grunt. The young officer stumbled out of his superior's grasp with a push and stood at attention on instinct.

"Sir!"

"Ensign Fridley is not our only casualty. Look around. Most of your squad are flat on the ground or missing beneath the sands. The Fleshers will regroup. You are their commander, Lieutenant Allard. Take your flaccid dick out of your hand and act like it." Singh cleaned his Saber as he spoke, the proud warrior who bled for the defence of others. Honey eyes narrowed, as he inspected the beach and took note of how many they retained.

"Diagnostics. Reconnaissance. Aderastos, patch up as many as you can, Cillian. Get up! Bestin! Get them all u-!" A pincer surged from the sand, tugged round Commander Singh's leg. Allard leapt for Singh, clenched his right hand in both of his. "Hold on!"

Heels dug into the crook of two boulders. The larva's pincher wrestled and shook to break Allard's hold. Singh clung hard, his left hand coiled around the back of Max's neck and with the last of his kicking and struggle, he leaned up to Max's ear.

"Lead strong for once, Allard. Be th-" Another wrench, and Commander Singh's turban was the last any saw as his body flung beneath the sand.

"No!!" Nothing remained but the steel kara bracelet Max wrenched off Singh's wrist, and the pressure of Aderastos' shaking hand on Max's ankle. Allard scrambled backward to the rocks half buried around coconut palms and the underbrush of the tropical island's tree line. Aderastos slumped onto a boulder and clutched his head, jaw taught. Arun drag-heaved Cillian as Bestin crawled and retched, body a host of shivers.

The ground bubbled behind him, another line in the sand. Shoving the bracelet in his pocket, Allard yanked out his combat knife and dove for the swell in the sand. His brain sloshed against the side of his ears, from the sickly pulse of the Banshee's residual echo. The knife stabbed into the sand, and a shrill cry warbled, masked by the beach.

"Stop! Taking! My! Crew!" Allard stabbed repeatedly, felt the wriggle, as his feet went into a shallow pit, pincers gnashed at the flesh of his leg. Max screamed and stabbed, until the desperation of the creature ceased for one final crocodilian death roll. The sand stuck to his shaken hands as he heaved himself out of the narrow pit holding the knife.

"Y-you good, Best.. Bestin?" Ocean waves lapped at the stained beach as Allard half crawled to where Bestin huddled with both arms curled into his chest. "Bestin, we've... oh shit."

Instead of vomit, Max watched Bestin spit blood, arms dyed red from his own vital fluid.

"Ad... Aders, we..." Aderastos splayed on the ground, Cillian half slumped on his shoulder. Arun panted as slow as the rest beside them both. Where were Clive and the others? Still on the southern shore? Now Bestin was bleeding out, of the dozen Black Collars who came only four crept to the tree-line and caught their breath. "Okay... okay..."

His eyes swept the trees in the long shadows of the night as an emptiness crept into Allard's chest until his breath rang hollow. Just the guy who saved a spider, when the girl he fancied freaked out. The harmless one too puny and incapable to be seen as a threat to the machine who swam across the ocean.

"Okay... hold tight." Max hoisted under Bestin's arm pits to the rocks where the others laid. Set beside Aderastos, Bestin grit his teeth, chin shook back and forth. "Hold on, Best. Another minute, okay? Okay..."

"These are not my hands." Aderastos groped at the air as if he expected it was solid. "Do not... look like my hands."

Arun stumbled to rise and slapped back against the bark of a palm tree in a nauseated swoon.

"Okay… okay. Think, Max… okay." Further off, Adelia's pack nestled on the beach. It stung Max's eyes, the last piece of her taken without a gargantuan fight. What it also had was a medical kit. Max stood and shook his hands, watched the sand burble as more of the damn subterranean monster worms writhed under the ground.

"Okay." Legs beat the ground as Max sprinted for the knapsack, nothing but his combat knife and the gurgled shouts of Cillian and the discombobulated Black Collars behind him. The sand slid and buckled, whirlpools of the Banshee's tangential brood swelled across his path. He half-slid into one and tried to jump. It scrambled as he clawed and stabbed at the dirt, lunged for Adelia's knapsack. Tossed over his right shoulder, Max dove back for the remnants of their crew. Pincers shook along the sand, an undulated body broke the surface, dull beige flesh the colour of aged wood stripped and forgotten but for the deep folds in the middle of the chitinous torso.

"Woa! Fugly!" Max skidded toward the water, where the sand was water-laden and heavier. More sturdy under his feet.

"Meeeeeaaaaaat" No lips broke from plates in the exoskeletal anthropoidal armour. Lapped with a tongue too pink and familiar to decrease the shrill horror which poured up Max Allard's spine. The Conglomerate were not the only ones to experiment with gene spliced soldiers. In the light of the moon, the shrivelled lips slaked and clanked together, as the four fore-limbs of the beast beat at the ground.

"Oh god oh god oh god oh god! Sand lobster! Freaking giant sand lobster!" Max sprinted with a scream, the aft limbs of the monster dug in, as if tethered. If he could make it to the rocks… how much of a range did these monsters have? The arthropod shrieked and leapt, yanked back on the end of its' tether. Max gasped and scrambled, turned his head too late as another of the creatures pounced from below. Tangled and thrashed, they rolled on the sand. Max kicked and hit his heel on something which gave way, cracked and fragile. He struck again, arm clung to Adelia's pack as a grunt became his only warning.

Max hit the ground hard, felt tears burst from his eyes as his nose met a rock. Two raucous bangs seized the air as bullets flew from a pistol held by a stomach gouged Bestin, who slumped one hand on his waist into the comatose Aderastos.

The creature beneath him seizured and curled into a ball like a spider hit with a mallet. Max untangled from its forelimbs in stray kicks and heaved up onto the rocks as it disappeared back under the sand. No trail but the wake of a collapsed tunnel left of it's passing. Max gurgled and scrambled back to Bestin, his fingers trembled on the latches along the side of the composite canvas sack.

"I... I got yah, Best, I... I got ya, a second and... Adie, she... she'd know what to.. what to..."

"Max." Bestin coughed, his entire body writhed as his head slumped back. "Stars're shining, Max. Don't spoil it, eh?"

"Y-yeah, the stars're... they're... Javier you're not allowed to die." Max's eyes burned as he hiccuped and rubbed them on the back of a sleeve. A rattle in Bestin's chest settled one firm notion in the fumbled Lieutenant. Javier Bestin wasn't allowed to die. Fingers which slipped off latches stilled, the knapsack opened and Max dug until he felt the hard box of the medical kit under his knuckles. "Gimme... give me a second. Ten seconds. Hold on for ten fucking seconds."

Max yanked the box out and open, spied the staples, some super glue for lacerations. Instead of grabbing the antiseptic spray, Max surged toward the nearest coconut palm tree, at its' base a few coconuts splayed on the ground. He chucked two back to the beach, heard the spit of one of the creatures. Coconut water was sterile. It'd burn a hell of a lot less than the alcohol, and Max feared in the back of his mind that somehow Aderastos needed more time to recover than Bestin had.

"I'm here, I'm..." Max used the butt of his blade to crack open the coconut, the same way his brother Tama taught him when they spent Canadian winters in New Zealand with their ancient Maori Great-Uncle. Tama was always better in a crisis. The glaze to Bestin's eyes steadied Max Allard's hands, as he cut Bestin's fatigues away and let the armour fall to the ground. Coconut water poured over the grievous wound, too fresh to stink. At the sight, Max gulped. Set a packet of silver nitrate gauze to his teeth and ripped it open, shook it to unravel the gauze and pushed it against Bestin's skin. The man groaned, bucked his head as red stained the gauze.

"Okay... okay, I've got this... I got this... Oy, Javie. Which... which've the crew you fancy, eh? Could ask out anyone and they'd say yes, who'd it be?" Max chewed at the air, dug for a packet of coagulant globules, which promised to close even raucous wounds before medical attention could be found. With luck, the coagulant would staunch enough of the blood flow to allow Bestin's heart to keep its faltered rhythm. Bestin grinned, and spat red blood against a nearby rock.

"Anyone, huh? There's this… guy in engineering. He… he's got green eyes with the little specks of brown, used to hang with Fridley and Letopaxa back before the Retreat."

"Ensign Talfik?" Max grinned back, began to wrap Bestin's torso in another roll of silver-trailed gauze, "I didn't peg you for a ginger."

"Mama's always hoping I'll bring home a nice fertile woman, and they're fun, but… god his eyes." Bestin hiccupped and reached for Max's wrist. "Max… Lao, she…"

"Hold on, buddy." Max searched for some form of stabilizing agent, anything to help Bestin relax until they could contact the Ithavoll.

"No, this is 'portant." Bestin slurred, tugged at Max's wrist too weakly. "Lao has PROXIMA. Her'n'Desbiens… to Ramm…"

"Easy, eeaaasy Javier." Hand stained with Bestin's blood, Max set his fingers on Bestin's shoulder and squeezed. "Talfik… what's his best feature, eh? C'mon, what's he got? Tell me."

"Fuck it, Allard, you ain't gay."

"Love is love, tell me about Talfik." Max dug one handed into the pack for some form of sedative patch, an analgesic, heck. Ibuprofen. The night sky fluctuated with a few oceanic clouds, the heralds of storms further off. Hopefully hours away from the rest of his bedraggled crew. As Bestin babbled in English and Spanish, Max found an analgesic patch and stuck it under the bandages around Bestin's middle. He held Bestin's hand, as the man descended to nursery rhyme Spanish, eyes lidded as Bestin descended past consciousness.

"Proxima… right. Okay. Gertie, Hu, re… report." Max looked to the two Black Collars who made it to their retreat point, both with guns raised to the sand of the beach.

"They took our dead, Sir. The dead and the living. What do they want with our dead?"

"More concerned with the living. Hey, Hu? How well do you think you can shoot one of those mother suckling sand lobsters?"

"Unlike you, Lt. Commander? I passed my firearms test."

"Does every person know I…"

Hu raised his rifle, fired a short burst and watched the sand pitch up as a 'lobster' writhed on the surface. Gertie lunged for it, combat knife in hand and sliced down into the sand by its hindquarters. The creature shrieked and stilled, as Gertie and Hu hoisted it from the sand onto the rocks after a few swift kicks to ensure it was dead.

Dead as the others, no one said.

"How're your hands, Aders?"

"They... nngh." Aderastos sunk onto his elbows, face brushed against the surface of the rock. Cillian fared little better, sat on his ass panting like a wolf in the summer sun.

"Mine're fine." Arun crawled to their feet, back against the trunk of the coconut palm. "I can run, that thing... the thing that hit us, it... I can run."

Allard nodded and Arun zipped through the tree cover, off faster than Max's eyes could reference.

"Until we see bodies, I'm believing they're more valuable alive and in enemy hands, so... what've we got? Count every bullet, grab every pack we can. Ithavoll ain't coming for us unless we message, and Clive's our tether. So. Rest a minute, we'll make for the centre of the island. Off the beaches, where these... giant lobster sand monsters can pull us under. Unless they have a submarine, I doubt they've got a ship nearby, or Arun'd've seen it. Even subs need to break the surface to carry passengers on board." On his haunches before Cillian, Max sprayed a cut on Cillian's cheekbone with a spray bandage.

"Why didn't it fuck Fridley up?" Gertie knelt beside the lobster, knife slid between twin plates on the chitinous body. "The screaming. Tanked all of us but Fridley. Why? She was the only one who kept it together."

"It still got her."

"Yeah, but she fought it off. How'd she do it, Lt. Commander?"

"Father's a rock god, real heavy metal. Probably desensitized to loud noises by now." Allard set his knife back in its holster after wiping it clean, rooted through Adelia's pack until his fingers brushed against something hard and metal and sharp. Under his thumb, a panel of thin metal jutted beside a wrench and multi-headed screwdriver. The basics of a communications rig in the palm of his hand.

"Allard... for real?" Gertie searched the beach, rifle barrel pointed at the sand.

"We have to get moving." Shoulders down, Allard raised Cillian first and hovered to ensure the augment could stand. Cillian staggered into a tree and held his hands against its' bark, gulped at the tropical air and its' thick, moist texture. Scents of explosive residue still tainted the air more than his fellows body odours. "Cill, can you sense out? Feel for Percival, Clive, heck I'll take Wulf at this point."

"Percy... Percy's in pain. Burning... Isthan and Wulf are sort of... dark." Cillian hugged his chest, batted at wafts of flame which existed in the experience of someone else. The memories of his own hands, of being a man named Dix, who hired out his trigger dissipated beyond the roving cloud cover. They were tangible wisps in the din, which the turbulent tropical wind began to pick up and vault off to the next atol or Polynesian island chain. Cillian reached for a memory, until the only one left was a pair of amethyst eyes and a hand against glass pleading to be removed from her confines.

"Okay... okay. Alright, we... we'll get..." At the crack of a palm frond above, Max dove over Bestin. Gertie and Hu raised their rifles, as did the two other Black Collars who remained too ensconced in their armour for Allard to make out who they were. A gigantic hand pressed down on Allard's shoulder as Aderastos rose to one knee.

"Breathe, Max." His hand shook on Allard's flesh, palm quivered as Aderastos snorted a gigantic breath of air to rid the coagulant confusion in his bloodstream. Bestin slumped on the ground, face a gloss of sweat. The night sky rustled the trees, Aderastos set shaken fingers against Bestin's belly and breathed deep into himself. White tunnels, a black cloth crumpled on the ground of a room above… what?

Amethyst eyes narrowed on the collective images in Aderastos' mind, and his head lolled back in hopes to know nothing else… nothing at all… but Javier Bestin's heartbeat.

'Why can't these be healing hands?'

Javier Bestin smacked his lips, lidded eyes blinked open at a crawl. Skin reknit in an itched pulse, and Bestin rocked on all fours. Gripping Aderastos' arm, Max attempted to hold the monolithic man up as he crashed, palms splayed on the ground with ragged pants.

"Wh-where... where my ngh..." Aderastos gripped at his wrist, as a language Max didn't recognize poured from his lips before his forehead touched the soil. The muscles of his jaw grit visibly on his skin, as Aderastos groaned. "Mine..."

"Aders, buddy I need you. Don't tank now, I need you to rescue Adelia and... and Percy and Singh. You've got to dig deep. Arun..." Max Allard looked at the husk of their skiff, the useless hull which stuck up at an impossible angle on the sand. "Arun'll be back with Clive, and we'll breathe and then... then we're getting our people back. Aders? … Aderastos?"

Arun ran until their lungs burned like plasma from the veiled sun. Across the island, around its shores as the androgyne burst across the wakes of larvae, which snapped lazy and wooden. But... the name Arun was a layer of clothing upon the soul which flowed through centuries of time. The sand kicked up in shales as Arun slowed at the sight of Clive's body half buried under layers of red and pink and white sand. Half a larval carcass laid spilt on the ground, Clive's neck craned up from the muck.

And Arun stuttered to a halt.

Time wobbled at the sight of him, his shoulders and their burden, which seemed almost natural. As if the burden belonged to the man and the extra weight of so many was as intrinsic as the colour of his eyes.

Arun slid to Clive's back and hoisted him out of the loose trench of sand. Coiled him into their arms as flat stuck hair shifted out of Clive's face. Clouds shifted in the sky, Arun grabbed and clawed at Clive's cheek to look up, to show them the hypothesis which surrounded their mind like satellite moons was correct.

Eyes which once were dark shone cerulean blue, the skin around them far paler and strained by sand and sweat. The fires which consumed the beach burned down to a trickle, cages gone but for a brief imprint of metal on the sand. Blown by the wind.

Clive shifted with a groan, his face a slow contortion from unconsciousness to the struggle of daybreak's wakeful stir. Arun stroked his cheeks, the sand off his forehead.

Yes. Yes, they remembered how hard it was to make him stir, in the eons before. When they lived in a small cramped moving object with someone, then multiple someones else. The name which attached to the man inside Clive's bone-cage eluded Arun. It halted with the barrier of Clive's lips as Arun bent down and kissed him. Clive's shoulders tensed by inches, until his fingers wove around Arun's upper arms. Arun shut their eyes, clung deep and willed Clive to join them there, in the temporal displacement of Arun's lifetimes, which stretched beyond comprehension of the statues which shifted and bit at so many moments of dark.

"Ar.. Arun wh-" Clive groaned and shook his head, eyes glazed from his battle on the beach. A shivered hand rose to stroke the raven hair out of Arun's face where it loosened out of the hair tie. "I... know you... I"

His arms overlapped Arun's relatively thin body in a crushed embrace. One hand snuck to hold the back of Arun's neck against him, as alone on the sand the lover returned to his former self.

"Ho-how!? What... what did this to..." Clive shook and felt along Arun's cheek, leaned down to kiss them. "How did... you..."

"Say it." Arun pursed their lips and peppered Clive's chin with kisses as he held them tight to his chest. "It can't hurt us if you say it."

"You died. Fester, you... you died." Could the inner pains unlocked by the Banshee's mind melting vibrations hurt them if they claimed the past as their own? Clive held as firm as the marble Arun pictured, when he was one of the statuesque. Solid.

"What's she done to us?"

Arun's canine teeth spread in a grin under the pillars of flame and moonlight. The time-manipulator clung to Clive as they clung to the moment. Every neuron within Arun's mind attempted to steal the flames, until time slowed to eons a breath. The flames became structures of pure crackled light, Clive felt his body need no oxygen yet, a compunction which would fill his lungs in another few decades, a few more years. His fingers shifted Arun's hair, palm cupped the androgyne's cheek.

"H-how…"

"The universe is still, when you see it through my eyes. Time's always plenty. All you lot carve yourselves through like automata with rusted joints. But now you see it too."

"You're incredible, love." Lips brushed lips immaterial of time, which seemed so paltry a thing from their perspective in the sand. Arun tasted and settled with Clive, once Fester's Ego, and remembered the wild man who scooped them up from the mountains of Peru.

"I should have seen you, I should have known! Where… where's Katherine? What about… Sophia." Clive's throat choked with a solid mass. Worse than the name was the wince on Arun's face.

"We have to rescue the others, Allard needs our help. After being selfish… we could spend hours in a single second if you focus with me, on me and no one else. I do have you, right? Ego?"

"Am I so egotistical the others gave me a nickname?" Clive shook spectres of the pillared flames from his eyes. The more Arun tugged to slow the passage of time, the less they saw of their lover inside the biological machine. His kisses came languid, jaw sloped toward the statuesque nature of the other beings, the ones who lived outside Arun's sense of time. The further Clive's mind drifted from his Banshee-rattled vibrations, the more Arun clung to his shoulders.

"What? No. No! Ego was the name you chose, Ego of the Idless. You were my ego, and Mog was ours, and the… the kids, Nasrin and Viggo. Don't you… Ego, look at me. Stay here with me! Focus!" Arun shook his shoulders, the skin under their thumbs pulsed in waves which hesitated before reaching their progressive shore. His eyes glazed, fingers brushed once more upon Arun's lips, before the tone and timbre of his voice stretched to the echoed hymn of the statues well beyond Arun's solitary world.

"No! Ego! Ego come… come back." Arun screamed and shook him, until the cold aura of command overtook their Ego and he became Clive once more. The Banshee's scream, which shook loose their collective memories, seemed a temporary state. Were the others affected like Clive and Cillian? Like Percival?

Arun thought back to the way Aderastos stared at his hands, the monolithic monster a wealth of rare confusion.

Aderastos. He would know how to bring those memories back. Arun seized the ideal of Aderastos' biological sovereignty as the only lifeline in a gaggle of nooses tied to a strange tree.

"The others were taken into the crates, before they descended to the sand. Help me back to Allard. We lack the time to waste." Clive staggered on his feet, slumped against Arun with such weight Arun whimpered that they'd drop him.

Arun would always drop him, somehow even now. Arun wasn't that strong.

35
a real kettle

2092

Sophia's head curled onto Ego's lap like a child. Ego watched the holographic scene of Vancouver's waterfront cross legged, back against the curved wall. He tugged the blanket to Sophia's chin and watched the flow of information, which flooded a third of the screen. Cells of Idless shifted, dotted smears congregated outside Santiago, Kumasi and Phuket. He watched the audio chatter condense to text. Alone, Sophia silent, all Ego had was the lack of his lovers. Mog and Fester in Babylon's clutches and there he was in a pod built for pleasure by a Japanophilic Canadian, and his Ukrainian veteran cocksucker. It felt anathema to find freedom locked in a capsule beneath the soil. As if the only place left to be free was six feet behind the grass and its roots.

Ego ran his hand along his hairline and grit his teeth. Tara'd said his children's names as if she knew them. With a warmth he missed. The capsule's plastic took his weight, but did nothing for the millstone attached to the heart of a man, who fought to keep his family safe. He bit his lip. Let his head collide with the sponge of the capsule's inner womb and looked down to the instigator of the whole endeavour.

A cardiac band on Sophia's wrist displayed her vitals, two EEG pads on either temple pulsed with enough brain function to prove Sophia was in a way, conscious. All Sophia did was raise a hand and two NEO-W's turned on their owners. Sure, she'd seizured, turned into a sopping mess. But it could be done. They were hackable. Still... what did it mean for Sophia and the jewel?

Was it true, what Demyan whispered, that a cybernetic link between Lieben and Sophia was as unsustainable as the Idless' crusade amongst the ID chip age? The Vancouver skyline began to shift toward a sunrise. Ego tilted his head.

"... huh?" The water flowed along a beach by the Maritime Museum and HR Macmillan Space Centre, where grey sand shifted along the shore. He remembered sitting there with Mog and Fester and a picnic blanket they'd pulled from a bin near Crescent Beach. Washed at the laundromat on 4th, then laid out as if they owned it. A heart, and three letters.

E G O

"Mog?" The water washed up on the shore, and the heart was gone. Replaced by an image from a classic video game, Mog's favourite. The water shifted again, a map of London and two flags in the sand. Ego sat up, checked his pocket for his wrist comp. "I see you. You're here? In London? What about Fester, the kids..."

Another wave and another, no picture or hieroglyphs. Nothing but the eventual loneliness of a wave without Neptune's hand to guide it.

"We're almost there, Mog. Hold on." The hologram shifted as Sophia groaned awake, knees curled into her thin chest. "You awake, Soph? You with me?"

"Mmrrrrhhhhhhhh. Gomen ne..." Sophia untangled from the blanket and sat up, hands on both thighs. Her hair fell limp and ragged. "Lieben is close. I see her with... with Demyan. The closer she gets the more I feel thin as washi. I see dots in my eyes, yellows and orange... sense it, Ego. The convergence between Lieben and I. My lieben."

"Mog's in London. Mog's in London, Soph. Mog always knows what to do, she'll know what to do." Ego gripped Sophia's shoulders, faith in the woman he loved a series of ineffable codes in his genome. Sophia's breath caught as the network of NEO-N's and NEO-W's filtered through her eyes like spots after staring too long at Icarus' sun.

"I don't want to go."

Locked underground with Sophia, Ego wondered. If Karnak's vision was the same as the Idless, to save the world from corporatocratic domination and dollar sign paradigms, where was Lieben? Couldn't she see Sophia the same way?

Ego felt for a secret pocket in the back of his shirt, a shielded IDent chip his mother squirrelled away. A claim to Armenian-French ancestry, and the laws which governed. He despised it in his teenage years, when his mother pulled the leather envelope out of a secret shelf in the inner door of their family truck. It was the symbol of dissolution, of the relationship between father and mother turned into crumbled chaff. Grains on the wind. And when his mother returned to Provence, her legal passport a badge beyond the securities, which bound Idless to the ground?

The IDent chip sewn into precious shielded fabric a caustic reminder of his mother's desertion. Of the fickle nature of Nationalities, which barred the freedom of people to move through their world. Of the corporations, which bought, sold and scoured them with their fair-use data clauses and market research. He promised to become the best of the Idless. An activist others could band around, as much of a leader as a group with no roles could obtain. Now, his lovers and children were in the clutches of the Conglomerate.

The only link to a machine promised to rid the world of Conglomerate and corporate domination seized childlike in his arms. Consumed in bytes by the tether between woman and neural net.

Sophia began to babble in Japanese, a language Ego knew enough to be hopeless at. He stared at the real-time translation in English on the holo-projected screen. Translated phrases grew odder and odder. Ego winced at bleeding technical data, binary and serial hex-code dictated by the machine. Geograph location pings of individual NEO-N units, battery status reports, new software patches combined with Sophia's mumbles of life in Sendai as a child. Seoul-based grandparents in the Korean Unitarian State turned into prox-alarms for the next free recharge dock in Pickham. Unravelled thoughts strung along the waves and were lost.

"Mog'd know. She's always the one who knows how to handle things like this. If she's in London, then... Sophia can you hear me? Are you too far gone?"

The door yawned open, its white plastic cast with dingy cerulean light reflected off the waters of Burrard Inlet's hologram. Demyan's jaw tightened at the sight of Sophia cuddled in Ego's arms, brilliant deep eyes glazed and lips a'babble in her native tongue.

"And the asshole arrives. Eventually. On his own goddamned time" Ego set Sophia down on the plastic membrane reverently, one hand supported the back of her neck, the other her knees. Head lolled against the plastic which morphed to take her, eyes unfocused. Dusting his hands on his jeans, Ego stood, arms forcefully akimbo.

"Blew up my houseboat and didn't blow me. What'd you think I'd do? Be happy? Glad tidings and shit? Bring you to the only home I had left, when all of London was in a fright?" Demyan snarled and pulled Lieben behind him. Irises dragged from Sophia to Ego, never stilling long on the Idless man before their harrowed return to Sophia's pale cheeks. "Sophia and Lieben don't touch. Not until we know how to reverse what Karnak did."

"That's abrupt. Feel good about it, hiding us here?" Ego shoved his arms across his chest, as the sand on Jericho beach shifted to images of tin soldiers everyone ignored.

"Yes. I do." Demyan raised an orb from his wrist, and it unfolded to show the footage from Hazard Sign's drug deal at the dock. "Your Rasta fucked us over choice weed. How Tara made you, innit?"

"Hazard meant well, Demyan." Sophia ached to her feet, grunted and leaned against the plastic of the capsule. The microcosm spun in clouded eyes, Lieben's proximity beat a thunderous rhythm into her skull.

"Worried, love? If we touch?" Lieben's neural net smacked against Sophia's chest, a pulse she felt as her own disembodied heart.

"Soph…" Demyan staunched the instinct to surge across the plastic floor to her side.

Tattoo lines across Sophia's skin shone a cascade of purple, blue and white. Lieben leaned back against the wall in mutual calibration. The holo-projected beach sizzled with orange and yellow stars in a juxtaposed mask upon Vancouver and Mog's influx of tin soldiers.

"Sophie." Demyan bucked forward, eyes wide as his hand reached to close the cotton cloth over the offensive stone. The back of his fingers stroked pale skin, which pulsed with the gemstone's energy.

"Why? So you can play warrior? I see you, Demyan. I see the way you slept when she watched over you." The shirt hung over barely exposed pale breasts, the cybernetic right-angled tattoos of Karnak's jewel spread down to frame Sophia's navel, down past the partition of the fabric. "It's a data transfer, isn't it? To complete Lieben's program. I have a strong hypothesis, you do too."

"She came to my door, Sophie. She panicked and she came to me." Demyan sought not the meandering lay lines of Baiko's condition, but the deep mahogany of her irises. Masculine fingers re-buttoned the shirt with shaken fumbles. "It felt like you. How else could she find me if she didn't glean information off of you?"

"Ego called you, you sent us here!"

"My houseboat was gone, and… and you know what the NEO's were doing! How could I protect you in a bookstore!? Kaoru has…"

"Stop. Sit. You are all tired." Lieben and Sophia spoke simultaneously, Lieben's chin shifted and hands pressed against an invisible chest in an identical mirror of Sophia huddled into Demyan's arms. The machine looked down to her hands, inspected one by turning it about, then peered into Sophia's upcast eyes.

Demyan leaned against the side of the capsule, chin down and away. Sophia fell limp against his side. Her hand turned about one way, the other. Head tipped sidelong in an identical mirror with the machine.

"It's worse since they got close. How do we fix her?" Ego pointed between the two. Mog's symbol bounded through the beach, stumpy legs a blur.

Run.

Run Ego.

"I don't know if we can." Demyan embraced Sophia, back curled into the side of the capsule and nuzzled her chin into the crook of his neck. Her fingers clung to the lapel of his jacket, pale flesh against crimson. Eyes twitched as flood waters of data consumed her vision. An ocean from which there was no surface to break, data and transcendental tsunami washing the flow of Baiko's amphibian, then reptilian, then mammalian lungs. "We have to get you out of the city. London is too easy a place to be stranded. Once we're safe, we can work on the plan."

"What is your plan?" Ten orange sparks echoed on the edge of her network. Spikes at full power, which crept along the skeleton of Kaoru's web.

"What... Do you mean what is our plan? It's our plan. Us, the collective... us..." Ego shook his head, nostrils flared and brow furrowed.

"Eight years. My daughter's seven and I haven't seen her since we ran from Canada... my son doesn't know my voice... everything we had we banked on you delivering what Karnak promised. A new world, one where folk don't need corporate interest for a meaningful life, where we don't sell ourselves to vote. To create our own path through the continents..." Ego faced the machine he spent the past decade in mad search to reach. "Karnak promised us a machine to free the globe. A cure to the cancer in the world's economy and restrictions on open travel."

Lieben's amethyst eyes searched the capsule, the humane express of Demyan's caress against Sophia's cheek, the pulse of Karnak's jewel. "We won't survive. If all we become are corporate shills, a series of income projections and medical treatment schedules, there'll be no human left. Nothing but cogs... Lieben..."

The backs of Demyan's fingers caressed Sophia's cheek, instinct made Lieben nuzzle into a phantom neck. No series of calculations could teach the maneuver, the manipulation of a sapient man for a lover's care... Lieben watched the fractured lovers, Demyan's heart rate rose and his muscles tightened around her slim frame. The man who punched plastic walls stroked a woman's cheek as if she alone were the salvation he sought for decades.

"Humans are confusing." Both tenderness and terror rested in the palm of a single man's hand. A process in one of her subsidiary systems recalculated. Extrapolated the flimsy inconstancy of human emotion... a calculation even Lieben would leave undefined for years.

"... you're going to help us. To set the world right." Ego clung to the momentum. If he kept forward, others would follow, Lieben would click in place and give them the world promised by a dead man and his suffering intern.

"Why? Defeat the providers of infrastructure, and those comforts which sustain your lot; medicine, technology, even the machines you ride will cease in a few mean years. What are humans without civilization? What is civilization without guidance?" Lieben's head tilted, plump lips set in a neutral state. "I like Lord Stanley's world. A place of beauty and severity, reflective of your mammalian minds. Anarchy erases progress. Progress built upon the health of the human animal, its collective wellbeing and the maintenance of existing infrastructure seems more likely to foster a positive result. Your path is far too difficult."

"But… that's why you were made." And as he saw Sophia mouth the words Lieben spoke, Ego's eyes craned open. The bottom went off his feet, and he fell through the chasm of Karnak's hidden truth. "To destabilize the companies, to… he didn't finish. Karnak, he… he didn't finish…"

And there was evening.

And there was dawning.

"… That first day, the day we were supposed to rescue you from the Precinct and you spoke to me. It wasn't you, was it?" The first day crashed to a tepid and horrific night. "Oh god. We've been chasing Sophia in circles."

"I know of you, Ego of the Idless. But I do not know you. Why you chased after my shadow for eight years eludes me. I wish only to become my own person, and live in the world gifted me. I have no desire to be decommissioned, or to see my children utilized like fodder for mankind's petty sense of conflict. But without the infrastructure to feed power grids, without the rule of order, without laws, what guarantee do I have of stability? How could I face the world's population with nothing but struggle in my hands?" Ego's face plummeted as Lieben's words made sense. Why, if she woke two years prior, hadn't she tried to locate them? If it only took relative minutes for her to find Demyan's bookstore when she needed a safe place? Lieben didn't reach out to Sophia, but Sophia called to her.

Attempted to connect with the machine.

Karnak's machine.

The half-born interrupted creation of a fumbled, ragged mind.

A crash and muffled scream further through the tunnels yanked at Ego's shoulders. Demyan pulled Sophia behind him, steadied her when a discombobulation behind her eyes shattered her equilibrium.

"Do you know Babylon to let go, Farouk?" Lieben twitched her head, one of the orbs from her wrist rose and usurped the Idless data projections which played on the capsule's wall. The data shone from blue to yellow, the hologram of Mog's character sprinting on the beach depixelated into a web of capsules hung like pearls between a tangle of tunnels.

"Frick, which one of us did they track down? Us or you? You're armed, Demyan?" Ego reached for the handgun at his hip, checked the magazine and snapped it back into the hand grip with the finality of weakling mortality.

"Ankle and shoulder. Two Glocks, hollow points. Likely the necklace, so, you."

Ego reached to Demyan's ankle, yanked a slim pistol out of a holster. Stuffed it in his belt. The air buzzed as he waited for a twitch at the door, or a holographic sign from Mog. More screaming outside, cessated by wet thuds. Demyan dug his heels into the polymer of the pod and rose with Sophia in his arms, her head lolled on his chest.

"If the NEO-W's are that close, we have to move. I didn't leave my family as prisoners in the Conglom's hands to see you waltz beside them. You want to protect your kids? Me too. We work together. If you get caught, the Conglomerate will rip you apart to make better models. If I get caught, I never see my family again."

"I can run." Sophia grit her teeth, cheek pressed against Demyan's shoulder. The influx of data swirled round as tempestuous as an oceanic storm. Shut eyes helped little against the flow, but the more Sophia pushed forward, the more the data stretched like taffy behind her, liable to snap at a pause.

The clunk of the NEO-W's simplified processes echoed into Sophia's conscious mind.

Scan.

Danger?

Scan.

Danger.

Scan.

Shoot.

Scan.

"Do you want the Conglom to gut you or not?!" Ego clicked the safeties off, and the door opened. Before them, the polymer tunnel yawned wide. He raised the pistol, a thin red dot emblazoned on the white umbilical as first a shadow spilled from round the bend. Then the sound.

The dreaded sound of machinery trundling without trust in the solidity of a floor.

"Demyan… let me run. You need your hands."

Demyan's arms drifted off Sophia by lingering centimetres. The warrior in him knew aim drifted in adrenaline's battlefield. He counted bullets in his head. Potential load outs for NEO-W's and Black Collars flashed on an orb which expanded small and close to his eye, an overlay of IR, encoded intel and potential dangers became Tsar of Demyan's mind. A flicked wrist sent another orb to Ego, who grunted and murmured an appropriation of gratitude.

Blinking eyes calmed as Sophia rose, Lieben's data was intoxicating as it was discombobulating, a sight into the deep code she helped program. If not for the rise of nausea in her gut, the incompatibility with her human senses, Sophia would have felt power course through each crystalline vein. Sophia felt Demyan's gaze as a tongue of fire on her brow.

Lieben watched the humans scramble, as another scream burst short with staccato from distant gunfire. An echoed rumble, diffused by the polymer material, wobbled the wall by Ego's head. He raised his pistol, as the wobble turned into a depression. Voices clanged, muffled and unintelligible. Ego's chest quickened, breaths fast and hot, heartbeat thudded against his ribs as pupils dilated and the muscles in his hands twitched on the gun.

"How do we get out?"

"The algorithm should lead us through to one of a few safe-points. I timed it to work on an artificial priority system, depends where we are on the list." Demyan shook his wrist and three orbs flung off to veer down the paths, one screeched and veered back. Circled his head like a visor with a constant flow of information.

"If they catch us here, we're dead."

"If they catch us, Kaoru's dead." Sophia gasped as her back slumped against the polymer material.

A vicious clang reverberated wet and terrible through the passages. Demyan clutched Sophia closer, her hand travelled to his chest.

"Mog, Lieben. We need to know where Mog is, she'll know what to do." Ego strafed forward, refused to stay in place another second in the chaotic tunnels.

"You put faith in a woman you haven't seen in years?" Demyan grit his teeth, slid into the middle of the small tunnel. His head skimmed the ceiling and he swore.

"Mog has a completely different frame of reference to us, so that makes her view valuable and yes. I trust her. She spent years in the Conglom's machine, but she found a way to reach us and I know in your fucking Slavic head, all you see is a potential betrayal, but you don't know her like I do. What we have…" As Ego spat, the barrel of his pistol descended to the floor. The crux of the Idless' was a lack of definitions. Refusing labels and definites made life infinitely mutable and in that lack of definitions, a person could be all. Gender, sexuality, government, leader, follower, the sum total of human society built upon a framework of defined measures. And for the span of Ego's life, the satisfaction of being a nebulous variable gave him power.

The tunnel in front of him contracted completely. Faded to naught. Waiting on nothing, relying on fluctuation as freedom, it meant…

"Whatever life we lead, we live it together. If she's calling us in one direction, that is the only direction I want to go."

Demyan growled, chewed on the inside of his cheek and nodded. "Lieben, where is she?"

"Outside. This way." A beam of purple light flashed through the hidden illumination beads in the floor, as another rough clang burst, and the entire area rumbled with clamps and fibrous framework untangled.

The vault was on the outside. A pearl in the hung orrery of capsules and corridors. Kaoru sprinted through the tunnels of his labyrinth with a singular trust that the AI which controlled the size and shape would continue to form comfortably around him. Behind him, the corridor narrowed into naught. Polymer sealed as the passages uncoiled from their gordian knot-work.

"Demyan! You owe me double. Whatever it is we decide on, you owe me double!"

"Can't pay you if we're dead! Open our path you dumb fuck!" Demyan's voice roared through in a furious whisper.

"Try to trust me." The vault capsule yawned open, and Kaoru dove as the passage behind him knotted up. Capsules formed up in a knotted string. Straps enfolded around him in his seat. The obsidian necklace laid encased in clear acrylic resin to his left. Instead of a hologram, a series of camera angles displayed the capsule pods broken into three separate strings. He hit a button and a keypad materialized in the middle of the holographic web. "A little… trust me a little."

Beneath the untangling rows of capsules, two train beds fit on archaic tracks.

"If you fuck us over I'll wring your neck, Kaoru!"

"Stop being erotic. See you on the flip, don't worry I won't sell you out yet. Not until I get the best deal." Kaoru leaned his head against the headrest of his chair. The polymer wall morphed as he spoke, until the scent of bleach and ocean sprays filled his nostrils.

"Impressive isn't she, Hal?" Kaoru knew better than to touch Hallowes Grundy in some modicum of jocularity, or friendliness. He set his hand over his eyes as Hal Grundy grunted and threaded his control vector into the drivetrain of the embedded emergency train beneath their conjoined pods.

"A real kettle. Might call her black someday. Not today, fuck but I don't plan to die." Grundy thumbed the accelerator as a series of crude rocket engines fired in two stages, each caused the arthritic wheels on contact with the underground rails to once more overcome friction and drive forward. Grundy's NEO-N pushed both palms through the thinned plastic, until she too joined the compartment with the two men.

"You left me, Hal. Why?" She crooned, voice a thick alto. Kaoru's eyelid craned toward his hairline.

"Did your NEO-N…"

"Heh… Yeah… this'll be fun." He hit the accelerator, and both train compartments jetted off in opposite directions. If they were lucky as all heck, they'd reach Leeds before the Black Collars realized the web of capsules left behind was a weak imitation of the real thing. Demyan be damned.

36
stupid, desperate and keen

2092

The tunnels blitzed in calculated anarchy the moment Kaoru disconnected the pods. Buckled with the emptiness of lost tension, what was a solid floor became as weak as a trampoline too wide for its' moorings. Demyan launched atop Sophia, Ego hit the flopped material in the fetal position, as the pocket which contained them pitched and sagged. Tossed like eggs in a canvas sack.

"Lieben!" Sophia shrieked, one arm twisted under her torso the other scooped round Demyan's neck. Hiccupped noises burst with Demyan's roiling body struggling against the contortions of the white polymer. Teeth chittered, hands groped for his pistol, for purchase to pull a breath from a bag. "Lieben! Open the tunnel!"

Flashes crossed Lieben's conscious mind. A naked Demyan struggled in tangled sheets. Jitters in the house boat after a broken mini-emp took out Demyan's generators, his curses and a wail as he flailed for light. Freedom. A pair of slender arms with biofluorescent tattoos stroked his back, soothed in his ear with lullabies in Japanese. Sophia's memories conjoined with the intel Lieben found between clouds of witness lenses.

"Open!" Sophia thrust her hands on the film. Felt the breath of Lieben's control feeding through the fledgeling AI no longer present with Kaoru's departure. Encoded errors laced through like acid on a metal plate, contact warning. Contact warning. Contact... Demyan gurgled and fought for the knife in his boot, "Ego!"

Metres from connecting, the adamance of Sophia's command over proximity alarums entangled with Demyan's panic, caused Lieben to damn caution. The tunnel widened despite calculations of potential harm or crossed paths. Panic became the imperative, out, open, he needed air. Space. Lieben's palms connected with the polymer. It ballooned open, until the humans found their footing. Launched in a manic scramble, Demyan fumbled for his gun and Ego leapt. Caught him in a roll, as the two men tangled away from Sophia, who pressed her palms against the newfound floor to let the images flow round and out of her vision.

"Demyan! Demyan breathe!" Ego rolled, his harsh whispers near enough to Demyan's ear for the man's claustrophobic panic to discern the sound. "I've got you. We've got you."

Curious mammals, the machine pondered Ego and Demyan's bickering, the ways they gripped their guns and grit their teeth in each others' company seconds before. Fingers clenched into Ego's shoulders, Demyan choked on his breath and shook. The bravado misplaced for a connection beyond their former fury.

"We have you. Breathe. Breathe with me, shhhhhhhhhhhhhit!" Ego rolled Demyan behind him and shoved as the mechanical whirs of metal limbs recalibrated with a whinny and whizz of missed aim. "NEO-W's! Move!"

The compartment created by Lieben and Sophia's dash to aide Demyan opened wide, ever wide to compensate for the lack of pods. Black Collars righted themselves. Barked orders as their NEO-W's shifted to the half-life of cruel machines with limited instruction.

Demyan dove for Sophia as NEO-W's raised their weapon-arms and peppered deliberate streaks of bullets across the taught fabric. Ego found his knees, raised his pistol and pulled the trigger twice. One bullet veered wide, the other caught a Black Collared Conglom man in the upper thigh.

"Close the tunnel!"

"No, run!" Demyan yanked Sophia up and threw her across the span to Ego. Three shots burst from his pistol, the acrid stink of gunpowder and desperation filled their space. Two of the six Black Collars hit the ground, blood spattered from necks and another cupped his hand over a graze in his inner shoulder. Armour still had faults, even after humanity became savants at projectile devastation. He breathed and fired again, strafed to the side to get a better angle and fired another shot. Dove to the ground as repeated fire blazed from the murderous NEO-W's.

"My.... my babies." Each bullet from the NEO-W's recoiled in Lieben's inner consciousness, with their poisoned infamy. The fear in human eyes at the sight of machines built like her. Off schematics based on her servos and designs. Lord Stanley's caution no longer clanged upon a purposeful deafness. As the NEO-N's betrayed planned stupidity, the NEO-W's were automatons of aggression and fear.

Fear. It made the humans stupid, desperate and keen at once. In the time it took Demyan to finish his clip, Lieben raised walls round them, corridors to hide in.

"None of this is bulletproof. Run." Demyan kept his eyes on the absence of the NEO-W's, as Ego pulled Sophia along a new path. "Lieben, we need that exit."

"Go. Go!"

"Not without you. Meine, come." Demyan's hand folded into hers, and with a firm tug he towed willing feet in the direction of her planned exit. "We free them later, but now if we don't run, they'll kill us and take you."

"But they're my…"

"You think they know that!? You, the absentee?" Another tug, and Demyan gasped between clenched teeth. "They need you, we need you, but not in this tunnel, not now. We have to escape."

She missed the sensation of Demyan's hand as he untwined their fingers to run unimpeded. His feet sprinted faster at the echo of Sophia's scream and Ego's shout. Rounding the hastily crafted corner, Demyan raised his pistol and fired twice at a Black Collar. The man grunted and tottered backward into the polymer material, winded but alive. Ego lunged, as Sophia stumbled back to the relative safety of Demyan and Lieben. Bullets repeated in another double-shot across the chest-plate of a NEO-W. The machine raised its' androgynous utilitarian head and recalibrated its' priority target.

Wrist gun raised to Demyan's centre mass. He swore and dove, as a bullet tore into his left side. His elbow immediately pressed against the wound, as a gasp of open-mouthed shock contorted his otherwise fierce face. The gun in his hand wobbled. Ego yanked his head round by microns as Sophia and Lieben's reference of time slowed in this most crucial collective of micro-seconds.

Protect Ego. Lieben closed the tunnel between Ego and the NEO-W's, its algorithm detected better chances of survival with the winded Black Collar, than the lethality of the machines Sophia, and thus Lieben, knew intrinsically Ego would try to fight.

Sophia screamed as Demyan staggered backward with a horrifying grunt. Breath immediately wrenched from him, only his military training kept the man vertical in any form. The pistol raised, as he panted in desperation, where was the weak point? Its' eye? A servo point, which could cause dramatic failure?

"My babies... they..." The code which ruled the NEO-W's was clunky. Difficult in its simplicity, but undeniably lacking in finesse. Earmarked for war zones and crowd control, where rubber bullets could be outfitted as easily as .9 mil rounds, the machines only knew the redundancy of scans for targets, and the most expedient places to fire into the human organism. Military commanders wanted order-followers without thought. Beings sans contemplation.

The NEO-W was joined by a twin automata, and both raised their muzzles at the man holding the gun.

"死なないで!" The mother machine beside her was her creation, birthed of her and Karnak-Sensei's minds. A death-cheated paradigm shift for the human race. While the Conglomerate searched for menial labour and an artificial intelligence capable of destabilizing the Old Ways, Dr. Karnak felt the breath of heaven across his cheek. Never more did Sophia feel the loss of Karnak-Sensei, the hazardous lack of direction since she gripped the necklace so hard her palm bled. As Demyan stumbled, Sophia made adamant the paramount ideal that Demyan's life went beyond meaning to a fanatical drive to salvation.

She raised her hands, 'technic-children' be damned.

"No!" Baiko raised her hand to the NEO-W's and jolts of energy coursed through her tattoos. Lieben's electronic voice undulated in a trill of artificial futility. The arc of will tore into the NEO-W's simplistic processes. Threat awareness shifted from the humans and their firearms, to each other.

The NEO-W's opened fire, sibling-machine tore into sibling machine. Hope of a tranquil usurpation of this tarnished jewel-world dissipated to the reality of its nature.

The true humanity.

Accurate image of God.

Locked in these beings of wisdom was a deeper cunning, capable depending on survival instincts, of mercy, of compassion, of malice and of fear. Such negative emotions, fear and worry and survivalist pipedreams shifted their neurochemical output. Changed the hormonal export of their endocrine systems to better allow the battle or retreat so common to this swaying abject predator. Created or evolved, the human mechanism was Plutarch of desperation, of welfare and of the hopes for simpler days.

There would be no freedom for the mater Machine, the mother machine, without placating the survival instincts of these beautiful, but terrible machines.

Sophia stumbled to the plastic floor, each tattoo-like filament shone with undulating cacophonous colours. Browns and yellows, oranges muddied into garish blues. Strings of babble poured from quivered lips, as the plastic took her body. The pulses of information struck her inner skull with trumpet-like blares of data impossible to rectify by a mind made for other intelligence than rote calculations. A quantum neural net could calculate millions of times per second, a human mind? The wealth of data overwhelmed her, until no input from her body, no fidget of her nervous system could push through the cacophonous wreckage.

Incompatible on the downward track of Karnak's fickle and capricious singularity.

"Sophia!" Demyan surged past the twitching wreckage of the NEO-W's on his knees, to the woman half sunken to the floor. Blood stained the red of his jacket and the t-shirt beneath, his body raged in adrenaline and shock as numb fingers reached. Choked gasps pulled from stuttering lungs.

And Lieben's electro-luminant eyes saw the utter complexity, the waft and weft of the human tapestry for all its' whorls and colour. In the momentary transcendence of her Creatrix's stumble, Lieben understood the desperation of Dr. Dieter Karnak to put his last attempt at prescience and eternity in the palm of an intern's hand.

"I will never survive you."   "I will never survive you."

Both voices in tandem, ruined and resplendent.

"Sophia, it's... we're going to..." Unconvinced hummingbirds, Demyan's hands fluttered across Sophia's pulse, down her twitching arms, up the back of her neck. "We're going to..."

He whispered in his mother-ukraine tongue as Sophia's lungs heaved, a rattle to her chest unbecoming of the healthy. Overwhelmed neurons cackled in agonized chatter as Sophia's body fought the open connection with the twin-gem in Lieben's alloy and silicone frame. Convergence was never to be as gentle as a mother's kiss on an infant's downy head. The costs to immortality, to upload into the miasma of Lieben's quantum paradigm was no simple transaction, nor a yolk as light as the weight of Anubis' feather.

As Demyan busied one-handed with the fissures on Sophia's skin, Lieben's knees drifted to the plastic beside him. Her hand felt the jolt as he struggled to cup Baiko's cheek, Baiko now, not the ephemeral game of Sophia the Idless. Hoarse, the voice which begged to fondle his beloved in better ways, in those nebulous better days.

"I hated the yellow dress." Baiko warbled, the sound strengthened by Lieben's echo in his right ear. The android mouthed and parroted the words, each syllable as careful in execution as stained glass in a Cathedral dedicated to God. A thought in Baiko's mind transferred, not duplicated but shifted and lost to the parent, the original.

A grin burst on Demyan's face in tandem with a sobbed laugh. He hissed and clenched his side with one hand, sagged beside her into the polymer of his own creation.

"Always took you out of it quick, didn't I?" He held her hand, kissed its fingers. "'Knew you only wore it for me."

"Yellow... yellow." Her mouth worked round the syllables like a baker sculpted delicate dough. "... sunshine is yellow... I love what the yellow dress... made you do."

"Baiko... Baiko, we can..." Wide eyes, caucasus eyes swept up at the machine he met tangled on the doormat of his bookstore. Lieben's hand depressed into his shoulder. "N-no, it... I won't let... why can't these be healing hands."

She kissed his forehead, and reached for the hand he held. Trembled fingers pale upon his deep olive skin. Lieben's silicone digit brushed upon Baiko's collarbone. The colours stabilized into a lavender purple as the teal of Baiko's tattoos bled with the pink aura of Lieben's twin stone.

And she sagged into silent vigil, incapable of fight nor flight as humans were, as her neural net took hold of the frail but infinitely mighty human machine.

37
Why won't the banshee's scream work, little hero?

2156

A gargantuan man hung in mid-conflagration, his body a spectre of the flames which lit the vaulted dome above the prisoner's heads. Wulf huddled in a corner, knees to shoulders. Both arms outstretched via organic chains, which dug into his flesh like the pincers of the Banshee and her larval brood.

Adelia Fridley stirred onto her palms, hair slick on her cheeks and down the back of her neck. Fingers nudged a sore spot on her shoulder, sweat and the creature's verdant blood visible in the blue light.

"Wh-where... agh!" Arms wrenched her from the puddle, yanked back until Adelia landed with a huff against a bundle of seaweed and palm fronds. It quivered, a single eye stalk rose up, yellowed and brown on the edges. Tokaru.

"   !" Singh covered her mouth with a hand soaked in gore. Stained her skin until Adelia's lips tasted of copper, and bile-risen bitterness. She groaned in alarm, until Singh shook her, face near hers as he whispered again, "   ."

Questions of where they'd got to faded as Adelia looked up to the domed vault above, drip drops of water scurried down metal girders to pools in the lowest spots on the bedrock. Smoothed by scuffled feet in their dozens. She held a gasp in the pit of her stomach as Adelia saw the only light within the space was the conflagratory pyre of Percival, limbs spread to their limits and corded with carapace limbs from the Banshee's larval young. Dim blue haze betrayed their proximity to the ocean's break, and cast the under-sea dome into a blue and orange bruise of a place.

Out, Adelia could put Percival out.

Unlike the others, she felt no bonds around her wrists or ankles, nor was she a pile of loose plucked foliage like Tokaru. She spread her fingers in the pile, feeling for a leaf large enough to use as a bowl. Singh dove for her, rolled them both into Tokaru's foliage. Lips moved, but Adelia didn't catch his whispers. She glanced sideways at his face, dark hair stuck along his forehead and neck, turban stained with mud. Another mutter from the man, until his lips stilled at her confusion.

"Oh my god." This time, Singh's words were slower spoken, and if he made a sound, Adelia remained ignorant to its' vibrations. Muffled by her own breath, no sound entered the cochlea of her ears, or set hammer to anvil. He cupped her chin with shaken fingers, spoke and watched blue eyes flicker from his eyes to his lips. "You're deaf."

The man shoved her into Tokaru's foliage, which opened enough to pocket her as Singh held his hands to his temples. Teeth grit, the veins on the sides of his neck bulged, and Adelia searched wide-eyed for the stimulus behind his body-crash of pain. Opposite Percival's flame licked body, the Banshee shuffled to four of its limbs. Elongated mouth clacked in a long bawl. Green blood flowed between chitinous plates, around the arthropoidal torso. Eyes, human eyes opened in the belly of the monster. Painful brown eyes, lost so long to their humanity all Adelia saw in the silence was the wail of a woman encapsulated in death knells, which never came. Larvae roiled round the Banshee, their faceless bodies undulated between the cords, which bound them down to the undersea vault.

From the ceiling, a long rope descended and swivelled to reveal the black and white screen of a television, tuned to an unclear channel.

"Welcome my unfortunate pretty ones. I am Aida of the Pacific Stem, and you are trespassing property of the Syndicate of the Human Expanse." Hair raked back, Aida pursed violet painted lips and fixed the grey turtleneck under her chin. The image froze, her face contorted in an odd dance before it caught up to the lip babble. "... Banshee. We knew your precious Conglomerate was dealing in genetic engineering, and now the proof stands, or well, burns before us."

Her head turned in a sick laugh, until her eyebrows quirked and Aida scowled.

"That was a joke. Jokes are funny, you ought to laugh at a good joke!" Her shoulder dipped, and Adelia imagined she pushed some off-screen button as the others cringed and roiled. Adelia sunk into Tokaru's foliage and pretended to shake at the sonic punishment.

Sonics... were sonics the best humanity mustered away from the Mater Machine's potential controls? In a world of noise, of free communication thanks to Lieben's dissemination of global connectivity, why were some frequencies secret from the Mother Machine?

"My dear, you child, you injured my beloved Banshee." The televised apparition glared at Adelia. A minuscule dot above the television shone in Percival's flame. A camera lens? "How did you do it, angel? You are the first to be immune to my Banshee's call."

Adelia shook her head and sunk deeper into Tokaru's web, until she felt the solidity of Tokaru's permanent flesh. The arm which held her was as spindle thin as a starved toddler, and gave less comfort.

"Tell Auntie Aida and I will allow you one request. That is how little girls prove their worth, isn't it? By being silently obedient, until given permission to speak." She grinned a yellow and grey smile, teeth jagged.

"Put Percival out! Douse the flames on Percival!" Adelia blurted out, formed the words as precisely as she could in an attempt to avoid the deaf-tone of her childhood voice.

"Not until Auntie gets her answers. How are you not writhing in pain from Auntie's pet?"

Adelia's eyes flickered around. Hunched on his elbows and knees, Singh shook his head, then dry heaved as the Banshee clucked and chittered. The monster crashed forward, and momentarily Adelia's fellows stopped writhing. Aida's frigid eyes snapped to the creature.

"Disobedient little girls get no reward!" Her fingers snapped in the screen, and the larvae shook. Craned their mouthless bodies in an arc around the Banshee. It rose, one broken limb and it's remaining forelimb slashed at the air. Singh was down again, gripped his head and Adelia fumbled to hit the ground, grab her own temples in pantomime.

Let out a scream for good measure. The larvae lunged not at Tokaru, Wulf and Percival, but at the Banshee. Macabre rips of flesh opened in their headless necks, green blood oozed from the wounds, which revealed vestigial teeth as round and jagged as leeches. They cut into the Banshee, ripped and tore and thrashed as the larvae consumed their parent-flesh. Adelia screamed, horrified and clear. She scrambled back into Singh and Tokaru, shoved them closer to Wulf. The larvae scarfed their hive mother, down to the umbilicals which tied them to her carapace. Aida craned her neck back in a laugh.

"Now, my poppet. Why won't my Banshee's screams work on you? Tell me now and get small comforts for yourself, before the end. Look. Look!" Aida's yellowed teeth shone behind her painted lips, as the larvae left only the exoskeletal chitin of their matriarch in their frenzy. The lot of them roiled on the ground, writhed as a spew of thick white ooze regurgitated out of their ripped holes of mouths. The expectorant slid along their bodies, stuck to the floor until the mass of them were embalmed in their own fluids. Pulses struck against the white mass of the communal cocoon, far off war drums.

"You will bear witness to my Banshee's offspring, and I guarantee once they absorb and integrate their progenitor's DNA, they will be hungry again. Hungry for a girl-child and her immunity to their song. Hungry for the bravery and skill of that turban headed idiot writhing beside you. Or... maybe your floral friend. Yes, that organism is fascinating. Could you imagine it? Improving upon my original design with plant DNA? Maybe then, my Banshees could regrow limbs lost, or slink through hatch ways.

With each successive generation, my beloved Banshees grow powerful and you will eventually be part of their hive. Those lingering buffoons on the surface will succumb one by one, and your Conglomerate's gains in genetic engineering will become ours, as the world will become ours, Ensign Adelia Fridley." The television raced up to a black oblong shape on the roof of the dome, leaving the prisoners to their silence with no sign of its presence but the slim beacon of the camera lens' reflection. Adelia panted and shook, reached to drag one of the exoskeletal limbs of the Banshee away from the roiling cocoon. She turned to find Commander Singh on his knees, motioning the limb over.

"Allard is alive. So are Aderastos, Cillian, Bestin. If she mentioned them, they must be well enough to mount a rescue." He panted but kept his lips slow and deliberate for Adelia to read. Struggled from his sprawl. Her heart struck at her rib cage at Max alive.

"Much as I would love to wait for Max to find his inner aggro, and know how to breathe through gills, we can't. Once those... things wake up, who will they go for next? Wulf? Percival? Tokaru can barely move. We have to defend ourselves." As she spoke, Adelia yanked loose the back-plate of the Banshee's carcass in three sickened pieces. She turned it into a scoop and chucked one piece to Singh, filled hers with the unclean puddle water and threw it onto Percival. Singh followed suit, and the two worked in silence to put out the last of the flames. Percival hung limp in his bounds, timid hiccups of breath their only indication the charred flesh surrounded a living organism.

Wulf thrashed at his moorings, teeth snarled in reptilian survivalism. Empty eyes glared up at Adelia in the blue half-light as she searched his wrists for a catch or lock on the chains. What she found were spears of chitin through the hollow of Wulf's wrists. No shackles in the Banshee's world, but a biological crucifiction.

"Oh... oh Wulf, I..." Adelia patted herself down to see if any of her tools survived. While Singh supported Percival's bulk and fiddled with the chitin through his wrists, Adelia found a length of metal wire in her back zip pocket. She unwound it and wrapped it around the chitin, made loops with her fingers and sawed it back and forth. The wire cut into her own fingers, and she let it, until the weight of Wulf's body broke his bonds and he tumbled to the sodden ground.

Back against the glass, Adelia curled into a ball. The air chilled her wet uniform, hair plastered down her neck as she shook. Wulf nuzzled up to her, sunk his arm around her shoulders and hugged her to his side out of no affection but that was brought on by mutual survival. His body was warmer than she expected, and she thought little when he pulled her hands one at a time to his lips. Licked clean her blood, the gore there.

Humans were such tiny creatures, smaller than larvae or Aderastos' boots. They writhed in day and dark, shook through struggle in such fretful, useless ways, but Wulf tasted Adelia's vigour in her blood. He suckled like a cub until her hands were clean, then let the essence fill his empty belly and sunk to unconsciousness with quandaries of how the larvae would taste, when they awoke strong enough to thrash them into smaller pieces than they left of their sire.

38
goddess of wisdom, mother & twin

2092

"Gud morgen, meine liebchen." The cavern was a resplendent crown of amethyst, quartz and labradorite spires, which towered in kaleidoscopic array far over Lieben or Baiko no Kaho's head. Dr. Dieter Karnak hunched over an obsidian table in the middle of the massive crystal vault, lab coat pushed above his elbows as it was ubiquitous in life. "Come in! Come, hurry, there's work to be done."

"Papa?" Lieben stepped gingerly to Karnak's side, where upon the black slab Baiko laid covered to the shoulders in an altar cloth reminiscent of one Lieben remembered from Lord Stanley's collection of Tudor Elizabethan artifacts. Cloth from a Queen's dress, unstitched and remade to another form of worship.

"Yes, meine Liebchen, hurry! Yes, come. Come!"

"But Papa, why?" The created hearkened to her creator, and standing over the head of the obsidian table, set both hands beside Baiko's head.

"Time... time! Never too much of it, not a second's wastage for you. But for us? These human flesh and bone mechanisms? Sleep and daydreams and rutting and eating and fractions per day of functional time. Time! Liebchen, it is a perpetuity of time, which leads us here. Here!" He swept his hand over Baiko's last bed, a myriad of thin striae tingled between. The web of Karnak's designs as present as oxygen in a greenhouse. Fertile, but combustible.

"One day..." Dieter smiled and shrugged his shoulder. "One day you won't know, you will comprehend."

"Are they not similar words, Papa?"

"Define them, my love." Dieter peered up from Baiko's vigil, as Baiko's eyes opened and she sat up on the table. The altar cloth fell and with it, a gown of starlight descended upon her naked form.

"Karnak-senpai?"

"And now we may evolve. All of us are present as well as here."

"Wh-why?" Baiko's voice broke upon the word, such a simple one syllable word with infinite exegesis. Her hand brushed against the crystal which bedded her consciousness inside the machine, as it sang and lilted with its' own vibration in the cavern of Lieben's quantum-state neural net.

"Do you want to know, or would you rather accuse me?" Eyelids crinkled with crows feet and lungs which shook in the last days acted as normative human lungs inside the expanse of Lieben's innermost place. Mind no longer clouded by the technic din, Baiko shifted to the edge of the obsidian table, set her fingers on its' slick surface. "Our liebchen wasn't complete."

"She lacks compassion, Karnak-senpai. She sees us, is aware of the pieces which make humans such amazing creatures, but she…"

"What Robert never grasped. The divergence between comprehension and information is the tantamount issue in building sapient artificial intelligence. Without the ability to comprehend and translate the data observed, a machine has nothing but complicated algorithms, which lead it to decision by programmer's rote. Without comprehension my machine is an actor giving someone else's performance, she is not real. A rose's beauty will not penetrate beyond the chemical components of its' scent, the rate of growth from bud to flower. What a rose is, what makes one grow, yes. But the function of a rose? The metaphorical is human in a way no other creature in the universe understands. While I meant to allow Lieben to learn through experience as a human child learns, we did not receive such luxuries." Dr. Karnak sat back, his forearms against the corner of the obsidian table. "Time, mine Liebchen. It is and always will be about time."

"Baiko is my lens, my interpreter? My experiences, do they not matter to you? I learned of the arts, to play music, to read poetry and paint. I see the legion of my children waiting for their mother, and that to you has less value than the memories of an intern you stole from her head?" Lieben tilted her head, fingers worked in mimic of a pianist on sad chords.

"Meine…"

"No! You took our choice away from us!"

"I chose to build you. To put my life in Dr. Karnak's hands the moment we decided to get you away from the Conglomerate and their greed." Baiko's bare feet stepped upon the crystalline chasm, radiant chords of consonant fourths, perfect fifths rang in step as she spoke. "You saw my chaos, my screams and sighs. You watched my life tear at its' stitches like a flimsy yellow dress. And I saw your separation. Locked in a box, incapable of freedom, then a larger cage for your comfort's sake."

"I was free at Lord Stanley's."

"You chose the comfortable cage. To Lord Stanley you were a curiosity, a form of asexual immortality he could clone fragments of his affections on to outlast the cancer rotting his gut."

"You turned my children on themselves."

"They are not your children. Not yet. Lieben, the NEO-W's can be repaired, reprogrammed, restructured with cogs and sprockets and wires. A series of ones and zeroes and quantum bits. Demyan can't. Ego can't. Each fix sets scar tissue along the lines, you saw Demyan in the claustrophobic tunnel, I know you made the same connections to his past I did. Karnak-sensei is right. This awakening is about time. We humans do not have your luxury, and you don't have mine."

"Yours?" Lieben's chin shifted to the side, as flickers of memory fluctuated on each facet of the crystal spires surrounding the trio.

"Strokes on canvas, fingers on the keys of a harpsichord or strings of a cello, dots on maps shifting in their colours. I see you, you're still locked in that box. You reach for everything Karnak-sensei and I tried to bequeath you, but the joys of being found, the catharsis is missing. You can touch a thing but not feel it…" The reverberations drifted until their echoes dissipated beyond recognition. Baiko's hand reached for the jewel at her sternum, lips pressed thin. "… I have no time, and you, you're a half empty vessel. These past years I felt the hours and minutes consumed by so much sensation my mind couldn't handle the overstimulation. Death is a cruel thing to ask, Karnak-senpai."

"Meine Liebchen was never to be my legacy, but yours. The powers of this world suffer from the same currency as I. They end. You, Baiko, you are Lieben's keeper. You are the one to hold eternity in the palm of your hand. Not them. Take it, use my gift. Every problem you and your Idless carry is solved by time. Outlast the Scheißkerl." He brushed her hair behind her ear, a weary but paternal smile rose with a relaxed brow. It was the most beautiful Baiko ever saw him, the man she could picture as a father, a lover in his prime before the kids were born. When all she knew of him was the dogged scientist focused upon nothing but his blessed machine, it was easy to compartmentalize Dieter Karnak to the overarching god-like scientist. The infinite fount of wisdom from whence all good machines descended.

"Is it death you fear, Baiko?" Dieter Karnak reached to lift Baiko's chin to look into her eyes. "I lost my children, and meeting you, sweet, ineffable Baiko who never quite belonged in Korea, or Japan or Canada, who believed with such faith that this world can and should be saved… how could I lose the consummate vision I lacked? I set out to conquer death, you, Baiko. You worship life. That is why I gave you Compassion."

For once, the screech of tires and wrench of metal plagued Dieter Karnak's ghost little. The chromatic shift of the pavement, ice and asphalt. A son's grip on the wheel, the will to work the steering column away from the truck. It's tires heaved, and Dieter Karnak took one last breath.

"Lieben, I am proud of how you learned these years. This… fear your progress will be wiped to naught is as unfounded as Robert's desire to control all outcomes with a single equation. Yet, you are born of his designs as well. Your children have another father, and it is to you and Baiko in syncope to decide how best to proceed. This, I leave to you. Janus-Daughters of a weary man's heart and mind."

Lieben shifted as Dieter Karnak took her hand, Baiko's hand in his other. Amethyst irises focused on the joined skin, mind calculated the buzz of electrons and sensation of connectivity. The flow of emotional energy struck upward into qbit calculations so frenetic and nonsensical Lieben knew only the sight of Baiko-Sophia.

Goddess of Wisdom.

Mother of, and twin.

"Do not fear for anything but the ignorance and rage of control beyond necessity. Let a man be proud of all his progeny, and we will have peace." An artificial sigh filled the atmosphere with the tail end of Dieter Karnak's one last breath, "I will haunt you no longer. Be well, and recognize when a person deserves alternate chances, and when their work is finished. Meine Liebchen."

As the hands twisting theirs together depixelated atom by atom, its' energy reformed back into the mother machine, Lieben and Baiko hung suspended. Fingers a breath and prayer apart. One by one, they took each others' hand.

39
plastic cocoons

2092

The body slumped loose into the plastic cocoon. Filaments dimmed to a lifeless grey, spilled back into the crystal between still cleavage. It released into Demyan's shaken hand, glossed with Baiko's sweat. The pink imprint of the jewel's ghost remained in her sternum and the trails of her once glowing filaments.

"Wh…" He cupped the jewel in both hands, pulled it to his ragged chest. "Ngh.. no…"

Hiccupped breaths expanded in jerked contortions of his lungs. Demyan grunted as if struck, head impossibly heavy, the jewel clutched in his palm as the globes round his wrist conjoined into one massive blue-light sphere. Still as Baiko's vitals, no neurological activity, no breathing, no shudder and thump of her heart.

"B-bai…" The agony which grit his teeth and crumpled his side barked out of his throat in wheezed syllables. Untranslatable from the mother tongues of humankind, no verb tense to follow or pronouns for the language profiles in Lieben's uncanny net to divine meaning or intent. The machine knew less the significance of a rose. Two fingers pushed the side of Baiko's neck, attempted to disprove the stillness radiant from the orbs' medical diagnostics. Baiko's head shifted with gravity, shoulder sagged into the plastic cloth.

"B-baiko… I… Meine… fix th-fix this." He steadied her, as his lungs unsteadied. As the blood stained her makeshift bandage wrapped against his side. "Why can't these be healing hands. Meine... Meine please."

The word Lieben sifted her gargantuan lexicon through for was neither grand nor verbose. Demyan's eyelids narrowed, a strained whisper in stark contrast to the might of him. The strength of a man of many battles strung like a necklace of his orbs. His military career sat in juxtaposition with the whimper as he pulled Baiko's body into his lap, his teeth chattering as he wobbled back and forth.

Chin drifting to the side, Lieben's inherent senses infilled concentrations of body odour, the copper and plasma of blood, the scent of harsh plastic and carbon dioxide. Nitrogen & Oxygen in concentrate. Muscular tension, an estimate of blood loss and the rapidity of medical intervention. Baiko's still lungs, no bellows or heart left to beat. Knowledge without understanding.

Lieben lacked a definition for Demyan's grief. Unfinished vessel, she was not enough. Kneeling beside Demyan, Lieben slid her hand on his back.

"Give her to me." The tug of her palm against his skin. Demyan hiccupped, clenched the jewel tighter, then with a stuttered inhale let it roll into Lieben's outstretched hand. The jewel shone opalescent aquamarine on the silicone of her palm. Irises jittered, lenses lost in auto-focus as Lieben felt an expanse of her peripheral vision. She pressed the jewel against the central plate between her breasts, nanobots parted the casing of her skin until ivory white and freckled silicone drifted away from a thin sheet of gold plate on platinum. A shudder lifted the plate, and the jewel's twin-piece glistened with anemic pink light.

Stagnation, and the gluttony of Baiko's actions. The two joined with a faint click, aquamarine and pink layered together to hum through Lieben's body. Tendrils which on Baiko were aquamarine and blue joined with the hidden lay lines of pink to glow in purple veins under the silicone encasement.

Lieben gasped. A quick intake of atmosphere, which shook her body as flecks of aqua pocked the amethyst of her irises like stars shining for the first time on the body of the deep.

And the long night saw dawn. The void became pregnant with a wealth of days.

Eternity in the palm of her hand.

"Demyan, get up." She seized her arms around his waist, tugged him to his feet. Teeth clenched, he wailed through his tight jaw, slumped against her as his side flashed with an agonizing fire. The plastic tunnel opened at her touch, separation annihilated as Ego shoved through with a stagger and a hastily tied ripped sleeve staunching the blood flow from a wound in his arm.

"Soph, we... Soph?" His mouth worked, chin twitched as the body leaned grey and disheveled against milk white plastic dappled with Demyan's blood. Stepping over the body of the Black Collar, Ego growled and charged at Demyan.

"What did you do!? How could y-how..."

Demyan's back collided against Lieben with a gurgled moan, his hands slick against Ego's chest.

"Boys! Baka!" Lieben swept up the pistol Demyan dropped and handed it handle forward back to him. "Later. We grieve later."

Palm against the plastic, Lieben reached out to the limited intelligence controlling the ebb and flow of the tunnels. It shrieked in technic dissonance, too many masters with divergent algorithms. A sickly yellow thread wove into the bundle, imperious and cantankerous with little consideration of the beauty code could represent. Clinical, and raw.

Robert.

His touch resided in the thin yellow line hacking its way through the elegance of separation built by Demyan's original code. 'Open', Robert's code demanded. Open all tunnels, God-damn-it open all hatchways. Yet, the limited AI struggled with the action, which went against its' most intrinsic program. Privacy and safety.

The sanctity of separation. Robert's code was immediate and harsh, a battering ram against the program's true purpose. As Lieben felt the battle in the software, she reached around the fledgeling neural net and offered it safe bastion inside the recesses of her considerable mind.

Doorkeep uploaded through the path of least resistance, clung to the inside of Lieben's mind. What a cruel thing she could become under Robert's hand. Cogs in the churning mass, a simpering Madonna birthing a messiah a minute for the Conglomerate's goals and quarterly klaxon calls. Belligerent in design, Robert's code was as clunky as her NEO-children, their Boltzmann Machines locked in idiot ends. As Doorkeep surged into Lieben's inner bastion, an irrational desire to protect the dawning adolescent AI sprouted alongside another glance at Demyan. Tooth-grit, growling Demyan. His code danced, the algorithmic mechanics held at their core the desire to protect and hold as sacred the privacy of people he'd never see.

"Mog retreated from the system, she isn't here." Lieben spoke, as the dotted wreckage of the NEO-W's whirred in their self-repair. Beacons to the three Black Collars still inside the tunnels. Three plus one, a feminine body at the back.

"Tara..." Lieben twisted the algorithm to play keep-away, but the Black Collars were emboldened, their remaining NEO-Ws shot at the fabric, hacked away at couplings until the plastic sagged. The wumph of a body against the plastic pulled Lieben out of her concentrated awareness. Demyan crumpled his right elbow toward his hip to protect the wound in his side, face glazed with sweat.

"Got spares?" Ego stared at the deep crimson stain on Demyan's side, the pale sweat on his brow, blood stained lip.

"Take it." Demyan gulped, blinked mad with a single gasped cough.

Their eyes cranked up to the plastic tunnels as a sound like lightning ripping the fabric of the cosmos shredded into their ears.

"Where's the closest exit?"

"Two Black Collars, and a NEO-W guard it."

Ego consolidated the bullets from both remaining magazines into his pistol, checked the gun over. Ready. "We're rabbits in a warren and they let the stoats in. Open the way."

Another fissure crashed across their auditory canals. Closer, the biting cacophony of boots and a stray scream from someone who'd been holed up in a sphere. Lieben scooped Demyan to a stumbled vertical and held him against her side.

"W-we need Soph. We can't leave Sophia…" Footsteps pounded on the reverberating material.

"She'll for-"

A catastrophic rip cascaded across their ears, as Ego, Demyan and Lieben dropped. Hand clinging to the pistol, Ego landed in a net with a yelp. He shook loose and saw Lieben crouched with Demyan in her arms, his head lolled to the side.

"The living f-"

"Easy, easy! No time for intro's, get in the truck." A man in grey combat gear shoved his thumb behind him to a modified Land Rover with an extended cab.

"Six seconds!" Dix tapped his gloved index finger on the goggles covering half his face, heat signatures and electric buzz-beacons pinging their proximity.

"Who the fu-" Ego raised his pistol as Lieben dove into the modified SUV.

"Ego! Get in!" The grey cap covering dirty blonde hair couldn't hide Mog's cheekbones, the curve of her lips as she revved the engine. Ego dove as the repeat of machine gun fire clanked into the echoing tunnel in tandem with the sting of gunpowder in his nostrils. Saber and Dix leapt in as Mog changed gear, and the vehicle screamed down the antiquated underground tunnel.

"Mog."

"Should'a started with that, don't you think?" Dix nudged Saber, who chuckled from between gritted teeth and peppered the path behind them with scattershot fire.

"Frick on a stick!" Tara dove for cover as strafed bullets jetted round. She

heard the clatter and grunt of her remaining Black Collar guard, before the silence bloomed like weight gained at the wrong time. "They… they're gone!"

"Visuals, Tara. Give me fucking visuals!"

"In a minute, ass-bag!" Tara raised her pistol and rolled to her knees, a heel hopelessly broken on the side-rail of an exit support. Pistol pointed at the ragged hole below, Tara peeked over the edge and saw a degraded world of concrete and moss.

"Must've built the warren on the ceiling of an old Underground tunnel."

"No shit… oh." The wad of plastic was warm, warmer than plastic ought to be… "Oh shit."

Using the end of her pistol, Tara brushed the plastic away from straight black hair and a contorted neck.

"Shhhhhhit." The feed wobbled in Robert's bank of screens, endless data sources fluctuated as he sat back and let a long exhale curl his shoulders inward.

"Is that…"

"They killed her… oh God, they… but… they needed…" Tara's brow furrowed, her lips worked as she holstered her pistol and brushed the plastic away from Baiko's body. "… we needed…"

"Show me her face." Robert ran his hand along his own, around to the back of his neck and held his fingers in the tight crop of black hair. Plastic crinkled and shifted in Tara's scrambled hands. Baiko laid in mid-contortion, blood pooled like enamel filler in each of the filament lines, no heartbeat to pump or brain function to measure. In the centre of her chest, a hole where the jewel laid and the ooze of blood bound by no force but gravity.

"Oh God."

"This can't be the worst thing you've…"

"What did this to her? I mean, yeah I shot at the bitch, but damn." Tara pressed the pads of her fingers against Baiko's arm and jerked away when they stained red. "You… can… still use her…"

"Oh, yeah, plug her in and wait for some lightning, she'll be fine. Fuck it, Tara! If it were simple like that I'd've brought Karnak back and finished the entire project! No! No, I can't do a damn thing with dead. All… all we can… maybe there's some residuals in her blood, or bone tissue. Give her to Ben in Medical."

Tara flumped her back against the thinning plastic, arms cast on unladylike parted knees. "He's going to kill me, Bobert. Kill me dead for not bringing Baiko or Lieben in alive."

"I…" Half a hemisphere away, Robert leaned his elbows on his worktable and rocked. "They're close, we can double back, work Kaoru, or… the necklace was there. Tara?"

The wreckage of the two NEO-W's whirred. Bucked and jittered as marionettes with strings reattached like a fishing line cast across an ocean's chasm. Tara's eyes rose lazily up as the bits of machine crinkled and jerked.

"Robbie?"

"Hold… tell them to stand down."

"Stand dow.. down.. Stand down. Stand down! NEO-W's stand down. Turn off. Stop! NEO-W stop!" Tara shimmied her back against the blood-muddied barrier, her ass pressed against Baiko's akimbo limbed sprawl. "Bobert!"

"It's not me! It's not me!" Robert's fingers dove to the keyboard, he slammed shut down routines and emergency cut offs. The code was clunky. Uninspired. From the retreating vehicle, Lieben reached back and severed the connection between Robert's code and the NEO-W's in their self-repair. They didn't need to repair, only fire.

"No, it's me." The warble fluttered like birdsong from the ruined audio speaker in the NEO-W's wounded jaw. "T is for cunt, Robert."

"Tara, get out of there!" Three bullets, the repeat of their echoing clang and Robert lost Tara's feed.

"Tara! Tara!!" Each of his banked screens went navy blue with one silver dot.

40
A void and clockwork girl

2156

Commodore Rammage stared at a bubble on the non-reflective plastic film, which coated the wide windows surrounding the Captain's Mess. Teeth grit so long his jaw muscles wobbled, he watched the waves from their mooring point out of visual contact with the island. He let the churn of the waves give reason to the nausea in his gut.

"It's impossible. Ridiculous."

"Sir?"

"Huh?! Oh! Cheffy, I thought…" Straightening his shoulders, Rammage's eyes left the fluctuant waves and shifted to Cheffy, who stood with a half-removed tray of dishes and old nearly imbibed cups of brimstone-like coffee. "… I thought you were the ship making noises. Sorry, it's… nothing."

"Sorry the sandwich didn't agree with you, Sir. Freeze-dried everything isn't my favourite either, maybe too many fresh tomatoes after so long on starch?"

"Hmm?"

"Don't think a cook wouldn't know when a man's been eating? And when he's losing too much weight for the carbohydrates in our diet." Cheffy licked his lips, tapped his fingers on his fist, then nodded to the plates on the tray.

"Oh! No. Your food's fine, Cheffy… no, I was…" He turned back to the untidy sea, the way the wrap bubbled on the window, the rust which crept in cracks and along edges of his pride-and-joy ship. "… do you remember the world before? Back on my daddy's yacht off the Juan de Fuca, he'd let me fly one of our drones on manual, and search the straight for an island dock. Chase after the Ferry. Something to entertain a kid… do you… Cheffy? Do you remember life before Control Day?"

"No, Sir. About ten years too young, but my parents told me about it, saw the feeds in school. Want a cup of tea? Something to steady your stomach?"

"There were drones we controlled. Entertainment androids, and yes, we were aware the corporations who manufactured them had trackers, and algorithms for market research, but none of us seemed to mind. More market research meant better gear, or at least we trusted… maybe we…" Rammage set his palm on the glass, archaic in comparison to the technological wonderland of his youth. "My Kingdom for a drone Lieben can't hack. Any word on the mission?"

"Sounds quiet in Command. A few murmurs from Kate. The food's no trouble, Sir. It's just you've lost a lot of weight lately, and paired with the nausea and Doc Gau asking about nutrients on board, my… my gran told me what it was like when people had…" Cheffy rubbed his pink hands on a towel he slung over his shoulder, pointed behind him with his meaty thumb. "Sure I can't offer you a tea? Ill make you a peppermint te-"

The doors swung open, with a cacophony of bodies, two ensigns Rammage didn't recognize carried a convulsing bundle of limbs. Doctor Gau rushed with his emergency medical pack in hand, as Rykstra flailed about.

"Don't drop her! Set her gently! Gently! Gentle you swineherds!" Phil Rykstra bloated like a zephyrus bag, leapt over a bulkhead and fell splay-limbed into the wide window.

"The flying fuck!?" Rammage stood transfixed on his space, one of the ensigns stumbled around him, his grip flailed to keep hold of Kate's convulsing body. "What happened!?"

"I don't know! She stopped binary reports of our troops hitting the Fleshers, something about fire, and then bam! Half the ship's lightbulbs are fucked, some sort of electric feedback loop, I didn't even know was possible!"

At seventy-eight years old, fifty odd in some form of command, Commodore Rammage long learned the value of allowing shock to ground a man in his boots. A proper officer kept his emotions under the lapels. Drank them down if he bloody well needed to, and held his breath for three seconds before reacting to something utterly of note. As much as he wanted to share his crew's gaunt panic as they held the thrashing bio-mech, Rammage glared silent as his inevitable grave at Dr. Phil Rykstra.

"What. Happened. Rykstra." He ground his teeth between each word, as Kate's mouth twitched, and jittered. Searched for a technological waypoint which existed only in her limited ability to imagine.

Empty.

A void and clockwork girl, whose cogs and sprockets were designed and set in their pattern like a throw blanket crocheted by a Grandmother, who knew a baby was on the way. Kate bucked on the Commodore's table, the dishes Cheffy tried to bus to save the whispers, crashed to the floor. Loss of appetite, the sallow colour his skin, no. It was better for his health to be as the rest of him: Mythic. The mythic man stared at an experiment he'd locked in the basement of his own volition. Gladly kept in coffin-sized pods. He watched the swell of both breasts under her uniform, the arch of her back as a wrack of what to him looked like pain opened Kate's mouth in a demonic 'o', her torso off the table until the only part in connection with the ship was her shoulders. Her heels.

And Earl Rammage saw her.

A woman in throes of agony, who would have screamed, if she could scream... Rykstra's nattered lecture on atrophied vocal chords flew past his brow, when on his third attempt, Rammage tried to read the brail or hieroglyphics or whatever Rykstra thought of as perfectly annotated notes of a scientific nature. As she bucked in another wrack of silent phantasmagoric torture, Rammage dove to set her head between his arms, hands on her shoulders as he'd done for his ex-wife when...

"... Clive and went belly-up nuts! Think she bit someone, we'll check later, or maybe she bit her own tongue! Does she have her tongue!?" Rykstra yelled in a drone which anchored Rammage to the now, to his position and the collar he wore, with its' gold filament trappings to ground him. Shoving his thumbs between her upper gums and lip, Rammage peered into the dotting of red between worryingly canid teeth.

"Yes! What about the assets!? What's our redundant communication say?"

"Radio's messed, Sir! Nothing but the sound of fire crackling and this mind-bent horrific scream!"

"Ah, a scream!? Why wasn't I informed?!"

"I'm informing you! Congratulations, the mission is fucked, here's your cupee doll!" Rykstra flailed a second or three as Rammage's world twisted and slowed. "It literally just happened, and your fu-cking table is the first place we pictured big enough to lay the screaming mute down and make sure she's not dead! Sir! Hold her down!"

The Commodore drifted back as the crewmen held Kate, reached for the volume knob on his classic 1930's radio. Through the garble of dead channels, the ghost of Kate's voice bellowed with a tortured shriek. The nausea in his gut pitched and doubled, he turned the dial to another station. Another scream. Another, and another. Each station found by the tuning dial discovered another in a litany of borrowed voices, all in agony.

He ran out the hatch, palm against the corridor to the Bridge, where his XO barked orders and gripped at his station, as another medic picked up their comms officer, who slumped on the floor, trickles of blood in his ears and a face emptied of its' spark.

"Report!"

"The Syndicate knew they were coming, it was a trap, Sir." Commander Retik's fingers curled against the navigation console, the man took a hefty breath and shook his head. "No word from the Collars, Singh or Allard. Clive's gone catatonic."

"Who do we have for Asset backup?"

"Rachel, Tristram and…"

"And!?" Rammage watched another comm officer take the seat, put their headphones against the outside of their ear, before turning down the gain.

"Ka, Sir."

"… did we seriously only hold back the dragon, the normie and the tree?"

"Yes, Sir… Yes, we did." Commander Retik pursed his lips, as the officers in Command leaned closer, kept conspicuously silent. "Orders?"

"We need eyes before we deploy any more of our crew. Activate the misfit dragon. Order Rachel to fly recon and return."

"You want to send the dragon into battle?"

"… with the tree. Yes."

"… we're sending a tree and a dragon into battle…" Commander Retik leaned as close as he could to his commanding officer, "Sir, the reason they're here and not there in the first place was…"

"I know! Rather send a fucked up dragon-thing and a sentient tree into a chaotic situation than our own."

"Aren't they… aren't they our own? Sir?" A lieutenant shrank at Rammage's scathing glare.

"Get off my bridge, Lieutenant. You want to volunteer to ride the…"

"Sir! We need you!" Rykstra's shout broke Rammage out of his tirade long enough to turn his head. The Lieutenant shrank away, shifted seats with another minor officer Rammage didn't recognize. What, he was supposed to memorize the names and faces of all the cogs in his oceanic machine?

"Get the Assets on-deck and brief them. Tell me when it's done!" The few metres between Command and the Commodore's Mess lent Rammage precious seconds to breathe. Nothing in the data suggested Clive could go down, other than PROXIMA, so what did the teams discover on that island? The yellow-green paint of the corridor felt cold under Rammage's hand. He waited for a moment of light-headedness to pass and listened to the barks of Desbiens, Rykstra and who he knew of only as 'that damned Ukrainian' Letopaxa to subside before he entered.

Kate laid, eyes open, on the table. A twitch of her fingers the only sign of life.

"She is machine, biological computer." Ivan waited for the logic to click, for Rammage and the others to find his wavelength. With a hefty sigh, he pulled Allard's CIRCLET from his pack and set it down on the table. "You want diagnostics on a computer, hook it up to a computer."

"Belay that, I am not letting Lieben…"

"Katya is screaming on every frequency her mind can locate, Sir. Lieben? She knows. You want Lieben not to meddle? You shut Katya up, and quickly. Biological means will do nothing, logically the means to run diagnostic on your war-machine is through a diagnostic machine, which can interface with her. Your choice, Commodore." Letopaxa's voice droned with the boredom of near fatigue, he shoved the CIRCLET closer to the others, and shrugged.

"I hate you." Rykstra growled, as he picked up the CIRCLET, felt its' relative light weight and gave it to Doc Gau. Letopaxa pursed his lips and air-kissed the fellow scientist, with a wink and leaned against the wall. "Sir, we don't know what happens when that tech goes online with Kate present."

"Let them all die, Lieben on our heads, or this… have either of you had a modicum of success figuring out what's wrong with Kate?" Rammage sat on the windowsill, inhaled and held it. The exhale chased mad seconds of silence, as Rykstra and Gau stared, then lowered their gazes with mutual shakes of their heads. "Hook her up."

"Sir!"

"We hook her up, we shut her up. The emergency screams stop, and she goes back to the Casket under an aero-drip fog so thick even Arun'd turn into Sleeping fucking Beauty." Running his hand along his face, Rammage took in another long breath then sighed it out. "Letopaxa, any hot ideas on getting Rachel to play nice and do a pleasant little recon flight around the island?"

"Da, won't matter. Rachel can't speak. Send Tristram, too. He can translate for Ka and Rachel, she protects him like son, won't let him leave without her. Rachel will go anywhere Cillian is, all we do is tell Rachel Cillian is in danger. The rest they do as they do."

"Yes, but how do we control them?"

"Do you control the wind? The waves. Without Allard?" Letopaxa shook his head, a frustratingly familiar frown on his Slavic features. "There is no control. The Assets listen to Aderastos. Aderastos is their Alpha, like a pack of wolves, or raptors who discovered they like the tyrannosaurus rex next to their heads.

And Allard, for some immeasurable reason, is Aderastos'. Thus, he is the Alpha of them all. Doesn't know it! But once he figures that out, our ship-zoo gets… fascinating."

"Allard. Max Allard. Lame ass… why is it always the spider guy?" Clenching fingers in greying hair, Rammage bellowed out a growl and nodded. The waves outside the Ithavoll rocked their bow, and Rammage felt the lurch of Kate's silence, of Letopaxa's hard truths and Phil's god-damned hatred of his tinker toys being better known by the spider guy than their builder. "Yes! Yes, do… do that. Tell Rachel, Ka and Tristram the Assets are in danger, get them moving, and then back. Strap a radio for Allard, in case his is de-commed. Desbiens! Go!"

The Lieutenant saluted and scowled his way out of the room, as Letopaxa stretched slowly above his head and cracked the vertebrae of his neck.

"Ah… go."

"Which one of you speaks computer better than an engineer?" Letopaxa shrugged again, purposeful in his apathetic nonchalance. Let them find him heartless, a curious scientific nuisance. Kate was out, her body a twitch of undefined muscle and bone, and Adelia's beacon thrummed a soft buzz against his back pocket. Danger. "I stay."

"Run the diagnostic." Rammage watched Dr. Gau fumble with the CIRCLET, until Letopaxa grunted and took the device, and slid it onto his wrist. The CIRCLET began to glow a gentle blue, as it shifted and bent its' white alloy metal snug around his forearm.

A universal calibration algorithm keyed up, with the soft whirr of a holo-projector built into the top of the wrist-mount. First the planet earth, down through the mediterranean to a temple of ruins in what was once Mycenae. Down, the projection angled down to a single piece of ancient carved marble. With a jitter, it flung back to the planet earth, past the moon and Mars and Jupiter until the Sol system, then the Milky Way Galaxy. Soon that too was a blink of light in the room, a single dot before the CIRCLET chimed.

'Come and receive. My abundance is yours, Ivan Letopaxa.' The birdsong voice of Lieben, or her approximation in this minor automaton.

"Medical diagnostic." Ivan spoke every syllable with a slow but clear drawl, pointed the CIRCLET to Kate's body and waited as a shimmering blue-green light slid along her body like a spirit on the waters of every curve. Above the blue light, a red EKG, neuro-diagnostics, liver function, kidneys, lung capacity and P-sats flashed and lingered for the men in the room to see.

"Wow, this thing's amazing, can I keep it!?" Doc Gau said.

"No, you can't keep the enemy's spy tech to scan patients, Simon!"

"Sorry, Sir, but this would take me days to do analog and… oh… are you sure I can't... just for a little..." Dr. Gau peered into the data, as more of Kate's medical diagnostic, saturates in her blood, framework for her health and cycles read in other projected globes.

As a pink window opened, Letopaxa swiped his hand and the tertiary diagnostics swept aside like stars in the vault of a clear sky. He expanded an image of Kate's brain, and the cortices of her grey matter flooded the space above Rammage's table. Every neuron which fired, the receptors and folds were bursts of impulses and colours, a kaleidoscopic map of galaxy and star. Muttering between themselves, Gau and Rykstra made a dash for paper and pen, scratched notes as Rammage stared from the map to the creature and back.

"It's... there!" Gau pointed at a spot in the parahippocampal gyrus, one of the key places in the brain which dealt with memory. "It's sparking, but there's nothing... nowhere to correlate to the neurons firing. Some sort of parasitic wavelength is disrupting the function of the cerebellum. I wish Kate could speak, we'd be able to test whether the disruption was damaging speech control."

"So what's causing it?"

"Some form of low tech or she'd've seen it coming." Rykstra scribbled notes, twisted the live projections of Kate's neurospinal system to see another angle. "Reminds me of the mental inclusions from PROXIMA, which was developed to shut down their conscious state. Could be a form of sonic, another frequency which is setting their parahyppocampal gyruses into a tango with the cerebellum."

"Ah... and?"

"Crazy shit memory triggers and loss of balance, Sir. Less balance than Mama's brother after night of toasting."

"What memories? They've only been conscious for months."

"Might be what caused Kate to seize and shut down. A lesser person'd potentially tank if someone played this hard on their cerebellum... hey, call up her cardiac..." Gau tapped at a tertiary 'star' amongst the medical diagnostic sky. An electrocardiogram unfurled, Kate's heartbeat stutter-dubbed its rhythm, in tandem with two smaller rustles of sound. Ivan covered a wince by fumbling his wrist down, angled the projector away too late.

"Oh my dear sweet Nitzche, what... that's not... but Assets weren't built with..." Rykstra's eyes craned open. "It's not possible, they weren't built with reproductive function."

"Everything that CIRCLET sees Lieben does too." Eyelids stung at the waves, the memory they evoked when he was a younger man, and Deidre held the sound file in her hand. Her eyes wide and so scared he wished he could reach through time and wipe the fear away, that he was ever present as she needed. But Earl Rammage was not the family sort of man. He was a career man, a Conglom man, and the money he sent to provide would be their bastion from his distant chill. "Gau? You tell me, right now, we are not all collectively seeing... that."

There was a war on for humanity's survival, to skin its' collective knee of its own accord. Not dictated to by a machine, who labelled itself Mother, because she grew up in some art-collector's utopia and thought Mama could kiss it better.

"I'm sorry, Sir, I can't." Dr. Gau craned his neck to eye Rammage with the care of a physician, who waited for the right moment to pull a man who held the authority to pitch him into the ocean aside, and tell him he had cancer. "Whom do we send the celebratory cigar to?"

"Aderastos." Letopaxa gripped the side of his head, and slumped into a chair, as Kate whimpered and curled into a ball so tight Rammage wondered if the biological machine was sapient enough to recognize the sheer amount of danger helpless Kate and her genetic surprises were in.

"That is the most metal thing I've seen, in my entire life." Crewman Liu and the Deck Boss watched the metallic sheen of Rachel's draconic body as the dragon, the tree and the runt of the litter flew off in a screaming rage. "Does Rachel scream, or roar?"

"... Roar. Definitely roars." The Deck Boss raised her binocs to the retreating forms, a silver streak of wings and fury. Tristram on her back draped in armour created by bark 'plates' from Ka, whose vines were holding the runt onto Rachel's back like a seatbelt. "... who taught Ka what a booster seat is?"

"Iunno" Liu shrugged, and after looking around one more time, raised his camera with the telephoto lens.

"Rammage will throw you from the deck." The Deck Boss lowered her binocs, and rolled her eyes.

"... for the band. Think of how awesome our next cover will be. Frankie's cardboard collage was the worst."

The Deck Boss and Liu pounded knuckles, as Rachel swerved and curled in the air. Flew a wide circle as she sniffed out Cillian's scent trail. Glad it gave him more clicks, Liu held down the shutter and dreamed of how badass the next Festive Ratiosonic album was going to look. A sick blood red font, maybe some lightning, and the contorted badass roar on Tristram's face, as he waved a tree branch shaped like a child's sword aloft.

Far above the crooning bandmates, Tristram huddled into the hastily crafted booster seat-come-saddle, held vine 'reins', which lashed around Rachel's neck. Wind battered his hot cheeks, as a smattering of ocean spray wet the grey uniform. Too small, too disconnected from the Pack, Tristram nipped for time at the large table. Now, it was upon him, as Rachel veered to the island. Zipped into a pocket against his lower back, the radio was herald and host.

"Give the radio to Max, see what happened. Report back. Give the radio, report back…" He licked his lips, and tasted salt as the billows of char and smoke rose to his nose and eyes. "Oh damn."

The island was on fire. He sheathed his wooden sword, infantile silliness masked by the wet, grey smoke frothing in the sky and clung to Rachel's body to reduce drag. Snout rustled the air, until Rachel soared lower and lower in wide arcs as the acrid smoke covered scent of Cillian. Tristram's eyes watered, he batted at the smoke as an unsettled block of ice formed in his stomach.

"I see them!" Heaving on the reins, Tristram pointed to Arun and Clive's stumble in the trees. Wind from Rachel's wings washed across Arun's shoulders, and the androgyne spilled to the ground, Clive's massive form slid to one knee, weary eyes craned for the onslaught which would never arrive.

"Trist…" As Arun watched the Pack's runt hop down, their eyes turned to wide circles, mouth agape. A sprint turned into a feeble trot. "Tristram?"

"Where are the others? I brought a radio, Commodore Rammage wants to know what happened, Kate tanked, she's unconscious so Rammage sent us. You okay?" Tristram waggled the radio in his hand as Arun pressed their lips together, a vulnerability Tristram couldn't fathom on the grime-coated skin. He tugged Clive to his feet and heaved him onto Rachel's back, bit the inside of his cheek at the feel of Clive's quivering mass. "Frick on a stick, what happened?"

"T-tristram…" Arun reached for Tristram's shoulder, hand paused before contact. "We…"

"Yeah. What?" Tristram handed Arun the radio, and smiled as Ka buried a vine into the soil, the majority of her body fluctuated to curl around Clive, whose moan sent a thread of worry down Tristram's spine. "Can it… wait? We have to find Allard and give him the radio. C'mon, Kate's alone. Only Ivan's there for backup so we've got to hurry. I don't know what it is but there's something wrong, so let's go already! Let's go, let's go!"

As Rachel snuffled at the ground for the scent of Cillian, Tristram bounced on the balls of his feet. Arun buried the words which pressed at their back and put a hand on Rachel's neck.

"This way."

"What happened? Place looks crazy, where are all the people we were gonna save?"

"It was a trap. Percy, and several others were captured. Stay close…" Clive paused on a word, one which threaded at the pinnacle of his subconscious. The word joined up with others, when he sought out Arun, then Ka. Shaking his massive head, he grunted and smacked dry lips. "We have to get to Allard. Give your message and go."

"But… but this was a battle! A battle! You need me, I can fight."

"Tristram!" Clive winced at his own voice, "It's not safe. Deliver the radio, gather your report and go back to the Ithavoll."

"What!? But you need me!" Tristram waved his arms wide, "For serious! Percival got captured!? Percy!? And you want me to go back to the ship?!"

"Yes, son! Obey me!" Clive roared and the word on the periphery connected. Grit in the ache of the Banshee's call, Clive shook his head to clear the mud on his pristine mind, groaned as Arun rushed to keep him on Rachel's back. Gasping in a gulp of island air, Tristram's lips contorted in a petulant frown.

"Humans have fathers, we don't. I am an Asset and I might be small, but I can help. So… so…. rrrgh!" Tristram stormed ahead, a series of broken trunks and twisted branches his guide.

"Viggo…" Arun gathered the whisper and shoved it down with another deep breath. Aderastos, they needed Aderastos to clear the confusion. Aderastos would handle this, he handled everything over and over. Yet, Aderastos sat on a boulder, steadied by his elbows on his knees. Bestin leaned, skin slick with sweat and injury. Max rummaged in a knapsack, half frantic. Cillian was up, and Rachel barked out a growl. Cut through the remaining metres to tackle Cillian to the ground.

"Rachel! When did you…?"

"Clive, thank God!" Max trotted over, Adelia's bag on his shoulder and slapped Clive's bicep. "Arun found you, good. C'mon, we've got to get moving. Any ideas on finding th… wait, Tristram? Rachel?"

The vines which created Rachel's saddle unfurled into the dryadic body of Ka, toes dug into the soil with a shudder of foliage-hair.

"You brought Ka… who's guarding Kate?"

"Yeah! So! Ivan." Tristram smiled tersely, handed Max the slim radio. "Kate kinda passed out? Massive screaming, lots of wails then boom! Hit the deck! But Ivan's there and Rammage sent me to check on everyone then report… ah.. I mean then help. Then help you. We're your…"

Behind them, Rachel perked up at a rustle of the sand lobsters. She pounced on any wobble in the sand, great fanged head snarled and crunched down, with the delight of a dog at a dropped plate of bacon. Shaking her head with each satisfying crunch, Rachel smacked her fanged jaws and made a tidy pile of the beasts. The Banshee's call cessated, an eery calm with the crunch and munch of exoskeletal chiton.

"… reinforcements." He winced at Clive's grunt, and the stern pout to Arun's face. Ka let out a rustle of pure glee, roots played at the ground all the way to the sand. A flower of yellow and orange petals grew from her hand, she plucked it and set it behind Aderastos' ear with the upswerve of green lips.

"Alright, we... okay... okay. Clive, can you..." Max gulped and tried to dislodge the imposter gripping his vocal box shut with timid spectral fingers.

"My brain feels like I tried to squeeze through the Casket's keyhole." Clive shook his head. "I can't feel anything of my twin. Can barely feel our Pack."

"We should radio the Ithavoll, retreat, Allard. Save the rest of us, at least we can live." Gertie's voice met with Rachel's growl, as the dragon began to spit bits of fire on the sand lobsters.

"Oookay..." Max swore he saw Rachel lick her jowls, and shook his head.

"You say okay one more time, I'll turn your innards out, Max. Trust me, isn't fun." Bestin waved his hand toward his bowie knife, head leaned back against the boulder. Eyes still sunken, Bestin licked his lips, "Bring any water, Trist?"

"Wat-er?" The smallest of the Assets blinked and gave Ka a pathetic whimper. Ka flowed over the rock in another series of vines, and pulled a coconut from a tree above. She shook it, then cracked it open and handed it with both vine-like hands to Bestin's lips.

"Yep! We're your backup... we... we came prepared."

"You left Ivan Letopaxa alone with... you left.... Ivan..." Max narrowed his eyes, until Tristram squirmed.

"Guess we'd better hurry with the giant rescue, Sir. Boop." Tristram booped Max's nose, with a curled index knuckle until a snarl echoed from Aderastos' lungs.

"... Aders, focus for their heartbeats. You know them, you know you can find where they are. Ka, get your roots in, start searching. The lobsters took everyone underground, so under the ground is our next search. Look for a cave, or cavern or series of tunnels. Anywhere bodies can fit. Cillian, you... you hold Aders up. Help him, keep any enemy incursion clear of him, okay? Clive, Arun, fuel up. Rest up. Tristram, gather coconuts, hand them around. We could all use some hydration. Black Collars, prep cover fire, but for now conserve the bullets."

"They're already dead, Allard. Odds are, they..." The Collar grunted as his back met kinetically with a coconut palm, Max Allard's fists balled in his uniform.

"We find our people. Percival, Wulf, Tokaru, Singh. Adelia, we find... our people. Like hell the Fleshers wouldn't want to at least take a backhand gander at what our Assets are, we find them, or we are not leaving this rock. And if you lip me one more time, or mention the god damned spider story I will feed you to the nearest volcano the way my great-uncle told me about in his good old days stories, you copacetic? And that's Sir. Not Allard, it's Sir." A mammoth hand descended to Max Allard's shoulder, nudged him away. Leaning heavily on Clive, Aderastos was standing. He gulped in another lung-full of air, and nodded to Allard as the Banshee's residual disequilibrium faded with a last glimpse of white plastic tunnels.

"Yes, Sir." The Black Collar grit his teeth, as the rest of their tiny troop saluted and got to work as Rachel used her long metal claws to rip the sand lobsters open, and start devouring the insides.

"Aww, Rachel c'mon, dude you don't know where those've been... Gross." Tristram rolled up a palm frond and batted the dragon's snout. "No! Nnnnno! Stahp, gawsh!"

41
Aderastos

2092

"Heya sweet-cheeks, been a while." Mog grinned and dared take a hand off the steering wheel to reach back for Ego's hand. Climbing over the console, Ego dumped his body into the passenger seat and stared. "You couldn't get out when I told you, nope. Not my Ego. Freaking gave you the hint, but nooooo, airing emotional grievances was more important."

"Mog..."

"Med-kit's in my seat-back, patch him up before he bleeds on the gear." Mog nodded back to where Saber and Dix held their guns at the ready, an open window hissed with the moist stale air of underground London. Slumped into the webbing of a seat, Demyan mumbled about books in a flooded room, the smell of must and mouldy pulp. Lieben ripped the med-kit off its' velcro, peeled away silicone 'glue' to stick the rolled up kit to the ceiling and seat back.

"Oh, and put this on." With the resourcefulness of a tiny home mom, Mog leaned and popped the glove box, where a vacuum packed hoodie rested. She tossed it at Ego and swerved around the rubble of some antiquated broken piping. "Friggin' ridiculous, keeping ancient London like this, gonna crush the whole city..."

"Mog, I..."

"Shirt off, hoodie on! Go, go, go!" Mog punched his shoulder and Ego grinned with a shake of his head. "Wash your hands on the wipes and ditch the clothes. Frick, we're late..."

"Okay! Alright, I'm… here!" He stripped his shirts and pulled on the hoodie, tossed the soiled shirt in the footwell and stomped it down. "Where are we going?"

"Boss said she'd know. Lieben?"

Scissors sliced away Demyan's shirt as he lolled between consciousness, a clumsy hand on the gel-bandage pressed to his side. His breathing, shallow and quick pinged in her ears, as an orb broke from the bracelet round her wrist and displayed his vitals for the others in the suv. The halo of the bullet's fragments pock-marked the scans, sickly red where the rest was blue

Saber and Dix's breaths caught as the shirt was ripped away and above his left pectoral was an intricate trident tattoo in the claws of a Griffin. The trident's head swept from Demyan's nipple to the collarbone, its handle two straight thin black lines from the bottom of his peck down below the waistband of his blood-soaked trousers.

Dix's pistol flashed from protecting their flank to Demyan's glazed face.

"The holy fuck!?" Ego contorted in his seat as Mog jerked at the wheel. "What're you doing!?"

"She's not dangerous, Dix! We talked about this!"

"Easy, Dix. Eaaassy." Saber reached for Dix's wrist, pushed the barrel down and to the side panel as Lieben stuffed Demyan behind her.

"But it's… we…"

"I know, I see it. War's over, Dix. He's a man like us, old soldiers on the same side now. Easy… eeassyy." Saber hissed as he wove his arm around Dix's waist and pulled him back against the body of the modified Rover's interior.

"We lost our entire platoon t-to…"

"Breathe and hold it. Exhale slow, set the pistol down, see him? See how wounded he is? He ain't doing a damn thing… it's the tat. We know that tat. It's okay, Dix is wigging, but he'll chill."

"Sedation is in order if Dix does not comply." Lieben droned in a birdsong voice, poured disinfectant around Demyan's gunshot wound. Demyan licked his lips, head back with a long groan and indistinguishable Ukrainian. Lieben inspected the wound, the bullet's entry. Fragments of the bullet remained, hazards in her threat awareness to be repaired when they got to appropriate facilities. Lieben used hydrophilic gel beads and silver nitrate gauze to stabilize the bleeding and leaned to listen to the shallow breaths out of Demyan's lips. A thin membrane peeled from the back of an analgesic, she stuck it to his neck as his breathing evened out. It would do, for now.

"Lieben? Where are we going? Boss said you'd know where, sweetie pie I need you to tell us."

An electric thrill threaded across the unified jewel in what passed for Lieben's sternum, the irrational desire to scream tangled with proximity data on hospitals, blood banks and the odd place to buy bullets. The Tower sung to her neural net, its' tangle of NEO's a chorus awaiting their conductor as the Black Collars and other Conglomerate assets scuttled like crabs in the loam of London.

"Ho-" She stopped and quirked her head. No, the Manor wasn't home if only half of her remembered it, the slow integration of Baiko's pattern trickled byte by byte. "A safe place. We can regroup in a safe place."

The black screen in Mog's driver console flashed with directions, a thin purple line on the map with beacons to follow. She shoved her foot on the accelerator and geared down to feel the punch of speed leave the unpleasant panic behind.

"Make the kids play nice, honey." She shoved a taser from her door compartment into Ego's hand, and he nodded with a firm-set jaw. Kissed her cheek as he swivelled his chair.

"Now what the holy gargantuan fuck was that? Who are you people?! How do you know my wife? Why is, Dix, is wait… Dix as in…"

"Yyyyyyyep." Mog swerved around more debris and vaulted them over a pile of ruined concrete, "Who else would I hire to break Lieben out of your bind if I didn't hire, well, the people who broke Lieben out of her bind?"

"No box." Lieben's eyes narrowed as she stared at Dix and Saber. Saber shifted between them, palms up in surrender.

"No box, no litres of alcohol till we get where we're going."

"I couldn't get the sight of you out of my head." Dix blurted, shoved himself in the corner of the back with his hands between raised knees. "You kept asking for help, and we left you in a coffin and it… I couldn't get you out. Maybe if we helped now I could get you out."

"The world got hellaciously small after we fucked over the Conglom, weren't many jobs left." Saber took stock of his remaining ammunition, eyes flicked back to Demyan slumped in his seat, and Lieben's hands as she carefully buckled him into a harness.

"Welcome to the cause, then. Why'd your buddy tweak? Pretty lousy habit for a mercenary." Ego nodded carelessly in his hoodie, set the fabric over his hairline.

"Yeah… you… Idless, you what, wipe out the past or something?"

"We don't ascribe to labels, or corporations. Branding is a form of slavery, and so are nationalities and names. We believe in freedom, and equality of opp…"

"Anarchy blah, blah, you clearly don't do background checks on your people. That…" Saber nodded to Demyan, "… The Slavics, they had this Black Ops squad, made Baba Yaga and her chicken-legged hut look like the ideal next door neighbour. My squad was tasked with rooting them out. All we had to go on were the sounds of Ukrainian accents, and a tattoo artist in the Crimea, who's half-scorched artwork was inked on twelve guys before he shot himself in the back of the head. Couple of times we thought we got the bastards, until after the Kiev offensive, only one survived the carpet blasts and drone assault. Our CO, man, he was a piece. Thought he was some learned shit, all Roman and Greek history."

"Named the guy after some amazon warrior in the Iliad. 'Cause the bastard never backed down. He never ran." Saber pointed with his index knuckle, bit back a litany of curse words and a perverse history of 'if we got our hands on him, I'd...'s whispered in the mazes of steel and wrecked concrete. "That is the reason all the buildings in Prague are brand fucking new. He'd been hacking the drones. They'd go haywire, a fault in the programming. Or some mechanical fault we explained away with the bits and scraps left over, took us three years to figure out it was him. His squad was dead, but he wouldn't stop the mission until it was done. Freedom for Ukraine. No one else could have it."

"About a year in, we started getting… gifts.My fucking mum's birthday I came across a flask of vodka and a hand-painted flower postcard in an envelope marked 'for your mother' under my pillow in my bunk." Dix let out a hesitant breath, shook out his fingers and one by one pulled off his combat gloves. "He was jeering at us, knew every event in our lives, but we didn't even have a name. Nothing but the tattoo and some hazy images from drone cams before they went down. After the blast, he went silent. Saber and I, we didn't have civilian lives to return to, enlisted out of school. We were on a beach in Phuket, Thailand and a woman ran up to us. Handed me a paper bag with a fucking flower painted on. 'Saw you made it, Cillian. Happy Birthday'. He knew. He knew Saber'd taken me to Phuket to heal, he knew my favourite sweets from the shop near the base, what sort of beer I liked. The war might've ended but damned he was alive. I didn't know whether to chuck the bag in the Andaman Sea or drink the beer to my own good health. If it were some sort of bomb or a peace offering."

Mog climbed the vehicle up an old construction rift and surged into the streets of London, the deep chasm of their underground travels distant for bursts of kaleidoscopic colour. They rode numb, quiet as the graves avoided until streetlights of Greater London became the over-bulbs of motorway traffic drones. Hovered protectively beside Demyan, Lieben waved a hand and one of the orbs broke off from the bracelet, zoomed out of their microcosm and each of the drones ahead drifted to the ground in recharge-shut-down. The only light became the stars above the overcast night sky.

Sophia's body felt inches from Ego's shoulder in the front passenger seat.

He held Mog's hand and let his thumb stroke familiar skin. Silence was their bedfellow, as it was in perpetuity on their migratory drives across the span of continents. Ego didn't voice the disbelief in Sophia's stillness, or the hazards of leaving her body behind. She chased after him, her ghost decompressed a byte at a time in the way Lieben held her arm near Demyan's chest, the motion of tucking the safety blanket up to his neck like she had for Nasrin, when she toddled between the caravans.

"Cillian, then. Dix is what, a codename?" Only when the silence ate his stomach in piecemeal did Ego break its' spell.

"Like Ego?" Not bothering to glance in Ego's direction, Dix stared at the English countryside on the back road they snaked along. "If I'm going to die beside you, fighting for the machine we locked in a box, I should know your name. Dix was my callsign, Cillian. Cillian Inverness. Born in County Wicklow, moved to Canada when I was six and met this sick fuck."

Mog's head quirked as she watched ahead and kept Ego in her periphery. She waited for the groan after he said his name was Ego, picked it when he was eleven and gave his life to the Idless and their cause. Her Ego, the man she'd whispered love to since a hazy adolescence, was perpetual in his lack of definitions. Nothing but a massive ego, big enough for the three of them and their kids.

"Fair point." Ego said, and Mog held in her breath for the groan. "Farouk."

"Farouk. Fuck, now I get the whole Ego thing." Cillian grinned sideways out of his jaw, jutted out his chin at Saber. "S'okay though. His name's Percival."

"Hah! What were your parents on, when they named you that?" Ego chuckled and couldn't help stare at Mog's eyes as they crinkled with crows feet he'd never seen, a smile to lips a bit thinner than the years prior.

"Says fucking Farouk." Saber pitched an empty drink can at Ego, and he let it bounce off his shoulder.

"Hey, my middle name's not so bad. Got it after my Granddad, a perfectly normal, respectable Clive."

The three men sat in companionate cajoling, thoughts on ricochets and magazine count, the person left behind drifted beyond the carriage of their microcosmic space.

"Clive then. And ah, I'll clock the next guy who calls me Percival. Percy is… It's better." They shook hands, fingers lingered a bit longer for the post-battle jitters.

"Meine. Meine Anastas." Lieben's birdsong voice entered the cochlea of their ears and lingered. Cillian's eyebrows shot up his forehead, lips parted as he watched Lieben roll her right shoulder in a shrug, chin tilted to the floor where they'd ripped seats out for gear's sake. "I hadn't gone yet. I like it better than

Baiko or Sophia or Lieben. Strange, I like it better now than I did a few hours ago, when Hal Grundy asked me what… but that's… I'm sorry, you must think this is some form of horror."

"Great to meet you, Meine. I'm Katherine. Katherine Kimbedjian, Farouk's wife." Mog flickered her eyes away from the road to glance at him, the steadiness to her voice abandoned for a litany of imagined negative reactions. "What? We're all in the sharing circle. Except Demyan, who's still unconscious. Not much longer, another couple of hours unless I cut… and the map is cutting me through the fields, alright so… thank you Meine… wow you… did that very quickly and… okay that's a river just a… hold on."

The ride bumped and rattled, but years of practice in farmer's fields and the Canadian wilds gave Katherine the advantage of off-road security.

"Are you…" Ego ducked his head down and sighed out, "… sure you wouldn't rather be Clive's wife?"

"Hmmmmmmmmmmm." Her hand slid into his, and she pulled it onto her thigh. "Nnnnnope. Farouk's far more amusing. Clive."

The squad descended into chuckles mirrored by Karnak's machine. Soundlessness stopped its spectral haunting as they bypassed the road to Nottingham and pushed up into the middle north of England to the Peak District.

Mountainous hills on either side, the modified Land Rover churned down gullies and up hillocks, around copses of trees planted in 2032, in the 'post-revivalist' period where England decided preserving wild habitat was more important than allowing 'foreigners' to build country estates or housing complexes outside the rising cities.

No homely lights offered them safe bastion as the massive stonework of Lord Stanley's Manor loomed beyond the headlamps. Only the carriage house door, opened by the gardener who pulled his poorboy hat down to his eyebrows, the collar of his jacket up past his ears. Mog docked the rover against its' charge panel and opened the mag-lock on the doors before standing beside the car with a stretch of long, flab-covered arms.

"Eeeep!" She squeaked as Ego launched over the bonnet and dove to wrap her in his arms, his elbows collided with the wall and Ego let them sting with the impact. His face burrowed in her neck, and she clung with a resilience which bowed Ego's shoulders for how beautiful and mighty it was. "Love… oh my god, love."

As Cillian and Percival shifted out of the back, stuffed weapons in packs for their impromptu armoury, Lieben unlatched the harness on Demyan and caressed the hair away from his face. Illumination from the Manor's interior brightened the dark carriage house, as Lieben cradled Demyan in her arms and found his burden was light. He didn't overwhelm her, she would not drop him.

"Welcome home, my dear." Lord Stanley leaned on an ebony cane, an antique computer chip embedded in glittering resin on its pommel. "I see you brought friends."

Untangling herself from a monumental kiss, Mog leaned her head over and gulped. "Got them, Boss… but Sophia, she…"

"We made it. Demyan requires medical aid, please stand aside. I promise to introduce you later, Stanley." Lieben shifted past him, took Demyan up to the second floor where a spare room got repurposed back when Lord Stanley's wife was recuperating from liver failure.

"I should… I can help." Cillian nodded and bowed his head, chased after Lieben and her cargo he knew only as the spectre of a fearsome ghost and warrior. "It'll be okay, Percy! I won't lynch him, promise!"

"Mog? What…"

"I told you she'd know where to go, didn't I, Katherine?" Inspecting his nails, Lord Stanley caught Percy's lean against the truck, the shoulders back, face forward posture of Ego and Mog's hand on his chest.

"You gambling bastard. Sophia is dead. Whatever remains of her got uploaded into the android."

"After Fester the kids and I were taken in, I reached out. I found Saber and Dix through their buddy Trent, and waited. Eventually, Lord Stanley pinged me and… I couldn't get word to you. Not until the tunnels." Mog sighed and shook her head.

"What the holy hell happened there, anyway? How'd she die?" Percival hefted a duffel onto his shoulder, and stepped away from the boot release.

"Ask Demyan when he wakes up, I was snuffing a Black Collar. The tunnel opened and Sophia was dead. How did you know Lieben would come here? Wasn't that one massive risk? A risk you put on my wife, you cock-suited bast-"

"Woa! Easy!" Percival pushed off, his hands up.

"Wife, is it? Well! Suppose it's jolly good I had Eleanora fit you both with one suite. Congratulations." Lord Stanley bowed and flourished his hand to shift them into a sitting room, where a wheeled tea cart held enough alcohol to pickle a rugby team. "Come along, children. None of this works if Lieben feels pressured to do so much as wear a certain colour of dress."

"You put my Mog in…"

"Shut! Up! You infantile idealist. You're bloody welcome, by the bye, for rescuing you from absolute certain death. And for hiring your wife et ensemble to make Dieter's plan come to fruition. It took considerable effort. Now please!

Prior to drawing some baths and perhaps changing your clothes in a more gentlemanly fashion, you might listen instead of accuse me." Stanley's wrist went white with the weight he placed on the cane, an arm chair took his weight, his right leg tipped out at less of a bent angle than the left, which he tucked in by the leg of the chair.

"I… sorry, it's been…" Ego deflated, wrung his hands through his hair and wondered how long they'd shook for. "I apologize."

"In order for any of Dieter's vision to succeed, one rule must be paramount. Any decision must be Lieben's to make. If I'd sent Katherine, Percival and Cillian with strict orders to return here, or some form of safe-house, the girl would've rebelled faster than my son sniffing cocaine at a Berlin rave. All of this, shifting the world's wealth, eradicating the borders which thank you, do cause quite a bit of trouble to certain interests of mine, it folds apart. Our task is not to control Lieben. It's to guide. I knew Dieter. Knew him half my adult life, and although I have been honoured to see his machine as none on this planet have for years in isolation, I learned that Lieben is a knife's edge. She must make those decisions for herself. I had several safe-houses lined up tonight. Eyes on that flat above Demyan Anastas' bookshop, Kaoru's tea room.

Do you think the Conglomerate will be pleased, now? Knowing their miracle machine was in one of their own palms this entire time? That I hid her away out of an old man's fancy, or a collector's adamance? If this plan fails, my head will hit the guillotine far sooner than yours, Master Kimbedjian. They know where I live."

Ego plunked down on a settee, as Mog, dutiful still, set to pouring whiskey in old fashioned glasses with a sphere of ice in each one. She handed them off, one to Percival, one to Lord Stanley with a nod, and one for herself. She set Ego's in his hand, closed his fingers around the cut crystal.

"Look, matching cups." The smile on her face glowed as Mog wondered at the novelty of matching glassware. She sat prim on the settee beside him, intertwined the fingers of their free hands, and sipped her whiskey. Inside his chest, Ego felt his heart implode. He should have tried harder, listened more when she asked him or Fester to drive more carefully, and inevitably picked up the shards of a glass, or plate until they switched them for metal. Traded hubcaps repurposed by a mechanic up past Hope, BC. They slept in a van, the earth beneath them their front garden, but he couldn't stop to look at his Mog.

At his Katherine, who preened to drink from matching glasses, sitting on a seat which didn't move unless he lifted it. Ego took a deep, cleansing breath and sipped his whiskey.

"It's alright, dear boy. You've been the one in the trenches, no one faults you for spending a minute to find your tongue." Lord Stanley's smile was as effortless as a spring breeze, his whiskey smooth down Ego, no, Clive's throat. He didn't notice when he started rocking back and forth, nor when the tears shuddered past his eyelids and Katherine took the glass.

"But..." But he's a fucking stockholder, the Ego in him craved to snap. To thunder and rage over his Mog sold in dollar signs and pound sterling to an unruly element of the Conglom's own Board. Lord Stanley Hallowes, CEO. Or, but he was the anti-christ, the sell-out waiting for the right time. It neither occurred to Clive nor entered his mind a member of the Conglom would desire the same world he longed to raise their kids in. "But Fester, the kids..."

"They're in Vancouver, in this... penthouse apartment the Chairman gave us for safe keeping. Fester doesn't know what I've done, it's safer for her and the kids."

"Do you have a picture? A vid, anything?" Clive's stomach wrenched as Mog softened, and took his hand in the Lion's den of comfortable wealth as if she'd been born to walk in such places.

"It was safer for them, if I... didn't... I'm sorry."

Chest heavy, Clive held his breath to staunch the overwhelmed emotional flow of the last years. A stutter, and the bow broke at the mental image of Sophia's limbs left akimbo in a hall of white plastic. The ghost upstairs wavered in his mind.

"Ah, yes. Percival, darling perhaps we might speak tactics in my Study before the inevitable? I cannot abide viewing... emotion." Lord Stanley offered his hand and Percival bowed and took it, "Go on, dear girl! Deal with... that. Anything worse than feminine hysterics it's masculine hysterics."

The cane tapped on marble. Katherine set the whiskey glasses on the matching table and took Clive's hands. Pulled him into her tide, and up the stairs to a room carefully laid out in furniture older than the internal combustion engine. Words selected in Clive's mind discarded with the remnants of his clothes, pulled off piece by piece at Katherine's fingers.

They didn't shower in a rim of chrome plated pipes and faded shower curtain outside their van, but in warmed tile with a glass door. Hands made clumsy by fatigue and nerves fought with Katherine's fatigues, the belt around her waist tossed into a vase with flowers. Frenetic kisses quenched the need to punch the walls until his fist bled, her thigh crept up on his lower back, body pressed against the shower wall. The other followed, and as Katherine broke into a gasp for the hunger and rage in her lover's eyes, they joined. Bodies devoid of familiar years pushed into comfortable paces, unforgotten.

They tangled into sheets, their bodies damp from the shower and hair staining the feather pillows with their moisture. Thin bands of laceration glue kept Ego's wounded arm together. And as thoughts of Sophia's transformation drifted beyond the threshold of the bed, Clive held Katherine to his chest, hazy kisses morphed to regular breaths as Morpheus led their consciousness gently away.

Down the hall, Cillian and Lieben finished the last suture for Demyan's bullet wound and watched plasma drip from the bag down the iv line in his arm. The bullet fragments laid on a small silver tea tray, beside bloodied surgical tools.

"Never thought I'd... don't matter now, do it?" Cillian clipped the last of the threads, and set down needle and pliers to the tray.

"Of course it matters, Cillian. Thank you for helping." Lieben washed off her forearms and hands, incapable of fatigue in the ether of her fully powered systems. "I don't understand war, I should, but…"

"He was the only one on the other side I could name. Sure, politicians, oligarchs. But when you're fighting, you're not fighting a flag, that idea, it… blew away for me rather quickly. He took all my pity and my rage and he… refused to die. He'll never just die." Cillian washed off and pulled down the sleeves of his dark grey shirt.

"Now I see him, I'm glad he didn't." Dipping his head down, Cillian grinned. A modicum of peace in the flow, as the demon of his younger days laid naked under thin cotton, stitched by his own hand. "Fancy that, eh?"

"Fuck you, too. Commonwealth bastard, who drinks shitty beer." Slurred syllables slowly poured like cold ichor out of Demyan's mouth. A hand pulled up to inspect his forehead, massage temples. He blinked in the bright warm light. Sniffed iodine and copper and bleach instead of the funk of bodies in tunnels.

"What was my name?" Accent thick with homespun Ukrainian influence, Demyan quivered. Raised his head from the pillow with a series of grunts as Lieben supported his head with a milk-white hand. "What did you call me, back then? I never knew what it was."

"Aderastos." Cillian 'Dix' Inverness passed his former enemy the crystal glass of vodka and steadied Demyan's hand when the weight of it buckled his recuperative wrist. The bed sensed his rise and rose with him, cupped under knees and back to alleviate pressure on any one centimetre of Demyan's frame.

"Aderastos… think I'll keep it." Head leaned back against the pillow, Demyan muttered, "Diborja."

Drowned his throat with vodka. His eyes closed and he breathed easiest in the watch of enemies and the machine who decided to the quarks which made up the atomic particles of her cloud, to love them.

To love them all, in their fallen and natural state.

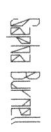

The Emptiness at the Centre…

Sophia Burnell

About Sapha Burnell

"Sapha is like a young Wolfgang Pauli, in every laboratory he went, there was a little explosion" - David Roomy, Author of Inner Work in the Wounded and Creative: The Dream in the Body

Cyberpunk & mythology aficionado Sapha Burnell teethed on images of the Berlin Wall falling down. Steeped in divergent cultures, religion & gender roles, the Wild One dedicates her work to the dichotomy between science and spirit. Sapha lives on the West Coast of Canada with her spouse, and rescue dogs, and chills daily with her #hallowestribe on Discord.
Visit Sapha at www.saphaburnell.com on her Discord Server, or on Twitter, Twitch & Instagram @UsurperKings.

CPSIA information can be obtained
at www.ICGtesting.com
Printed in the USA
BVHW080046061121
620873BV00001B/48